This is the first book to describe the scientific basis for the action of plant polyphenols in a wide range of technologically important phenomena.

Basic understanding of plant polyphenols (vegetable tannins) has increased dramatically over the past 30 years. This has opened the way to an understanding of practical polyphenolic phenomena. The book opens with a general summary of polyphenolic structure and a discussion of the physical and chemical basis of non-covalent molecular interactions (the process of molecular recognition) and the means by which polyphenols initiate the many and varied properties they display. There has been a big increase in interest in our ability to understand the importance of polyphenols in areas as diverse as agriculture, ecology and food selection, foodstuffs and nutrition, beverages (astringency and bitterness), natural medicines (the so-called 'French Paradox'), floral pigmentation, natural glues and varnishes and the age-old methods for the manufacture of leather. The role of polyphenols in all of these areas is discussed.

This book will be of value to a wide range of researchers in chemistry, agriculture, food science, ecology, oenology, plant physiology and pharmacology with an interest in polyphenolics.

PRACTICAL POLYPHENOLICS
From Structure to Molecular Recognition and Physiological Action

PRACTICAL POLYPHENOLICS
From Structure to Molecular Recognition and Physiological Action

EDWIN HASLAM

University of Sheffield

CAMBRIDGE UNIVERSITY PRESS
Cambridge, New York, Melbourne, Madrid, Cape Town, Singapore, São Paulo

Cambridge University Press
The Edinburgh Building, Cambridge CB2 2RU, UK

Published in the United States of America by Cambridge University Press, New York

www.cambridge.org
Information on this title: www.cambridge.org/9780521465137

First published 1998
This digitally printed first paperback version 2005

A catalogue record for this publication is available from the British Library

Library of Congress Cataloguing in Publication data
Haslam, Edwin.
Practical polyphenolics: from structure to molecular recognition
and physiological action.
p. cm.
Includes bibliographical references.
ISBN 0 521 46513 3 (hb)
1. Phenols. 2. Botanical chemistry. I. Title.
QK898.P57H37 1998
661'.82–dc21 9-8686 CIP

ISBN-13 978-0-521-46513-7 hardback
ISBN-10 0-521-46513-3 hardback

ISBN-13 978-0-521-67559-8 paperback
ISBN-10 0-521-67559-6 paperback

For M

Che faro senza M

Contents

Preface

Over the past three centuries Science has increasingly come to represent one of mankind's most formidable achievements. In areas as diverse as agriculture, materials and medicine its technological applications have transformed the lives of many throughout the world. Science conversely often exacts its own price, for benefits and risks are frequently found to be complementary aspects of the same technological advances; society has to judge between them. A popular misconception fostered by some writers and media pundits is that progress in Science, the accumulation of new and greater knowledge superseding previous discoveries, proceeds exponentially ever upwards and onwards. This is misleading. Science is a human endeavour and scientists are creative beings. Science is therefore spiced with pride and prejudice; it is often subject to personal folly, wrong turns and cul-de-sacs. False dawns, unfortunately not always constrained to the confines of the laboratory, are legion. The odyssey of the science of the vegetable tannins (plant polyphenolics) as its has journeyed from darkness into (comparative) light is littered with examples of the frustrations which underly scientific progress and discovery.

Records show that extracts of a plant origin (vegetable tannins) were employed in Mediterranean regions to convert raw animal hides and skins to durable leathers at least 2000 years ago. Investigations of the chemistry of such extracts, despite the outstanding contributions of the Nobel Laureate Emil Fischer at the turn of the century, merely served to show the complexity of the problems which they posed and by the 1950s this area had become one of the murky and unfashionable corners of organic chemistry. There is often an ebb and flow to scientific research and the advent of new techniques, new ideas and new personnel has, over the past forty years, not only transformed the state of the fundamental knowledge of these compounds but has also revealed their crucial importance to many other areas of science – from ecology and agriculture, nutrition and foodstuffs, to beverages, fruit and floral pigmentation. Thus particular attention has focused recently on the possible amelioration of long-term degenerative disorders using beverages (such as teas and red wines) and specific diets rich in anti-oxidants (natural phenolics and polyphenolics).

Guardians of the public purse increasingly demand from scientific research a greater poundage of flesh; the emphasis has shifted to one in which short term exploitation of its discoveries is pre-eminent. This text therefore summarises the established chemical knowledge appertaining to plant polyphenols and seeks to use this information to address and interpret the many, important and wide ranging technological and applied phenomena in which plant poly-phenols are intimately involved and which have been noted above. The ap-proach adopted is an all inclusive one and is based upon critical evaluation of the physical forces which underly the '*recognition*' by polyphenols of other molecules, often prior to further irreversible chemical reaction. There seems little doubt that this area and that of the genetic and biochemical control of polyphenol synthesis *in vivo* are the ones which, in the future, will come to dominate scientific research involving plant polyphenols. The period from 1950 to 1990 was one in which public attitudes towards the status of science helped to facilitate fundamental research in seemingly arcane areas, such as that of the plant polyphenols. As the turn of the twentieth century approaches and siren voices increasingly beckon a long past rustic Arcadia it is timely, however, to gently remind those same guardians of a seemingly diminished share of the public purse for pure science of the words of Louis Pasteur (whose father was incidentally a tanner) that 'there are no such things as applied sciences, *only applications of science*'.

We are like dwarfs on the shoulders of giants, so that we can see more than they ... not by virtue of any sharpness of sight on our part, or any physical distinction, but because we are carried high and raised up by their giant size.

As in most scientific fields the expansion in knowledge and ideas is due to the many, amongst whom some major figures stand out. During the past forty years the contributions to the chemistry of plant polyphenols of Schmidt, Mayer and Weinges in Heidelberg, of Okuda and Nishioka in Japan, of Haworth in Sheffield and Roux in Bloemfontein were outstanding. The science of the vegetable tannins also occupies a unique position which borders not only on chemistry, but also botany, food science, agriculture and animal physiology. Two persons, Tony Swain and E. C. Bate-Smith, had the vision to recognise this crucial pivotal relationship and to see the importance which vegetable tannins were destined to play in succeeding years in many areas of applied science. The author had the extreme good fortune to have glimpses of that cherished vision in a long correspondence with one of these figures, E. C.

Bate-Smith. The correspondence was not only of inestimable scientific value but it was also full of encouragement and not a little humour. A few extracts are reproduced herein, not only for their scientific but also their human insights. May they stand for all who follow as a reminder of one of the true pioneers and giants of the science of vegetable tannins.

The sixteenth/seventeenth century poet and later Dean of St. Paul's John Donne did not have scientists in mind when he wrote that 'No man is an Island, entire of it self'. However, one of the most satisfying facets of a career in scientific research is the opportunities which it gives to meet and collaborate with others world-wide; to test one's ideas with like-minds thousands of miles distant. Unlike governments, science does not recognise national boundaries, it thrives on the spirit of intellectual challenge which arises from differences of opinion and interpretation. It is entirely appropriate that the author should note therefore the special contributions made by others to his own travails in this field – most notably Professor T. H. Lilley and Dr. M. P. Williamson (Sheffield), Professor L. G. Butler (Purdue University, U.S.A.), Dr. L. J. Porter (New Zealand) and Dr. Zhaobang Shen (China); to numerous postgraduate students from the original eponymous 'Lord Tannin' to the 'Last of the Mohicans'; to friends and acquaintances in the Phytochemical Society of Europe (and the earlier Plant Phenolics Group), the Phytochemical Society of North America and the Groupe Polyphenols (France); to those in industry with whom it has been the author's privilege to collaborate – Mme Daniele Magnolato (Nestlé Suisse), Dr. M. Saltmarsh (Mars G.B.) and Dr. A. Davies and Dr. Y. Cai (Unilever).

The author would also like to express his thanks to those who have helped in the preparation of this text; to Dr. Simon Capelin, Beverley Lawrence and Ian Sherratt (Cambridge University Press) who have overseen the transformation of the manuscript to the printed text; to Dr. Christopher A. Hunter (University of Sheffield) who read and critically evaluated the two chapters on molecular recognition, and who also abstracted and organised the X-ray structures of simple phenols from the Cambridge X-ray data bank; to David Lodge and Reed Books who gave permission to reproduce an extract from the novel *Small World*; to Verlag Carl Ueberreuter (Vienna) for permission to use the illustration by Ulrik Schramm from *Gullivers Travels* and to the other authors who are specifically noted and who gave permission to use figures, at appropriate parts of the text, from their own scientific publications.

Sheffield, U.K.

Introduction

Et voyez ici (dans le cas du vin ...) comme la science pure, dans ce
qu'elle a de plus élevé, ne peut avancer d'un pas sans faire profiter
tôt ou tard les applications de ses preceiux résultats.

Louis Pasteur, 1862

... the future ... is to use the chemical information as the starting
point for questions that lie in the realms of biology.

T. A. Geissman, 1965

According to Albert Einstein we live in an era of perfect methods
and confused aims. For example in Organic Chemistry the known
synthetic methods allow us to prepare an astronomical number of
compounds; the gap between the **possible** and the **relevant** becomes
larger every day.

V. Prelog, 1986
18th Solvay Conference

Liebig's *Handbuch der Organische Chemie* of 1843 listed around 2000 substances derived from natural sources. Amongst these were several phenolic compounds ranging from gallic acid, isolated towards the end of the eighteenth century by Scheele from gall nuts, to morphine obtained from the opium poppy by Séquin in 1804 and a year later by the apothecary's assistant Serturner. The *raison d'être* for the growth of organic chemistry as a scientific discipline can be traced to this point at the beginning of the nineteenth century and to the desire of chemists to comprehend more fully these 'children of nature'. For the century and more which followed the elucidation of the chemistry of natural products – their structure, chemical properties and ultimately their synthesis – were the dominant themes and goals of organic chemists. In this same period the number of known natural products, including phenolic compounds, increased exponentially. Today this 'stamp collecting' emphasis has, quite rightly, changed – a fact which might well be considered appropriate in view of the percipient observation of Prelog made just over a decade ago.

In the case of plant polyphenols (*syn.* vegetable tannins) the surge in knowledge of a fundamental chemical and biochemical nature which has taken place over the past 40 years has now paved the way, as predicted by Geissman, to an

1

interest in questions which lie in the realms of biology (not least the questions of the purpose or function which these forms of metabolism serve in the organisms concerned). Equally it has given rise to a burgeoning concern with practical polyphenolic phenomena and an understanding of the importance of plant polyphenols in areas as diverse as agriculture, ecology and chemical defence in plants; foodstuffs, nutrition and beverages; fruit and floral pigmentation; natural glues, varnishes and exoskeletons; the efficacy of traditional and herbal medicines in the treatment of certain pathological conditions; traditional processes for the making of leather and the generation of important industrial chemicals. Whilst this change in emphasis might be considered timely it also perfectly illustrates the seminal observation of Pasteur that the proper understanding of complex applied scientific problems invariably follows periods when substantial advances in fundamentals have first been accomplished. Underlying a great number of the practical phenomena in which polyphenols are intimately involved is their distinctive ability to form intermolecular complexes with each other and with other molecules, large and small – the property of *molecular recognition*. In the case of polyphenols such non-covalent interactions are not only important in their own right but are often an indispensable prelude to further chemical reactions in which new chemical bonds are formed. The fundamental basis of molecular recognition is provided by the potential energy surface that represents the interaction of two or more molecules in a cluster as a function of their mutual separation and orientation. The number of independent variables upon which this intermolecular energy depends increases as the molecular size increases; for molecules the size of typical polyphenols in an aqueous medium the number of variables is astronomical and rationalisations are, necessarily, macroscopic and qualitative in their nature. Nevertheless understanding how this facet of the physical and chemical behaviour of polyphenols is achieved through their particular chemical structures has become a question of paramount importance in the understanding of the wider practical issues noted earlier and this particular question is given prominence in the ensuing discussion.

'E.C.' – A personal memoir

'E.C.' on his Eightieth Birthday
(reproduced by kind permission of Jeffery Harborne).

E. C. Bate-Smith ('E.C.' as he was almost universally known) was born at the
beginning of the century and died near its end in 1989. His academic training
was as a physiologist but it is probably for the period when he became Director
of the Low Temperature Research Laboratory (L.T.R.C) in Cambridge from
1947 to 1965 that his career will be best remembered. The immediate post-war
period was an exciting time in the plant sciences and 'E.C.' became one of the
great pioneers of comparative plant biochemistry. His studies of plant phenolics and his collaborations with Tony Swain were acknowledged world-wide

and were seminal in the development of this area of science. One of the characteristics of the truly great scientist is his or her ability to judge when the time is ripe and propitious to press ahead in a particular field, and equally when it is certainly not provident to do so. The demonstration by Partridge in the L.T.R.C. of the potential of paper chromatography for the separation of sugars stimulated him to show very quickly that it was an equally beautiful method with which to explore the chemistry and biochemistry of plant phenolics. There was an explosion of activity in this field, none more so than his own, much of it carried out by himself, often in his spare time; his surveys of the distribution of the commoner phenolic constituents from over 1500 species of plants from nearly one-half of the known plant families, published in the Journal of the Linnaean Society (Botany) in 1962 and 1968, is a monument to that endeavour. It remains a work for all ages.

'E.C.' always preferred his contributions to be referred to as Phytochemistry and, with Tony Swain, he was able to demonstrate that the field of vegetable tannins was now, and for the first time, open to serious and detailed biochemical and plant physiological study. Over a period of some thirty years his researches gave rise to a constant flow of ideas and theories. These he illuminated with a clarity and wit and presented with a boyish enthusiasm which bordered upon a passion, unequalled amongst most professional scientists of my acquaintance. He possessed a huge intellectual curiosity, a need to know and understand; his objectives were encapsulated in a letter written as late as September 4th 1985:

I myself started with the 'information' that tannins were waste products deposited in the wood of trees because the plants had nothing else they could do with them: a horrible thought! Convinced that they had a function I set out to try and find it by way of their systematic distribution linked to the idea of their astringency.

On a personal level the author owes 'E.C.' an everlasting debt for his unfailing kindness and thoughtfulness for he, amongst others, prompted me to participate in the investigation of the vegetable tannins in the 1950s, when it seemed a fathomless black hole, and moreover one which in its time had engulfed many eminent scientists and their endeavours. He had sensed that the time was now ripe to systematically examine these recalcitrant children of nature with the new armoury of methods which were becoming available. His shrewd comments and continual support were a constant theme to my own work and that of my many colleagues. It is exemplified by a correspondence which lasted some twenty five years until just before his death; a correspondence beginning 'Dear Haslam', it ended 'Dear Eddie'. He understood the need to constantly give encouragement to the young; how else might one judge the significance of one's own contributions? How could one's spirits fail to be lifted by the following letter (October 17th 1977) which commenced:

By one of those coincidences I have long since ceased to regard as remarkable the last thing I wrote as I went up to bed last night was ...
'There is now ample evidence that evolutionary dispersal of woody dicots is accompanied by progressive loss of hydroxyl functions from proanthocyanidins and flavonols and by many other transformations of these regular constituents.'... And behold your letter was waiting for me in the lab this morning!

Or again (January 8th 1979):

How immensely encouraging the chapter you've written for Eric Conn's book is! I miss having a kindred spirit to discuss my bits of research with and I often think how nice it would be to float ideas and pass on idle thoughts – some of which might even prove to be useful ones.

Thoughtful and frequently stimulating comments and criticisms flowed unceasingly from his pen and typewriter (now in the AFRC Institute in Babraham) well into his eighties. His continuing interest in Phytochemistry, Function and Taxonomy and his humorous asides upon life are illustrated by a brief selection of extracts from these letters.

Taxonomy, metabolism and function

Your letter and enclosure have given me a good deal of pleasure during an otherwise dismal week.... Your paper on Secondary Metabolism fits in oddly with this question. In your first paragraph you ask the question 'Why?' You unwittingly answer it yourself in your postscript; you quote Hoppy as saying living matter becomes dead when the chemist touches it and Robert Robinson as saying, virtually, chemistry comes alive when it touches living matter. I knew both of these people, the former very well. ... I think the truth is simply that neither chemistry nor biology can answer the question 'Why'; both in their particular way can answer the question 'How'. Between one and the other there is a bridge to be crossed, a metaphysical one. I call it 'The bridge over the River Why'.

But the two months have not been entirely wasted. I have written a long letter to Dahlgren (copy herewith) about his revised system, which I dont like so much as his original one. I regard the galloyl esters as the key components in the systematics of the dicots, and am relying on your work to bring some coherence to their distribution. By now you must have a good deal more than that belatedly published in Phytochemistry. I haven't written to you before because I wanted to get a reply from Dahlgren to my (Friendly!) criticism and I not only got this, but also a visit last week from Dahlgren himself. I hadn't previously met him, and we got on splendidly. I have also had a visit from Kubitzki, and have written to Thorne, who had sent me his latest phylogenetic tree, not yet published. It agrees more closely with the chemical data than his earlier one did.

Summing all this up, I think your speculations about the biosynthesis of the gallo-
and ellagitannins are well founded, especially the need to dispose of them by blocking
hydroxyl groups or by complete elimination when they are not needed. This can
happen at the fringe of evolving genera (e.g. Geranium) or when they are replaced by
more efficient deterrents. This seems to be what happens with the iridoids... An
interesting situation is when the galloyl esters occur with the deterrent which it is
assumed to be an improvement upon them and may eventually replace them. This
co-occurrence represents the kind of evidence which suggests that the iridoid-contain-
ing taxa have evolved from what were formerly galloyl ester-containing ancestors.

As I said I have been making good use of what otherwise would have been a tedious
break by reading your paper and pondering about it. ... The next question involves
function. This is my own particular province and was, in fact, the starting point of my
whole interest in phenolics. Their function as astringents has, I think never been
questioned, and so far as I know no other function has yet to be assigned to them.
'Astringency', of course includes their inhibition of enzyme action and other consequences of their association with proteins which you point out. As 'broad spectrum'
repellents one would not expect a high molar efficiency – such as one might get in an
alkaloid with a specific predator target. Your evidence of the wide variation in
structure and ability to associate has now to be seen in the context of systematic
distribution. My reaction is to try and identify a limited number of **groups** of structures
which as such can be treated as taxonomic characters, like, in fact you have done with
the *Rubus* and *Quercus* groups, and see how they organise systematically.

It is the significance of the distribution of the different classes of tannins that is the
ultimate reason for working on their chemistry. As I said there is immensely important
taxonomic material emerging from your present work and your future intentions.
Okuda's work too is important. I am glad to have the xerox copy of his paper which
you sent and which I would otherwise not have known about. It's interesting to reflect
that after the very first chemotaxonomic symposium in Paris in 1962 I suggested that
as an exercise people interested in different chemical fields should combine forces and
have a go at Casuarinaceae. The idea didn't catch on. My contribution would have
been the tannins – but what a hope in those days! Later, perhaps Seikel and Hillis
might have got the answers, with the advantage of access to the species where they are
indigenous. Until casuarinin turns up again there wont be any clue as to the possible
relationships of this family, which Melchior describes as 'Ein recht isolierten Formen-
kreis', which is why I wanted to have a go at it.

I have been reading Gottlieb's book which he kindly sent me, and he has a categorical
philosophy of micromolecular evolution from the simplest to the more complex,
mostly by oxidations and dehydrogenations. This may well be true of particular cases,
but those I know anything about seem to proceed from more complicated to less
complicated forms with evolutionary advance, e.g. the flavonoids, which lose B-ring
hydroxyls as they become more advanced (but they do of course acquire 6 and
8-hydroxyls). So in the case of the iridoids I prefer Robert Hegnauer's view of the
aucubin glycosides as occupying a central position (Naturwiss, 1971) evolutionary
changes taking place both by oxidations and reductions. The truth is, as you point out,
we don't know enough yet about the systematic distributions of most of these constitu-

ents, let alone their evolutionary history. ... The condensed tannins are an interesting case, and probably a special one, because they were presumably inherited as such from the gymnosperms, complete with myricetin and prodelphinidin. Evolution in the angiosperms seems to have involved loss of these constituents concomitant with loss of arboreal habit or by replacement with hydrolysable tannins, or, as many think, with more thrifty repellents. ... I haven't any reason to question the Schmidt and Mayer biosynthetic route to gallic and acid synthesis, but it had to begin somewhere and be confined to the dicots, and that is where I'm struggling to find the answer.

Tannins, proteins and haemanalysis

Your letter is an answer to a prayer! A thorough investigation of the 'galloyl esters' of fresh, green plant tissues is urgently called for, and I'm so glad you have been and are, doing this. I am not surprised you are finding none of the substances so laboriously studied by Schmidt and Mayer in commercial extracts. I have long been convinced that these are mostly artefacts of dead tissues.

My third point (and my last, so as not to make this letter too long), is to come back to using blood as the substrate for determination of astringency. To me its first advantage is the ease with which precipitation can be measured. This is equalled by its availability, if only a few measurements are to be made and people can bring themselves to pricking their finger occasionally. ... It is a protein omnibus! I think it is necessary, when considering the tannins generally in a biological context, to have simple **agreed** methods of evaluating astringency.

Beautiful!
Your results have certainly given me something to think about. ... What especially pleases me is that the RA of the procyanidins is much the same as I get by haemanalysis. ... I'm sorry that you can't bring yourself to shed a little blood in a good cause. The method's so simple and so very quick – 10 minutes from door to door!

As regards your paper on the association of proteins with polyphenols, I have to confess that I find up-to-date chemistry very difficult to cope with. (The last chemistry lesson I had was in 1922!). However I got some help with the Scatchard diagrams from Denis Sharman, and I can find nothing in your paper with which, translated into my own arcane idiom, I couldn't, from my own experience, agree. In fact you are saying, in more precise and elegant fashion, what I have been thinking ... and trying to say ... about 'haemanalysis' for several years.

Plant phenolics and the Phytochemical Society of Europe

Very modestly he acknowledged his role in the 1950s in setting up the Plant Phenolics Group which was eventually to become the now thriving Phytochemical Society of Europe; few of its members may be aware of the immense debt which they owe, amongst others, to 'E.C.''s foresight, enthusiasm and intuition.

Reading the names and the Institutes represented I have a sensation of 'déjà vue'. It recalls as I mentioned in a letter to Ron a week ago, the beginnings of our work on tannins and the formation of the Plant Phenolics Group. This started with an Inter-laboratory Symposium between Long Ashton and ourselves, the name being suggested by Wilson Baker. It developed several years later into the Phytochemical group and then into the Phytochemical Society. I wrote an article in 'Nature' entitled 'Plant Phenolics into Phytochemistry' of which I have no reprints and have lost the reference.

The group idea came from the Society of Chemical Industry which has a number of subject groups, ...

Life

I am enclosing a reprint of the paper which has the distinction of being the first 'Review Article' in Phytochemistry! It is a concession by the editors, in consideration of my senility, and is a compromise with my first proposal to publish the 1964 memorandum entire.

Incidentally I'm learning a lot of chemistry on Sunday mornings listening to the Open University. I shall soon be able to understand some of the papers you send me! Even some of Gottlieb's too.

I'm very frustrated by my inability to get to the Botany School Library. Without a car life is very difficult and I can only ask for transport when the situation is urgent. Otherwise my life as a villager is very pleasant and also is in the height of fashion!

It seems to me the time has come to withdraw (not retire) from phytochemistry. The message came with the last number of the Bot.J.Linn.Soc. which had a paper on computerised systematics which was written in a language which might have been Ancient Egyptian or Martian so far as I was concerned. I shall be leaving the board of Phytochemistry, but still getting my copy of the journal. I shall hope to keep up a lively correspondence with a select company of interesting people.

And finally and very sadly shortly before his death:

I had a slight stroke about a fortnight ago. My writing's not very good but it's improving. As regards mental activity I'm not aware of any deficiencies, but that's not for me to judge. I enjoy the usual occupations. I'm led to expect improvement. Hope to hear from you.

In today's high-powered business-like and business-oriented laboratories, manned by the proverbial 'big battalions' of researchers we shall be fortunate if we see again such an individualist and pioneer as 'E.C.'. He was unfailingly courteous and gentlemanly, thoughtful and erudite. His work truly belongs to the ages.

The abiding recollection which I have of 'E.C.' is from one of the very first meetings of the Plant Phenolics Group. It was at Englefield Green in the spring of 1960. 'E.C.' came into the opening session as chairman clutching in his arms a freshly picked, gorgeous bunch of pink, white and red camellias. As he wisely remarked they were the focus of many of our scientific interests and it was therefore highly apposite that they should also give, throughout the duration of the meeting, such aesthetic pleasure as well.

1

Polyphenols – structure and biosynthesis

1.1 Phenolic metabolism in plants

Studies of the chemical basis of flower colour variation comprised one of the first successful scientific excursions into phenolic metabolism in plants. Its success depended on the collaboration of geneticists, biochemists and chemists who established unequivocally that genetic variation in anthocyanin colour was mainly due to simple structural modifications (hydroxylation, methylation) of a basic pigment molecule – the anthocyanidin (Harborne, 1960; Scott-Moncrieff, 1981). It remained an isolated example for some time for although it was recognised that a very wide variety of phenolic compounds were to be found in plant tissues it was the advent of two techniques – chromatography and the use of radioisotopic labelling – which permitted wide-ranging and successful experimentation in this field. Thereafter interest in the biochemical and physiological aspects of phenolic metabolism literally mushroomed (scientific progress indeed hinges on the continuing discovery and extension of new laboratory techniques). Biosynthetic pathways to the principal classes of plant phenols were established and, more recently, enzymes which catalyse various steps in these pathways have been isolated and characterised. It seems likely that further progress in this field will follow the more widespread use of the techniques of molecular biology in this area of plant secondary metabolism.

The kind of substances that a plant contains depends on the kind of plant that it is.

Writing in 1964, E. C. Bate-Smith (a true pioneer in this field and a giant on whose shoulders we now stand) looked forward to a phytochemistry that increasingly recognised the biological relationships between the substances that plants elaborate. He continued:

10

It would be much more venturesome to say 'the kind of plant a plant is depends on the substances it contains'... The special biochemistry of a species must not only be characteristic of that species, but must inevitably be reflected in the exomorphic characters of that species. It is impossible to concieve of stable differences of morphology not being associated with equally stable physiological differences which in turn determine the processes which lead to a particular, and possibly unique, array of secondary products in the tissues of a plant.

One distinction, of the type alluded to above, and to which Bate-Smith repeatedly drew attention in his researches was that between 'woody and non-woody' plants (Bate-Smith, 1962; Bate-Smith and Metcalfe, 1957). He suggested that this distinction was sufficiently pronounced for it to be possible to speak of a typically 'woody' pattern of phenolic secondary metabolites in the leaves of certain plants. The same 'woody' phenolic constituents – apparent in a wide range of ferns, gymnosperms, monocotyledons and dicotyledons – are moreover not found in mosses, algae, lichens and fungi. Bate-Smith noted the structural analogies amongst these 'woody' phenolics to lignin and he postulated that they might well be concerned, in some way, with the development of a vascular system in those plants which contain them.

According to Bate-Smith, three classes of phenolic constituent overwhelmingly predominate in the leaves of vascular plants.

(i) *Leucoanthocyanins* (*syn. proanthocyanidins*) – principally those which yield cyanidin and delphinidin upon acid treatment. The nomenclature of leucoanthocyanin arose from a belief that these compounds were 'leuco' (colourless) forms of the parent anthocyanins. They are now correctly represented as indicated below.

The 'monomer' units of procyanidins (R = H) and prodelphinidins (R = OH) are phenolic flavan-3-ols which are linked primarily through respectively their 4 and 8 positions. The stereochemistry at C-2 is most commonly encountered as the 2R configuration.

(ii) *Flavonol glycosides* – principally those of kaempferol ($R^1 = R^2 = H$), quercetin ($R^1 = OH$; $R^2 = H$) and myricetin ($R^1 = R^2 = OH$).

(iii) *Esters, glycosides* and amides of the various *hydroxycinnamic acids* – principally
p-coumaric $(R^1 = R^2 = H)$, caffeic $(R^1 = OH; R^2 = H)$, ferulic $(R^1 = OMe; R^2 = H)$ and sinapic $(R^1 = R^2 = OMe)$ acids.

One of the most familiar examples of this class of phenolic metabolite is chlorogenic
acid – the 5-*O*-caffeoyl ester of quinic acid.

chlorogenic acid

It is generally assumed (as suggested by Bate-Smith) that the biosynthetic
origin of these three classes of phenolic metabolite (i, ii, iii) is intimately
associated with the development of a vascular character in plants and with the
development of the ability to synthesise the complex structural polymer lignin.
This is achieved principally by the diversion of the aromatic amino acid
L-phenylalanine (phe), and in the Gramineae L-tyrosine (tyr), from protein
synthesis to that of the phenolic polymer. In this particular context Kubitzki
(1987) and Kubitzki and Gottlieb (1984) have further suggested that the origin
and early evolution of land plants must have been linked to the expansion of
phenylpropanoid (shikimate) metabolism.

The three early steps in the conversion of L-phenylalanine to the various
hydroxycinnamic acids and their derivatives are common to all the major

pathways of phenylpropanoid metabolism and as a result this particular sequence has been termed '*general phenylpropanoid metabolism*'. The enzymes catalysing the individual steps in this sequence are respectively: (i) L-phenylalanine ammonia lyase (PAL; EC 4.3.1.5); (ii) cinnamate-4-hydroxylase (EC 1.14.13.11); and (iii) 4-coumarate: CoA ligase (EC 6.2.1.12), Figure 1.1.

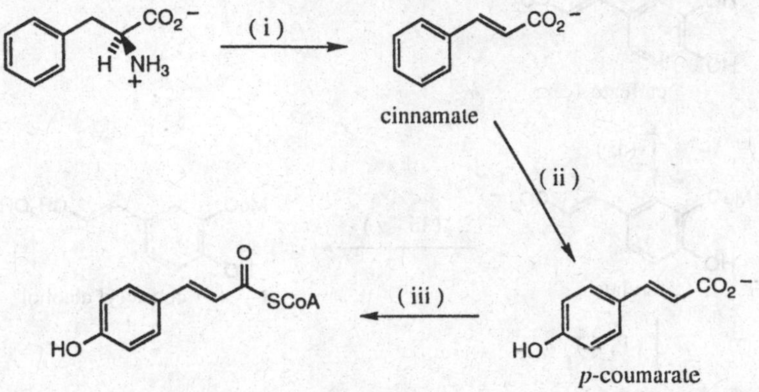

Figure 1.1. General phenylpropanoid metabolism. Enzymes: (i) L-*phenylalanine ammonia lyase* (PAL, EC 4.3.1.5); (ii) *cinnamate-4-hydroxylase* (EC 1.14.13.11); (iii) *4-coumarate: CoA ligase* (EC 6.2.1.12).

In the case of the synthesis of lignin and the soluble hydroxycinnamoyl esters other steps intervene after the formation of *p*-coumarate and prior to the CoA ligase reaction. These are transformations of the aromatic ring involving sequential hydroxylation and methylation to give successively caffeate, ferulate and sinapate, Figure 1.2. From these intermediates the monolignols – 4-hydroxycinnamyl, coniferyl and sinapyl alcohols – are then derived by reduction of the corresponding CoA esters and these act as substrates in the oxidative polymerisation to give lignin. The biosynthesis of soluble esters of the various hydroxycinnamic acids requires an activated form of the particular acid and the coenzyme A esters often function in this role. However, glucose esters may also act as acyl donors.

The flavonoids constitute one of the most common, widely distributed and characteristic groups of phenolic secondary plant metabolites. Their chemical structures are based upon a C_{15} ($C_6.C_3.C_6$) skeleton with a chroman ring (ring 'C') bearing a second aromatic ring (ring 'B') in position 2, 3 or 4. In some cases the six-membered heterocyclic ring is replaced by a five-membered one (aurone) or exists in an open-chain isomeric form (chalcone). The oxidation state of rings 'B' and 'C' are crucial in the classification of the various categories of flavonoids. Generally most of these types of flavonoid (excepting the phenolic flavan-3-ols such as epiafzelechin) are found *in vivo* in glycosylated forms.

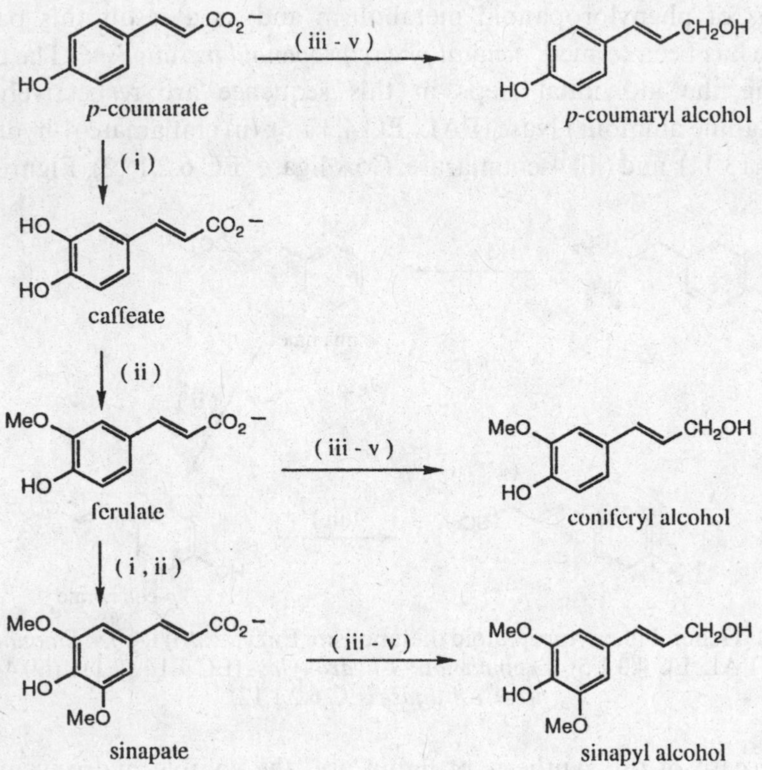

Figure 1.2. General phenylpropanoid metabolism: biosynthesis of the hydroxycinnamic acids and the monolignols. (i) Hydroxylation; (ii) methylation (*S*-adenosyl methionine); (iii) *cinnamoyl: CoA-ligase*, ATP, CoASH; (iv) *cinnamoyl CoA: NADPH oxidoreductase*; (v) *cinnamyl alcohol dehydrogenase*, NADPH.

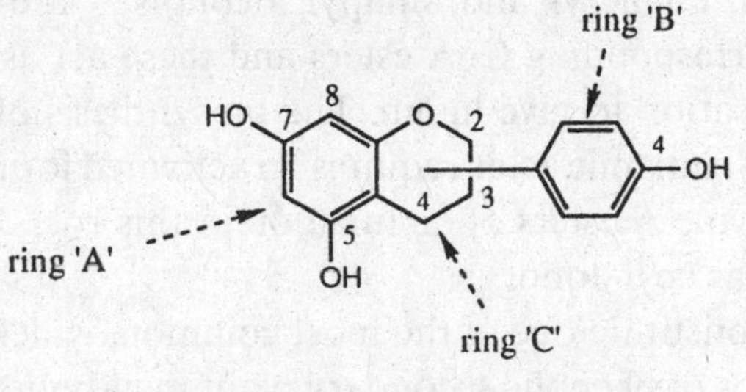

***general features of flavonoid skeleton
nomenclature , numeration***

Early experiments using standard isotopic tracer techniques established that ring 'A' of the flavonoids is derived from acetate (malonate) metabolism and that L-phenylalanine (and hence the shikimate pathway) gives rise to ring 'B' and carbon atoms 2, 3 and 4 of ring 'C'. In the formation of all the flavonoids a key central metabolic intermediate is the chalcone or its isomeric flavanone.

The pathways of biosynthesis of the various sub-groups of flavonoids are closely related and the early stages from 4-coumaroyl coenzyme A (*p*-coumaroyl coenzyme A) are depicted in Figure 1.3. More recent work at the enzymic level has largely confirmed earlier speculations concerning the sequence of steps in some of these pathways of biosynthesis. Plant cell cultures have proved to be a particularly useful vehicle with which to conduct these studies.

The enzymes involved in flavonoid biosynthesis may be sub-divided into two groups (I, II; Figure 1.3). The first group (I) comprises the reactions catalysed by the enzymes of general phenylpropanoid metabolism. The second group (II, ten or more enzymes depending upon the nature of the final flavonoid metabolite) are those specifically concerned with the synthesis of flavonoids. Studies with cell cultures of parsley (Hahlbrock and Grisebach, 1979; Hahlbrock and Scheel, 1989), show that the two groups of enzymes respond separately to different physical stimuli, such as light.

Since chalcone is the central C_{15} intermediate for the synthesis of all flavonoids, chalcone synthase can be regarded as the key enzyme of flavonoid biosynthesis (Heller and Forkman, 1988). The enzyme has no requirement for cofactors and its substrates are malonyl CoA and the CoA ester of a hydroxycinnamic acid. For all the chalcone synthases tested so far 4-coumaroyl CoA is the best substrate. However the substrate specificity of chalcone synthase in respect of the hydroxycinnamoyl CoA substrate has been a cause of some interest and speculation. This concern arises from the desire to know at which stage the hydroxylation/methoxylation pattern of ring 'B' of a particular flavonoid is established, i.e. prior to the synthesis of the chalcone or at the $C_6.C_3.C_6$ level once this carbon skeleton has been generated. Genetic evidence and the presence in some plants of specific flavonoid ring 'B' hydroxylases and methyl transferases also indicate that further substitution of ring 'B' occurs at the $C_6.C_3.C_6$ level. With few exceptions the various classes of flavonoids, such as the proanthocyanidins and the flavonols (i and ii above) are then derived from a flavanone intermediate, naringenin, Figure 1.4, (based upon the 4',5,7-pattern of phenolic hydroxylation, with a 2S absolute configuration). Hydroxylation, in particular of the 'B' ring of the flavonoid skeleton, methylation of free hydroxyl groups, *O*-acylation, *O*- and *C*-glycosylation and *O*- and *C*-isoprenylation are the principal additional reactions which then lead, within each class, to the various flavonoid structures.

Whilst the chemical nature of the steps from naringenin to the various types and classes of flavonoid are now well documented, it is eminently clear that in practice the experimental evidence which relates to some of the enzymes involved in these reactions requires further study.

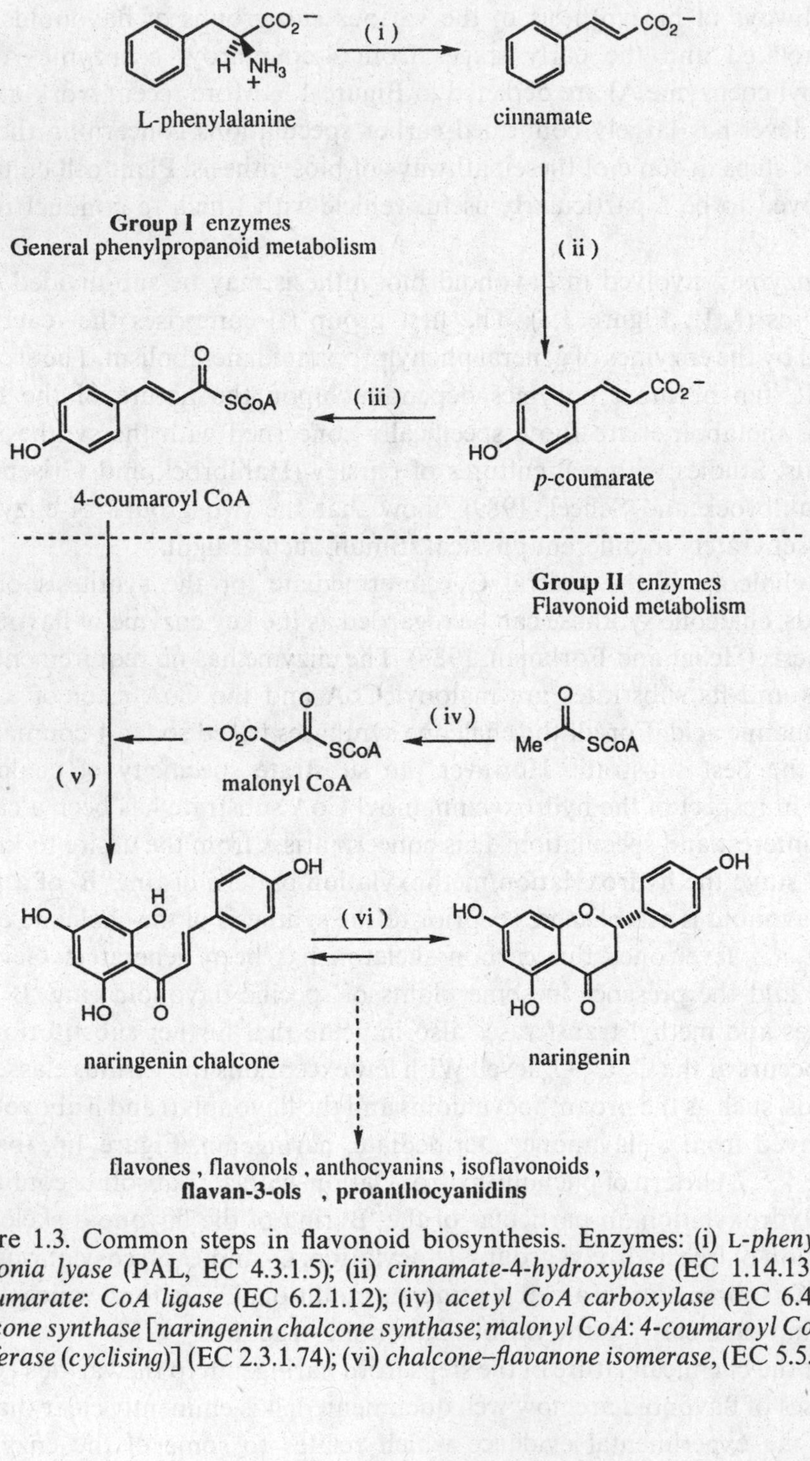

Figure 1.3. Common steps in flavonoid biosynthesis. Enzymes: (i) L-*phenylalanine ammonia lyase* (PAL, EC 4.3.1.5); (ii) *cinnamate-4-hydroxylase* (EC 1.14.13.11); (iii) *4-coumarate: CoA ligase* (EC 6.2.1.12); (iv) *acetyl CoA carboxylase* (EC 6.4.1.2); (v) *chalcone synthase* [*naringenin chalcone synthase; malonyl CoA: 4-coumaroyl CoA transferase (cyclising)*] (EC 2.3.1.74); (vi) *chalcone–flavanone isomerase*, (EC 5.5.1.6).

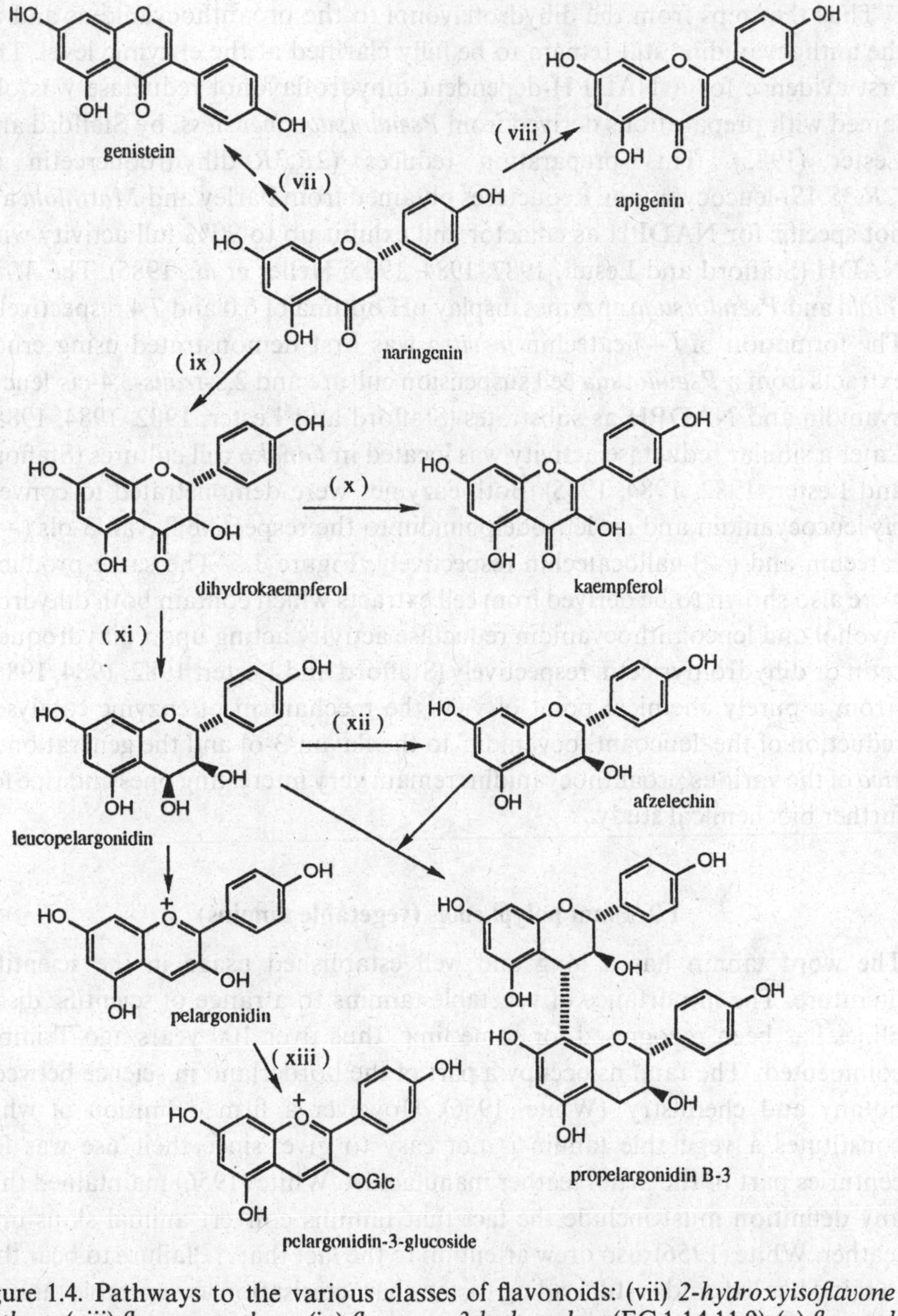

Figure 1.4. Pathways to the various classes of flavonoids: (vii) *2-hydroxyisoflavone synthase*; (viii) *flavone synthase*; (ix) *flavanone 3-hydroxylase* (EC 1.14.11.9); (x) *flavonol synthase*; (xi) *dihydroflavonol reductase*; (xii) *flavan-3,4-diol reductase*; (xiii) *anthocyanidin/flavonol 3-O-glucosyl transferase* (EC 2.4.1.91).

Thus the steps from the dihydroflavonol to the proanthocyanidins and to the anthocyanidins still remain to be fully clarified at the enzymic level. The first evidence for a NADPH-dependent dihydroflavonol reductase was obtained with preparations derived from *Pseudotsuga menziessi* by Stafford and Lester (1985). This preparation reduces (2R,3R)-dihydroquercetin to (2R,3S,4S)-leucocyanidin. Reductases obtained from barley and *Matthiola* are not specific for NADPH as cofactor and exhibit up to 90% full activity with NADH (Stafford and Lester, 1982, 1984, 1985; Heller *et al.*, 1985). The *Matthiola* and *Pseudotsuga* enzymes display pH optima of 6.0 and 7.4 respectively. The formation of (+)-catechin *in vitro* was first demonstrated using crude extracts from a *Pseudotuga* cell suspension culture and 2,3-*trans*-3,4-*cis*-leucocyanidin and NADPH as substrates (Stafford and Lester, 1982, 1984, 1985). Later a similar reductase activity was located in *Gingko* cell cultures (Stafford and Lester, 1982, 1984, 1985). Both enzymes were demonstrated to convert *cis*-leucocyanidin and *cis*-leucodelphinidin to the respective flavan-3-ols (+)-catechin and (+)-gallocatechin respectively, Figure 1.5. The same products were also shown to be derived from cell extracts which contain both dihydroflavonol and leucoanthocyanidin reductase activity acting upon dihydroquercetin or dihydromyricetin respectively (Stafford and Lester, 1982, 1984, 1985). From a purely chemical point of view the mechanism of enzyme catalysed reduction of the 'leucoanthocyanidin' to the flavan-3-ol and the generation *in vivo* of the various proanthocyanidins remain very interesting ones and ripe for further biochemical study.

1.2 Plant polyphenols (vegetable tannins)

The word tannin has a long and well established usage in the scientific literature. The importance of vegetable tannins to a range of scientific disciplines has been recognised for some time, thus over 100 years ago Trimble commented: 'The tannins occupy a part of the borderland in science between botany and chemistry' (White, 1956). However a firm definition of what constitutes a vegetable tannin is not easy to give; since their use was for centuries part of the art of leather manufacture White (1956) maintained that any definition must include the fact that tannins convert animal skins into leather. White (1956) also drew attention to the fact that... 'failure to bear this in mind has led, and is still leading to, the characterisation as vegetable tannins of many low molecular weight substances which have no tanning action.' Although the reproach that everything which is isolated from plants and which gives a blue colour with ferric chloride is a tannin is probably not entirely justified, failure to appreciate the distinctive characteristics of vegetable tan-

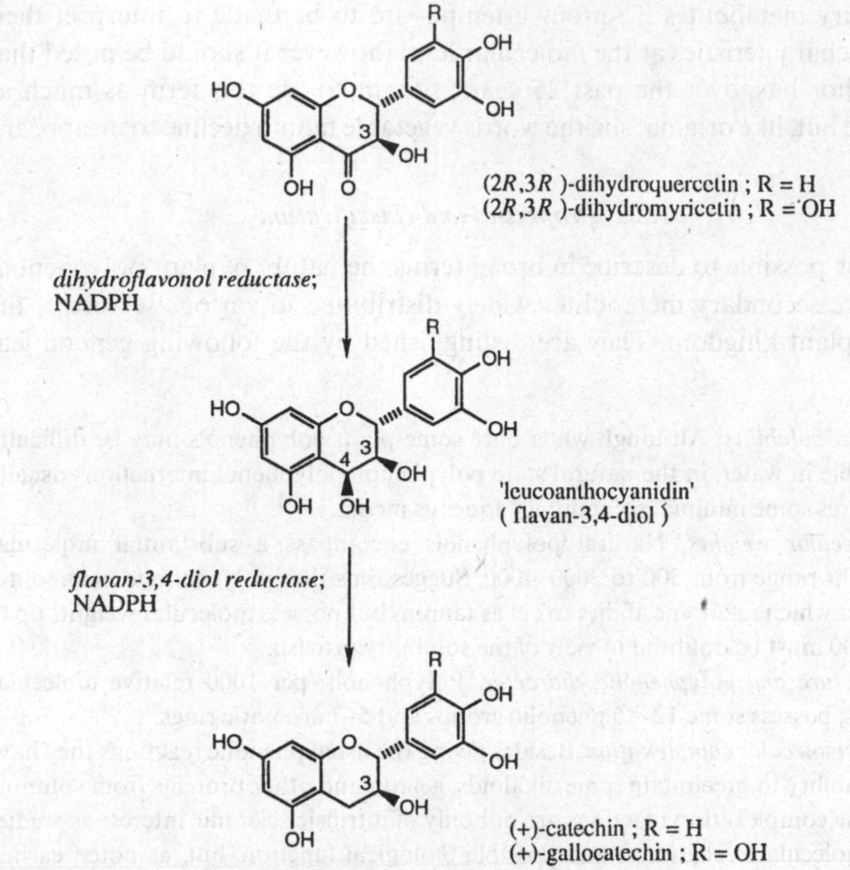

(2*R*,3*R*)-dihydroquercetin ; R = H
(2*R*,3*R*)-dihydromyricetin ; R = OH

dihydroflavonol reductase;
NADPH

'leucoanthocyanidin'
(flavan-3,4-diol)

flavan-3,4-diol reductase;
NADPH

(+)-catechin ; R = H
(+)-gallocatechin ; R = OH

Figure 1.5. Biosynthesis of phenolic flavan-3-ols from dihydroflavonols (Stafford and Lester, 1982, 1984, 1985; Heller *et al.*, 1985).

nins, as opposed to simple plant phenols, has led to some confusion in the scientific literature concerning the role which these compounds may play in a number of fields.

Probably the most acceptable concise and simple definition is still that of Bate-Smith and Swain (1962) who based their classification on the earlier ideas of White (1957).

Water soluble phenolic compounds having molecular weights between 500 and 3,000 and, besides giving the usual phenolic reactions, they have special properties such as the ability to precipitate alkaloids, gelatin and other proteins

Many still prefer the term vegetable tannin, which they find valuable simply because of its lack of precision. Scientifically and terminologically, ***plant polyphenols*** is to be preferred as a descriptor for this class of higher plant secondary metabolites if serious attempts are to be made to interpret their diverse characteristics at the molecular level (however it should be noted that the author has, over the past 25 years, sought to use this term as much as possible but, like original sin, the words vegetable tannin decline to disappear).

1.2.1 Properties and classification

It is now possible to describe in broad terms the nature of plant polyphenols. They are secondary metabolites widely distributed in various sectors of the higher plant kingdom. They are distinguished by the following general features.

(a) *Water solubility*. Although when pure some plant polyphenols may be difficulty soluble in water, in the natural state polyphenol–polyphenol interactions usually ensures some minimal solubility in aqueous media.

(b) *Molecular weights*. Natural polyphenols encompass a substantial molecular weight range from 500 to 3000–4000. Suggestions that polyphenolic metabolites occur which retain the ability to act as tannins but possess molecular weights up to 20 000 must be doubtful in view of the solubility proviso.

(c) *Structure and polyphenolic character*. Polyphenols, per 1000 relative molecular mass, possess some 12–16 phenolic groups and 5–7 aromatic rings.

(d) *Intermolecular complexation*. Besides giving the usual phenolic reactions they have the ability to precipitate some alkaloids, gelatin and other proteins from solution. These complexation reactions are not only of intrinsic scientific interest as studies in molecular recognition and possible biological function, but, as noted earlier, they have important and wide-ranging practical applications.

(e) *Structural characteristics*. Plant polyphenols are based upon two major and one minor structural theme, namely:

(1) ***condensed proanthocyanidins***. The fundamental structural unit in this group is the phenolic flavan-3-ol ('*catechin*') nucleus. Condensed proanthocyanidins exist as oligomers (soluble), containing two to five or six '*catechin*' units, and polymers (insoluble). The flavan-3-ol units are linked principally through the 4 and the 8 positions. In most plant tissues the polymers are of greatest quantitative significance but there is also usually found a range of soluble molecular species – monomers, dimers, trimers, etc. (Haslam, 1989; Porter, 1988, 1989).

The 'monomer' units of procyanidins (R = H) and prodelphinidins (R = OH) are phenolic flavan-3-ols which are linked primarily through respectively their 4 and 8 positions. The stereochemistry at C-2 is most commonly encountered as the 2*R* configuration.

Oligomeric condensed proanthocyanidins have been held (Bate-Smith and Metcalfe, 1957) to be most commonly responsible for the many distinctive properties of plants typically attributed to 'condensed tannins'. Mole (1994) has recently commented on some of this earlier work and has suggested that fewer plant families are characterised by the presence of 'tannins' than heretofore thought. In the context of the later discussion on the polymeric proan-thocyanidins it is pertinent to point out that, simply on the basis of solubility differences, Sir Robert and Lady Robinson in the 1930s (Robinson and Robinson, 1933, 1935) originally subdivided the leucoanthocyanins (condensed proanthocyanidins) into the three classes indicated below:

(i) those that are insoluble in water and the usual organic solvents or give only colloidal solutions,

(ii) those readily soluble in water but not readily extracted therefrom by means of ethyl acetate, and

(iii) those capable of extraction from aqueous solution by ethyl acetate.

In so far as the total complement of condensed proanthocyanidins (procyanidins and prodelphinidins) found in plant tissues is concerned the soluble oligomeric forms (monomers, dimers, trimers ...) are in metabolic terms but the 'tip of the iceberg'. According to the Robinson's classification they represent category (iii) above. For the generality of plants it is now quite clear that condensed proanthocyanidins which fall within the two other categories (i and ii) invariably strongly predominate over the more freely soluble forms. They are, metaphorically speaking, the base of the 'metabolic iceberg'. Indeed in the tissues of some plants such as ferns and fruit such as the persimmon (*Diospyros kaki*) there is an overwhelming preponderance of these forms. They are also of frequent occurrence in plant gums and exudates.

(2) *galloyl and hexahydroxydiphenoyl esters and their derivatives.* These metabolites are almost invariably found as multiple esters with D-glucose (Haslam, 1989; Okuda *et al.*, 1989, 1990; Haslam and Cai, 1994) and a great many can be envisaged as derived from the key biosynthetic intermediate β-1,2,3,4,6-pentagalloyl-D-glucose. Derivatives of hexahydroxydiphenic acid are assumed to be formed by oxidative coupling of vicinal galloyl ester groups in a galloyl D-glucose ester (Schmidt and Mayer, 1956); dependent upon the positions on the D-glucopyranose ring between which oxidative coupling of the galloyl ester groups takes place then a particular chirality is induced in the twisted biphenyl of the resultant hexahydroxydiphenoyl ester (Haslam *et al.*, 1982a,b). Acid hydrolysis of hexahydroxydiphenoyl esters gives rise to the formation of the bis-lactone of hexahydroxydiphenic acid, the planar and extremely difficultly soluble ellagic acid from which the characteristic nomenclature of ellagitannins is derived.

bis-galloyl ester   hexahydroxydiphenoyl ester

ellagic acid

Gallic acid is most frequently encountered in plants in ester form. These may be classified into several broad categories:

(i) simple esters;

(ii) depside metabolites (*syn* gallotannins);

(iii) hexahydroxydiphenoyl and dehydrohexahydroxydiphenoyl esters (*syn* ellagitannins) based upon:

    (a) $^4C_1$ conformation of D-glucose,

    (b) $^1C_4$ conformation of D-glucose,

    (c) 'open-chain' derivatives of D-glucose;

(iv) 'dimers' and 'higher oligomers' formed by oxidative coupling of 'monomers', principally those of class (iii) above.

(3) *phlorotannins*. More recently a third, and relatively minor class of natural polyphenol, has been recognised – the phlorotannins. They are isolated from several genera of red-brown algae and their structures are seen to be composed entirely of phloroglucinol sub-units linked by C–C and C–O chemical bonds. Presumably they are formed biosynthetically by oxidative C–C and C–O coupling of phloroglucinol nuclei. Chemical investigations of these polyphenols have revealed a variety of low, intermediate and high molecular weight compounds. The low molecular weight fraction can be separated into its individual components by chromatography; a typical low molecular weight phlorotannin is fucofureckol, obtained from *Eisenia arborea* (Glombitza and Gerstberger, 1985).

fucofureckol

Although these low molecular weight polyphenols have received the greatest attention they often constitute only a minor proportion of the polyphenolic fraction from red algae – the remainder are composed of oligomers ($M_R \sim 10^3$ – 10^4) and polymers ($M_R > 10^4$) based similarly upon the phloroglucinol building unit (Ragan, 1985; McInnes *et al.*, 1985). Detailed chemical and spectroscopic analysis of the polymer from the marine brown alga *Fucus vesiculosus* suggests (Ragan, 1985; McInnes *et al.*, 1985) that it is highly branched with $\sim 20\%$ of the constituent units being at chain termini. There is no evidence for the presence in the polymer of large rings of phloroglucinol nuclei. The interior backbone consists predominantly of ether-linked phloroglucinol units. Based upon a comparative survey of analogous phloroglucinol polymers derived from other red-brown algae, McInnes and his colleagues (1985) concluded that they are formed biosynthetically by a *non-random* assembly of sub-units.

1.3 Condensed proanthocyanidins

Real and substantial progress in the chemistry of the proanthocyanidins began to be made in the 1960s following the pioneering work of Klaus Weinges and his collaborators in Heidelberg (Weinges *et al.*, 1968, 1969, 1970). It is worth briefly underlining the seminal nature of the observations of the Heidelberg school, made some 25 years ago, particularly since one of the most common events in natural product chemistry today is probably the re-invention of the wheel! This group were the first to isolate and fully characterise the four principal procyanidin dimers (B-1, B-2, B-3 and B-4) as their peracetates. They assigned, largely from the ^{1}H NMR spectra, the correct structures and pointed out the very strong possibility that the procyanidins exhibited the phenomenon of restricted rotation about the interflavan bond.

As techniques and strategies for the separation and isolation of plant proanthocyanidins developed then so did work on their structure, chemistry and biosynthesis. These researches have been regularly reviewed particularly since 1975; most recently by Porter (1988, 1989) and Haslam (1989). These articles

give a detailed and comprehensive summary of flavans and proanthocyanidins
– isolation and detection, analysis, nomenclature, a comprehensive register of
plant sources and botanical distribution, biosynthesis, biomimetic synthesis,
chemistry and conformational characteristics. Other reviews which have recently been published deal with chemical transformations and conformational
features of this group of natural products (Ferriera *et al.*, 1992) and, in a text
designed principally to provide information and methods for the isolation and
analysis of plant phenolics for those involved in ecological work, Waterman
and Mole (1994) provide a very useful and comprehensive survey of the
'hands-on' methods appropriate for examining proanthocyanidins in plant
materials. Salient features of the chemistry and biochemistry are summarised
below; particular attention is also given to procyanidins from fresh plant
tissues and to the recent developments in this field.

1.3.1 Flavan-3-ols and dimeric procyanidins – structure and stereochemistry

Initial studies of the proanthocyanidins concentrated upon those found in
fresh plant tissues (leaves, fruit, fruit pods, seeds and the seed shells) and this led
to the isolation of the four principal dimeric procyanidins (B-1, B-2, B-3 and
B-4) as their peracetates by Weinges and his colleagues and as their free
phenolic forms by Haslam and his group (Thompson *et al.*, 1972). The simple
but trivial nomenclature introduced by these groups (B-1, B-2, B-3 and B-4 and
A-1 and A-2 etc.) has proved itself possessed of a very long shelf life. However
as investigations of proanthocyanidins developed over the past two decades
the isolation of further proanthocyanidin dimers, oligomers (often with mixed
oxidation patterns in the rings 'B' of the flavan-3-ol sub-unit) and polymers
underlined the need for an acceptable and usable but simple form of nomenclature for this group of compounds. Porter and his colleagues (Porter *et al.*, 1982)
met this requirement with a system which is at the same time both neat and
attractive. Details of this form of nomenclature are described in the original
reference and in articles by Porter (1988, 1989). The four principal procyanidin
dimers thus become:

> Procyanidin B-1; epicatechin $(4\beta \rightarrow 8)$ – catechin
> Procyanidin B-2; epicatechin $(4\beta \rightarrow 8)$ – epicatechin
> Procyanidin B-3; catechin $(4\alpha \rightarrow 8)$ – catechin
> Procyanidin B-4; catechin $(4\alpha \rightarrow 8)$ – epicatechin

Other proanthocyanidins are similarly based upon the structures of other
naturally occurring phenolic flavan-3-ols, such as gallocatechin and epigallocatechin, afzelechin and epiafzelechin, fisetinidol and robinetinidol (Haslam,

1989; Porter 1988) with which they almost invariably co-occur. The most widely distributed members of this class of phenolic flavan-3-ols are the diastereoisomeric pair (+)-catechin and (−) epicatechin. Their 3′,4′,5′-trihydroxy 'B' ring analogues, (+)-gallocatechin and (−)-epigallocatechin have a distribution which parallels that of myricetin and are found especially in more primitive plants such as the Coniferae. Resorcinol based 'A' ring flavan-3-ols, such as fisetinidol and robinetinidol are generally confined to the Leguminosae and Anacardiaceae.

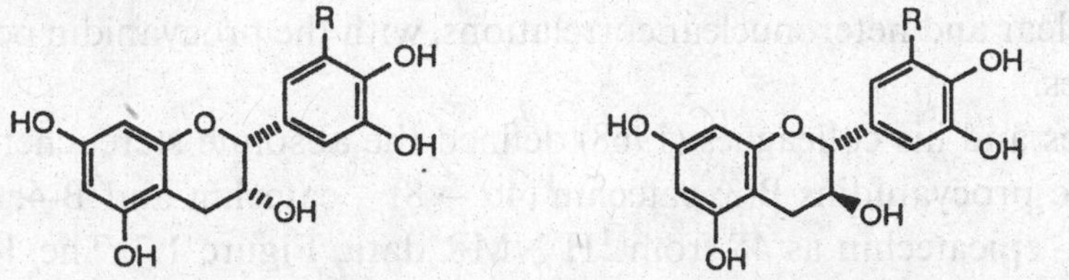

R = H ; (-)-epicatechin
R = OH ; (-)-epigallocatechin

R = H ; (+)-catechin
R = OH ; (+)-gallocatechin

Flavan-3-ols, such as ent-catechin and ent-epicatechin, with the 2*S* configuration, occur relatively rarely. However the presence of ent-epicatechin is widespread in the Palmae (Porter, 1988; Marini-Bettolo *et al.*, 1971, 1972). Nevertheless the particularly facile epimerisation at C-2 of phenolic flavan-3-ols in hot aqueous solution should always be critically borne in mind when considering the natural occurrence of these epimers. Freudenberg (Freudenberg *et al.*, 1922; Freudenberg and Purrmann, 1924) summarised the interconversions for (+)-catechin and (−)-epicatechin as indicated in Figure 1.6.

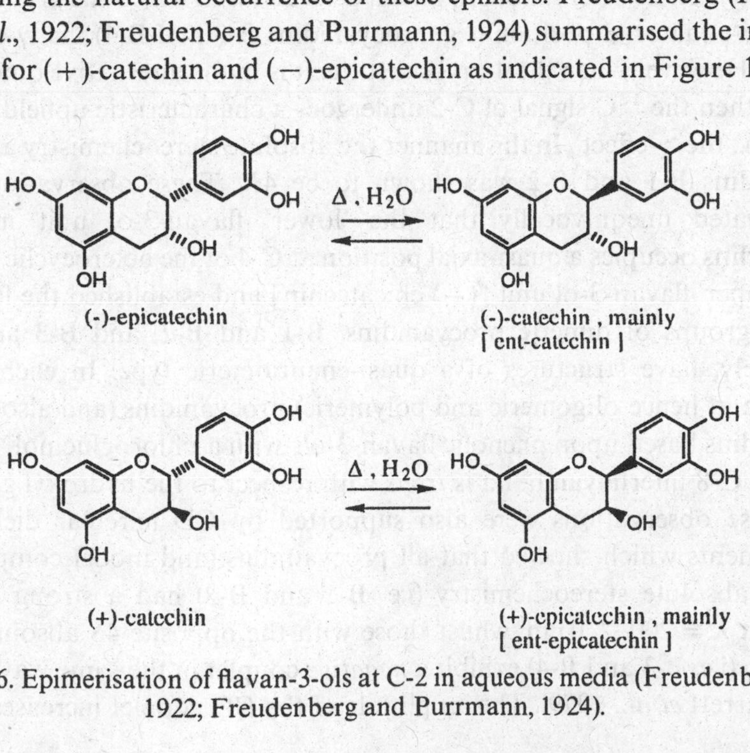

(-)-epicatechin

$\Delta . H_2O$

(-)-catechin , mainly
[ent-catechin]

(+)-catechin

Δ , H_2O

(+)-epicatechin , mainly
[ent-epicatechin]

Figure 1.6. Epimerisation of flavan-3-ols at C-2 in aqueous media (Freudenberg *et al.*, 1922; Freudenberg and Purrmann, 1924).

Although the structures originally proposed for the four principal procyanidin dimers were accepted for over 25 years, unequivocal proof that the interflavan linkage in these dimers joined C-4 of the 'upper flavan-3-ol unit' and C-8 of the 'lower flavan-3-ol unit' was not obtained until very recently. Many plausible chemically based arguments were deployed to support the C-4 to C-8 linkage but it was finally left to Balas and Vercauteren (1994) to elegantly demonstrate that this is in fact the case. This was achieved using high resolution NMR techniques currently available, in particular long range homonuclear and heteronuclear correlations, with the procyanidin peracetate derivatives.

Weinges and his colleagues (1968) defined the absolute stereochemistry at C-4 in the procyanidins B-3; catechin ($4\alpha \rightarrow 8$) – catechin and B-4; catechin ($4\alpha \rightarrow 8$) – epicatechin as $4S$ from ^{1}H NMR data, Figure 1.7. The 'lower' or 'terminal' flavan-3-ol unit in these dimeric procyanidins thus occupies a quasi-equatorial position at C-4 on the heterocyclic 'C' ring of the 'upper' flavan-3-ol [(+)-catechin] unit. The absolute stereochemistry in the procyanidins B-1; epicatechin ($4\beta \rightarrow 8$) – catechin and B-2; epicatechin ($4\beta \rightarrow 8$) – epicatechin was determined as $4R$ in later work using ^{13}C NMR (Haslam *et al.*, 1977), Figure 1.7. These observations were based on the so-called 'γ-effect'; when an aryl group is substituted in C-4 in a flavan its effect on the ^{13}C chemical shift of C-2 is dependent on its orientation. In flavan derivatives in which the aryl group is in a quasi-equatorial position (e.g. procyanidins B-3 and B-4) the 'γ-effect' is minimal. However when the 4-aryl substituent is in the alternative quasi-axial position then the ^{13}C signal of C-2 undergoes a characteristic upfield shift of 4–5 p.p.m., the 'γ-effect'. In this manner the absolute stereochemistry at C-4 in procyanidins B-1 and B-2 was shown to be $4R$. These observations thus demonstrated unequivocally that the 'lower' flavan-3-ol unit in these procyanidins occupies a quasi-axial position at C-4 of the heterocyclic 'C' ring of the 'upper' flavan-3-ol unit [(−)-epicatechin] and established the fact that the two groups of dimeric procyanidins, B-1 and B-2, and B-3 and B-4, respectively, have structures of a quasi-enantiomeric type. In each of the dimeric (and hence oligomeric and polymeric) procyanidins (and also proan-thocyanidins based upon phenolic flavan-3-ols with a phloroglucinol 'A' ring) the C-4 – C-8 interflavan bond is *trans* with respect to the hydroxyl group at C-3. These observations were also supported by CD (circular dichroism) measurements which showed that all procyanidins (and model compounds) with $4R$ absolute stereochemistry (i.e. B-1 and B-2) had a strong *positive* couplet at $\lambda = 200$–220 nm, whilst those with the opposite $4S$ absolute con-figuration (i.e. B-3 and B-4) exhibit a *negative* couplet in the same wavelength range (Barrett *et al.*, 1979). The amplitude of the CD couplet increases as the

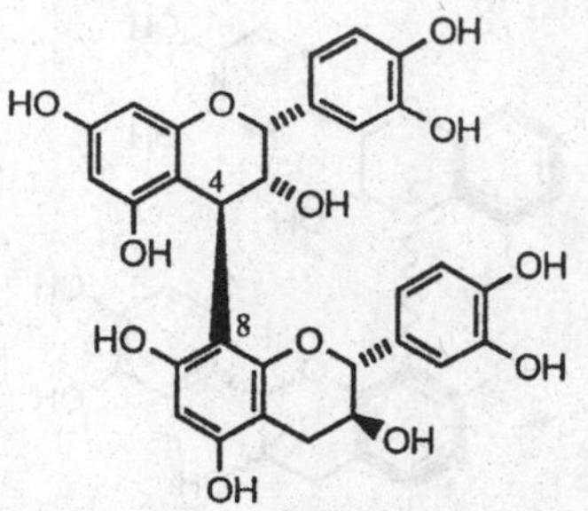

procyanidin B-1

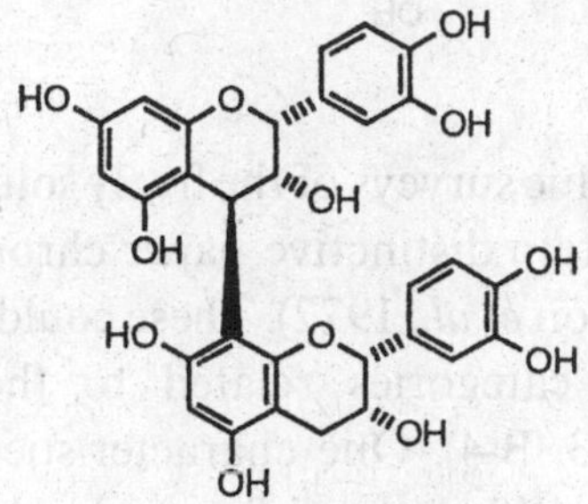

procyanidin B-2

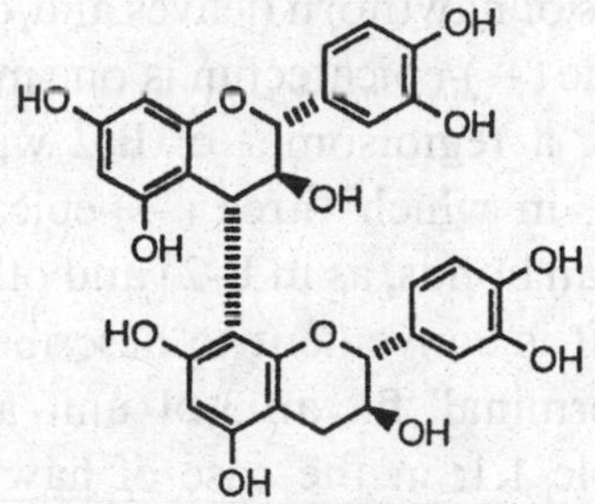

procyanidin B-3

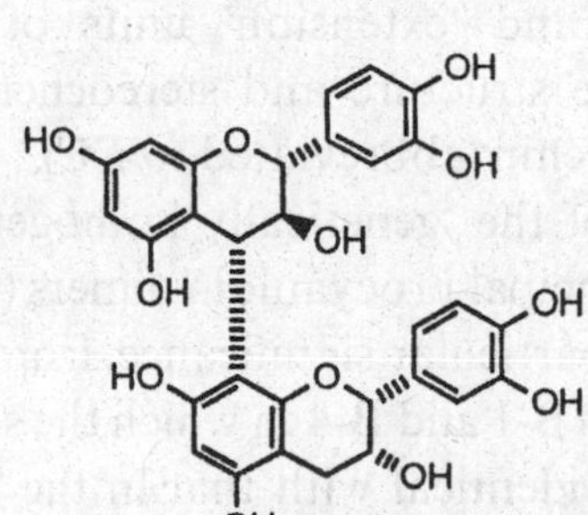

procyanidin B-4

Figure 1.7. Structure and absolute stereochemistry of principal procyanidin dimers found in plant tissues (Weinges *et al.*, 1968; Haslam *et al.*, 1977; Balas and Vercauteren, 1994).

degree of oligomerisation of the procyanidin increases. The origin of these distinctive CD couplets is believed to derive from the interaction of two aryl chromophores – the two 'phloroglucinol A' rings in the procyanidin dimers – and is a reflection of the absolute configuration at the terminus of the inter-flavan bond at C-4.

'A' rings

The initial paper chromatographic surveys of the freely soluble procyanidins in fresh plant tissues revealed certain distinctive paper chromatographic 'fingerprints' (Haslam, 1977; Thompson *et al.*, 1972). These could be classified into four 'genetically homogeneous' categories related to the four principal procyanidin dimers (B-1, B-2, B-3, B-4). One characteristic 'fingerprint' was that displayed by the procyanidins of hawthorn (leaves and/or fruit *Cratageus monogyna*). Co-occurring alongside (−)-epicatechin is one major procyanidin dimer (B-2), a minor dimer (B-5, a regioisomer of B-2 with a C-4 to C-6 interflavan bond), a trimer [C-1, in which three (−)-epicatechin units are linked by two C-4 to C-8 interflavan bonds, as in B-2] and other (quantitatively minor) oligomers, Figure 1.8. It is convenient to describe these and other procyanidins in terms of the 'terminal' flavan-3-ol unit and similarly the 'extension' flavan-3-ol units, Table 1.1; in the case of hawthorn (*Crataegus monogyna*) both the 'terminal' and 'extension' units of the constituent procyanidins thus have the same structure and stereochemistry as the co-occurring flavan-3-ol, (−)-epicatechin (abbreviated to **EC**).

Other characteristic features of the 'genetically homogeneous' categories related to the three remaining principal procyanidin dimers (B-1, B-3 and B-4) are summarised in Table 1.1. Of particular significance, from the biosynthetic point of view, are the procyanidins B-1 and B-4 in which the stereochemistry at C-3 of the 'extension' unit is not identical with that in the 'terminal' unit. It should also be noted that the observance of these highly characteristic procyanidin fingerprints in the plant kingdom is overwhelmingly the exception rather than the rule. The most commonly occurring situation which is likely to be encountered is that typified by the case of the fruit of the mountain cranberry (*Vaccinium vitis idaea*) – namely a variable ratio of the two diastereoisomeric flavan-3-ols [(+)-catechin and (−)-epicatechin] and similarly the co-occurrence of varying quantities of all four possible major dimeric procyanidins (B-1, B-2, B-3 and B-4), Table 1.1. The probable complexities of the procyanidin composition of such plants becomes readily apparent if one

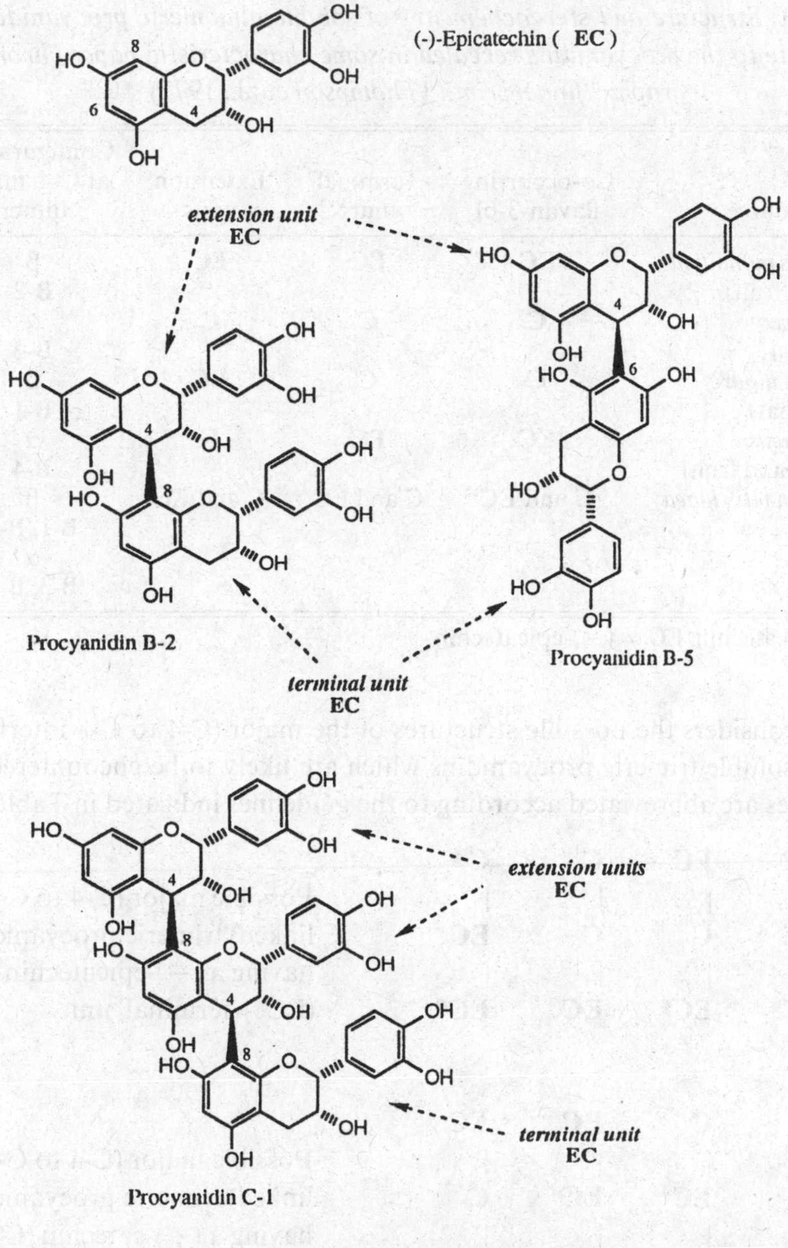

(-)-Epicatechin (EC)

extension unit
EC

extension unit
EC

Procyanidin B-2

Procyanidin B-5

terminal unit
EC

extension units
EC

terminal unit
EC

Procyanidin C-1

Figure 1.8. Structure and stereochemistry of major soluble oligomeric procyanidins of *Crataegus monogyna* (Thompson *et al.*, 1972).

Table 1.1. *Structure and stereochemistry of soluble oligomeric procyanidins as they relate to the procyanidins revealed in some characteristic paper chromatographic 'fingerprints'* (*Thompson* et al., 1972)

Plant source	Co-occurring flavan-3-ol	'Terminal' units	'Extension' units	Configuration at C-4: major dimer
Crataegus monogyna (leaves, fruit)	EC	EC	EC	β B-2
Salix caprea (catkins)	C	C	C	α B-3
Sorghum vulgare (seed coat)	C	C	EC	β B-1
Rubus idaeus (leaves and fruit)	EC	EC	C	α B-4
Vaccinium vitis idaea (fruit)	C and EC	C and EC	C and EC	β B-1, B-2 α B-3, B-4

C = (+)-catechin; EC = (−)-epicatechin.

simply considers the possible structures of the major (C-4 to C-8 interflavan linked) soluble trimeric procyanidins which are likely to be encountered (the structures are abbreviated according to the guidelines indicated in Table 1):

```
EC      EC      C       C
|       |       |       |          Possible major (C-4 to C-8
EC      C       C       EC         linked) trimeric procyanidins
|       |       |       |          having a (−)-epicatechin
EC*     EC*     EC*     EC*        (EC*) 'terminal' unit
```

```
C       C       EC      EC
|       |       |       |          Possible major (C-4 to C-8
C       EC      EC      C          linked) trimeric procyanidins
|       |       |       |          having a (+)-catechin (C*)
C*      C*      C*      C*         'terminal' unit
```

If one also then considers the additional isomers (albeit minor ones) which would ensue if one or more of the interflavan linkages in such trimers were of the C-4 to C-6 type then the complexity of the (soluble) procyanidin composition of such plants is, in principal, readily envisaged: **2** – flavan-3-ols, **8** –

procyanidin dimers and **32** – trimers! The analysis, separation and isolation, and identification of individual components of such extracts becomes a formidable task. In applied work involving such extracts it is important therefore, before embarking, to identify the needs and the goals of the investigation; in this way tedious and time consuming detailed separations may well be obviated.

1.3.2 Proanthocyanidins of the A group

A significant number of proanthocyanidins are now recognised in which there are two linkages (one C–C and one C–O) between two of the phenolic flavan-3-ol units. Typical of these compounds are proanthocyanidins A-1 and A-2 which were the first members of this class to be isolated. Proanthocyanidin A-2 was obtained originally from the seed shells of horse chestnut (*Aesculus hippocastanum, A × carnea*) by Mayer and his colleagues in Heidelberg (Mayer *et al.*, 1966, 1973) and later from avocado seed (*Persea gratissima*; Jacques *et al.*, 1974). It is clearly related formally to procyanidin B-2 [epicatechin $(4\beta \rightarrow 8)$ – epicatechin] but possesses a second interflavan bond between O-7 of the teminal unit and C-2 of the extension unit. It is therefore named as [epicatechin $(2\beta \rightarrow O7; 4\beta \rightarrow 8)$ – epicatechin]. Proanthocyanidin A-1 from mountain cranberry (*Vaccinium vitis idaea*) is similarly related to procyanidin B-1 [epicatechin $(4\beta \rightarrow 8)$ – catechin], and is named as [epicatechin $(2\beta \rightarrow O7; 4\beta \rightarrow 8)$ – catechin] (Weinges *et al.*, 1968). Trimers, tetramers and two pentamers of proanthocyanidins of the A-type have now been isolated from various plant sources (cf. Jacques *et al.*, 1974; Morimoto *et al.*, 1983, 1985, 1988). Doubly bridged structures such as these have much more rigid conformations than their counterparts, the procyanidins of the B class. One result of this relative inflexibility of shape is a much reduced solubility in water; proanthocyanidin A-2, for example, in the crude form is readily crystallised from water but once isolated in a crystalline form is virtually insoluble once again in the same solvent.

1.3.3 Higher oligomeric and polymeric proanthocyanidins

As Sir Robert and Lady Robinson first noted in the 1930s, by far the greater proportion of proanthocyanidins found in plant tissues is invariably in the form of higher oligomers and polymers. In the formal sense all the evidence presently available suggests that proanthocyanidin higher oligomers are built up in the same way as the dimers and trimers discussed above, namely the successive addition of phenolic flavan-3-ol 'extension' units through C-4 to C-8

proanthocyanidin A-1   proanthocyanidin A-2

or C-4 to C-6 interflavan linkages to the growing oligomer chain. However, proanthocyanidin polymers may also show chain branching by occasional and irregular *double* substitution on the flavan-3-ol 'A' ring; a typical polymer structure thus has the general form indicated, Figure 1.9. The structures of proanthocyanidin polymers from over 100 plants have been recorded by Porter (1988). From these data it can be seen that propelargonidin polymers (4'-hydroxyl substitution in the flavan 'B' ring) are exceedingly rare; procyanidin polymers occur widely as do mixed procyanidin/prodelphinidin polymers. In the latter co-polymers the procyanidin units (3',4'-hydroxyl substitution in the flavan 'B' ring) usually predominate. Polymers containing ent-epicatechin and/or ent-epigallocatechin units are relatively rare and are found mainly in monocotyledonous plants. The polymers are polydisperse and it has been suggested that they may contain molecules with degrees of polymerisation ranging from 2 to 50 or more.

Comprehensive studies by Porter and his group have significantly enhanced the level of fundamental knowledge of the soluble polymeric forms of proanthocyanidins which can be isolated from plant tissues (Porter, 1988, 1989; Foo and Porter, 1978, 1980; Czochanska *et al.*, 1980; Hemingway, Foo and Porter, 1982). Soluble proanthocyanidin polymers can be extracted from plants by water–acetone mixtures and refined by subsequent chromatography on Sephadex LH-20 (Jones *et al.*, 1976). The chemical composition and constitution may then be derived by a combination of chemical degradation (acid catalysed fission of the interflavan bonds in the presence of toluene-α-thiol or phloroglucinol), and by ^{13}C NMR spectroscopy (Czochanska *et al.*, 1980; Porter, 1988, 1989; Porter *et al.*, 1982; Haslam, 1989). This procedure, with various modifications as necessary, has been very widely employed to determine the structure of higher proanthocyanidin oligomers and polymers (Hsu,

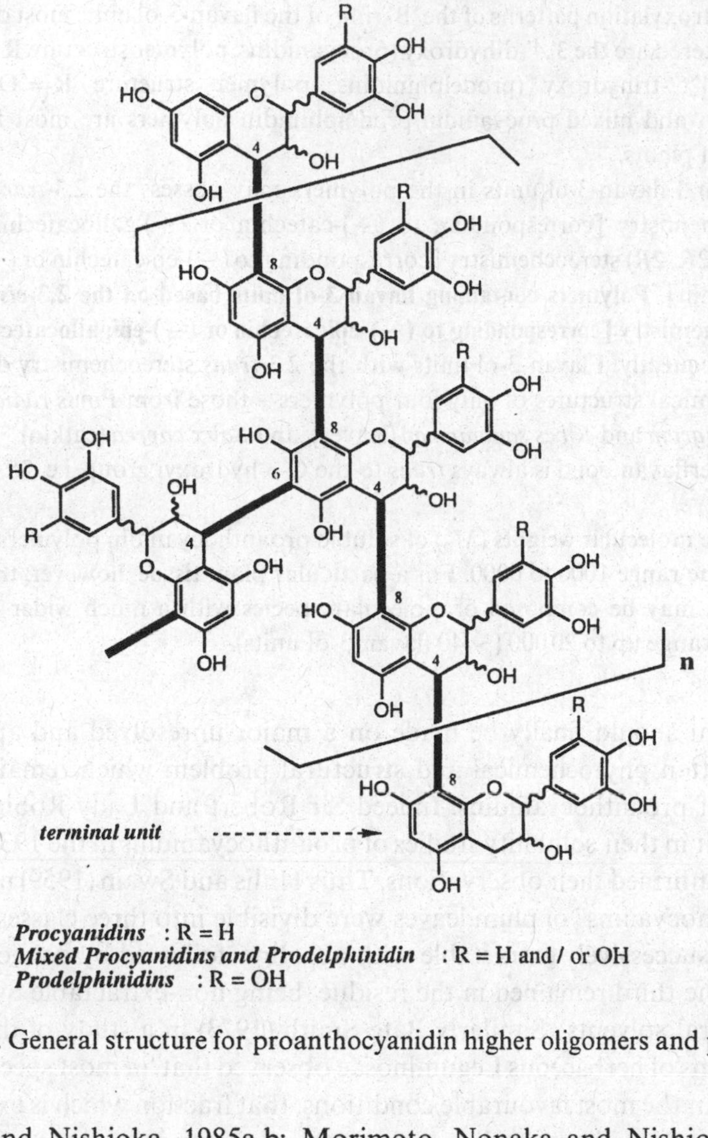

Procyanidins : R = H
Mixed Procyanidins and Prodelphinidin : R = H and / or OH
Prodelphinidins : R = OH

Figure 1.9. General structure for proanthocyanidin higher oligomers and polymers.

Nonaka and Nishioka, 1985a,b: Morimoto, Nonaka and Nishioka, 1985, 1986).

The broad general characteristics relating to the polymeric propelar-gonidins, procyanidins and prodelphinidins isolated from plants which emerge from these and related investigations may be summarised as follows.

(i) Polymeric proanthocyanidins which are soluble are composed entirely of flavan-3-ol units. The majority of interflavan bonds comprise the C-4 to C-8 type but some C-4 to C-6 linkages are probably present and these give rise to a degree of polymer branching.

(ii) The hydroxylation patterns of the 'B' ring of the flavan-3-ol units most commonly encountered are the 3′,4′-dihydroxy (procyanidins, polymer structure R = H) and the 3′,4′,5′-trihydroxy (prodelphinidins, polymer structure R = OH). Procyanidin and mixed procyanidin/prodelphinidin polymers are most frequently found in plants.

(iii) Individual flavan-3-ol units in the polymers may possess the 2,3-*trans* (2*R*, 2*S*) stereochemistry [corresponding to (+)-catechin or (+)-gallocatechin] or the 2,3-*cis* (2*R*, 2*R*) stereochemistry [corresponding to (−)-epicatechin or (−)-epigallocatechin]. **Polymers containing flavan-3-ol units based on the 2,3-*cis* (2*R*, 2*R*) stereochemistry [corresponding to (−)-epicatechin or (−)-epigallocatechin] occur most frequently**. Flavan-3-ol units with the 2,3-*trans* stereochemistry dominated the chemical structures of only four polymers – those from *Pinus radiata* (bark), *Ribes nigrum* and *Ribes sanguineum* (leaves), and *Salix caprea* (catkin).

(iv) The interflavan bond is always *trans* to the C-3 hydroxyl group, i.e. 3*S*, 4*S* or 3*R*, 4*R*.

(v) Average molecular weights (M_R) of soluble proanthocyanidin polymers generally fall in the range 1000 to 6000. For a particular plant tissue, however, the number average may be composed of molecular species with a much wider molecular weight range up to 20 000 ($\sim$ 40 flavan-3-ol units).

Comment should finally be made on a major unresolved and apparently long forgotten phytochemical and structural problem which remains in the province of proanthocyanidins. Indeed Sir Robert and Lady Robinson first alluded to it in their solubility studies of proanthocyanidins in the 1930s. Other workers confirmed their observations. Thus Hillis and Swain (1959) noted that '"leucoanthocyanins" of plum leaves were divisible into three classes, the first two being successively extractable with absolute, followed by aqueous methanol and the third remained in the residue, being non-extractable by these or other neutral solvents'. Similarly Bate-Smith (1973) in a study of the proanthocyanidins of herbaceous Leguminosae observed that 'in most species examined, even in the most favourable conditions, that fraction which is extractable is only a very small proportion of the total proanthocyanidins present'. The same observations have been made time and again in the author's own laboratory. Plant debris which results from repeated extractions of fresh proanthocyanidin-containing tissues with methanol, invariably retains far greater amounts of materials which give the various characteristic colour tests for proanthocyanidins (Porter, 1989) than that which has been extracted with other solvents. Moreover further extraction of the plant debris with boiling DMSO or DMF (both powerful solvating reagents) effects no further extraction of proanthocyanidins. This feature is particularly common in the proanthocyanidins of ferns and the herbaceous Leguminosae. One interpretation of these observations is that these forms of plant proanthocyanidins (just as is the case of the lignins) are covalently bound to an insoluble carbohydrate (or other polymer) matrix within the plant cell (Shen *et al.*, 1986). Heretofore it has been tacitly assumed that when a plant tissue is, for example tasted or sampled in some way, it is the soluble proanthocyanidins which are most important in the development of characteristics such as taste and flavour, which ensue. The role of the insoluble forms is far from clear. The nature, structure and properties of the 'insoluble' proanthocyanidins thus constitutes a major unresolved problem for phytochemists and ecologists in particular.

1.3.4 Biosynthesis of proanthocyanidins

Current lines of thought suggest that the proanthocyanidins are formed, in some way, as by-products of the processes in which the parent flavan-3-ols are biosynthesised (Haslam, 1989). An impressive body of knowledge has been assembled concerning the general pathways of flavonoid biosynthesis (Hahl-brock and Scheel, 1989). The chalcone–dihydroflavone pair (naringenin chalcone–naringenin) are the first recognised intermediates on this pathway but stages from naringenin to the individual classes of flavonoid, remain, in a number of instances, to be fully clarified at the enzymic level (Haslam, 1993). Although Stafford and her colleagues have considerably clarified some steps in the biosynthesis of the phenolic flavan-3-ols (Stafford and Lester, 1984, 1985; Stafford, Lester and Porter, 1985) using cell suspension cultures of Douglas fir needles (*Pseudotsuga menziesii*), Figure 1.10, the question of whether the final assembly of the proanthocyanidins is under enzymic control or not appears still undecided. Stafford and Lester demonstrated the presence of two sequential NADPH dependent reductases which catalyse the reduction of (+)-dihyd-roquercetin [and (+)-dihydromyricetin] to the 2,3-*trans*-flavan-3,4-*cis*-diol, leucocyanidin (and leucodelphinidin) and then to (+)-catechin (and (+)-gallocatechin), Figure 1.10.

However the nature of the final enzymic reduction step is, in chemical terms, problematical – particularly if one bears in mind Sir Robert Robinson's famous dictum that 'enzymes are unlikely to disregard the laws of chemistry'. Similarly the origins of the 2,3-*cis*-flavan-3-ol structure (corresponding to (−)-epicatechin and (−)-epigallocatechin and which is by far the most widely found structural unit in the natural proanthocyanidins), still remain largely shrouded in mystery because of the failure to detect 2,3-*cis*-dihydroquercetin (the required substrate analogue corresponding to (+)-dihydroquercetin) in nature. Various suggestions have been canvassed to circumvent these doubts and problems; that shown in Figure 1.11 speculates that the reaction formally

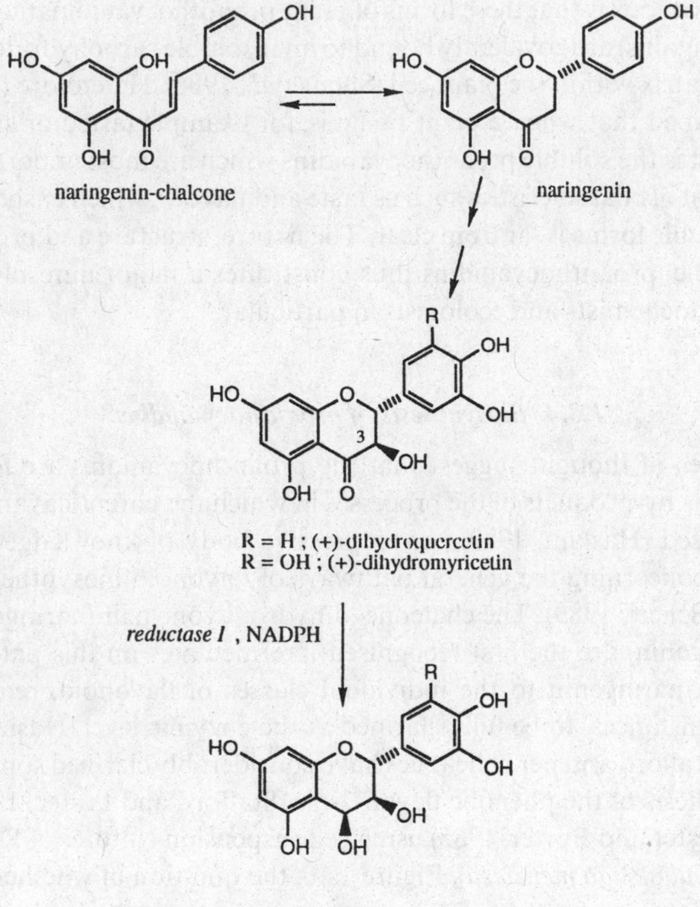

naringenin-chalcone

naringenin

R = H ; (+)-dihydroquercetin
R = OH ; (+)-dihydromyricetin

reductase I , NADPH

reductase II , NADPH

R = H ; (+)-catechin
R = OH ; (+)-gallocatechin

Figure 1.10. Biosynthesis of (+)-catechin and (+)-gallocatechin (Stafford and Lester, 1984, 1985; Stafford *et al.* 1985).

catalysed by the *reductase II* is in fact a two step process – loss of water, *followed by* reduction of the quinone methide (or the equivalent protonated carbocation). Facile inversion of the critical C-3 stereochemistry could then occur at the stage of the activated quinone methide (a vinylogous ketone) intermediate.

Any hypothesis put forward to explain the origins of the proanthocyanidins must, perforce, seek to rationalise a number of distinctive features; principal amongst which are:

(i) the biosynthetic nature and origins of the 'termination' and 'extension' units of proanthocyanidins,
(ii) the stereochemistry of the C-4 to C-8 (or C-4 to C-6) interflavan bond which, in the case of the procyanidins and prodelphinidins, is invariably *trans* in relation to the vicinal C-3 hydroxyl group,
(iii) the formation, in certain plants, of proanthocyanidins in which the nature of the 'extension' units is different in stereochemistry at C-3 in the heterocyclic ring, to the sole flavan-3-ol present, cf. Table 1.1 – *Sorghum vulgare, Rubus idaeus,* and
(iv) the relationship, if any, of proanthocyanidin synthesis to flavan-3-ol biosynthesis and/or anthocyanin biosynthesis.

Current lines of thought suggest that the synthesis of oligomeric proanthocyanidins in plants is intimately involved with the formation of the phenolic flavan-3-ols, Figure 1.12 (Haslam, 1974; Hemingway and Foo, 1983). It has been speculated (Haslam, 1977) that the proanthocyanidins are formed in an oligomerisation process in which the flavan-3-ol first forms a proanthocyanidin dimer by a stereospecific nucleophilic capture at C-4 of the putative quinone methide (or its protonated carbocation equivalent) intermediate in flavan-3-ol biosynthesis. The dimer then captures a further quinone methide (or its protonated carbocation equivalent) to generate a trimer and so progressively by the capture of further quinone methide species oligomers and finally polymers are formed, Figure 1.12. In this scheme the majority of the interflavan bonds are formed C-4 to C-8, although a minority have the C-4 to C-6 orientation; a flavan-3-ol molecule thus contributes to the 'terminal unit' of the proanthocyanidin and the extension units are derived from the intermediate quinone methide (or its protonated carbocation equivalent) species. Clearly in this scenario the flavan-3-ol and quinone methide precursors may have differing stereochemistry at C-3 in the heterocyclic ring and different hydroxyl substitution patterns in ring 'B' of the flavan-3-ol units if they are derived from several different biosynthetic sources, Figure 1.12. The picture presented thus encompasses the formation of dimers such as B-1 and B-4 and the biogenesis of prodelphinidins in which the 'extension units' usually contain a mixture of both tri- and di-hydroxyl substitution in ring 'B'.

This biosynthetic reaction forms the basis of biogenetically patterned chemical syntheses of proanthocyanidins which appear to mimic uniquely the proposed biosynthetic pathway, Figure 1.13. The required carbocation or quinone methide equivalent is thus released, by chemical means, into aqueous media in

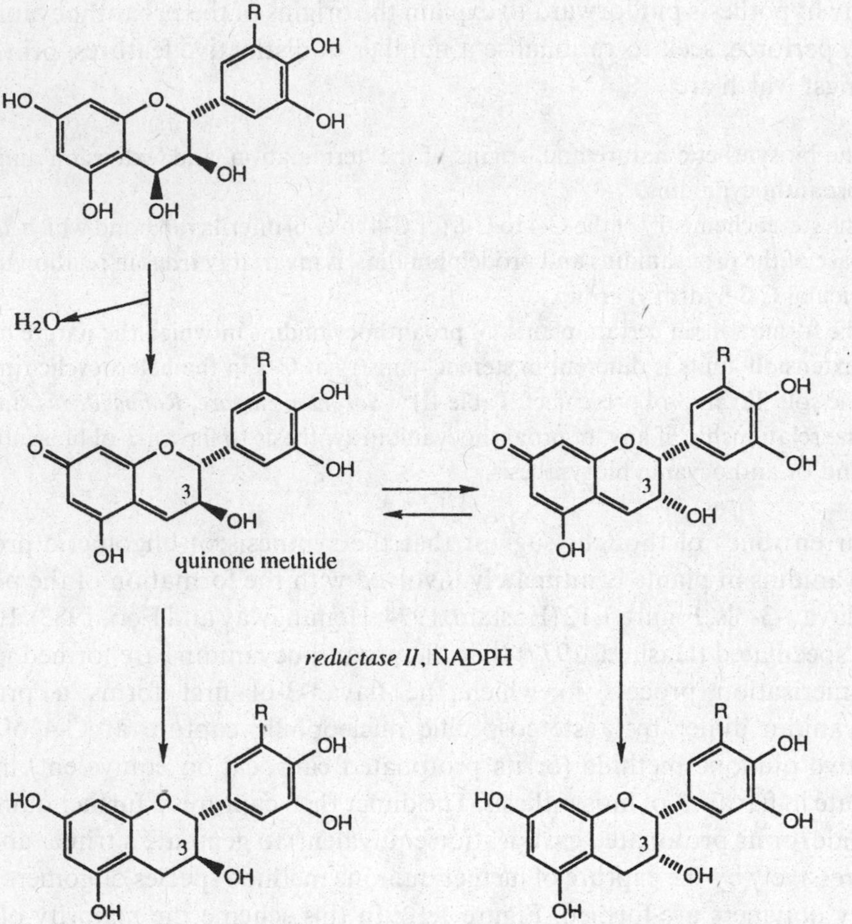

R = H ; (+)-catechin
R = OH ; (+)-gallocatechin

R = H ; (-)-epicatechin
R = OH ; (-)-epigallocatechin

Figure 1.11. Possible route of biogenesis of phenolic flavan-3-ols with 3*R* or 3*S* stereochemistry from a common flavan-3,4-diol intermediate.

the presence of the appropriate flavan-3-ol metabolite (Haslam, 1974). The soluble products derived from such biomimetic syntheses qualitatively and quantitatively match the procyanidins found in particular plant tissues. Thus, for example, reaction between the carbocation (or quinone methide) corresponding in stereochemistry at C-3 to (−)-epicatechin and (−)-epicatechin itself gives a product profile (procyanidins B-2, B-5 and procyanidin C-1, etc.) identical with that occurring in *Crataegus monogyna* (Haslam, 1974, 1977, 1989), Figure 1.8. The biomimetic syntheses are clearly under chemical and thermodynamic control and, in view of the criteria noted above, it is important to point out that, employing flavan-3-ols and quinone methide equivalents

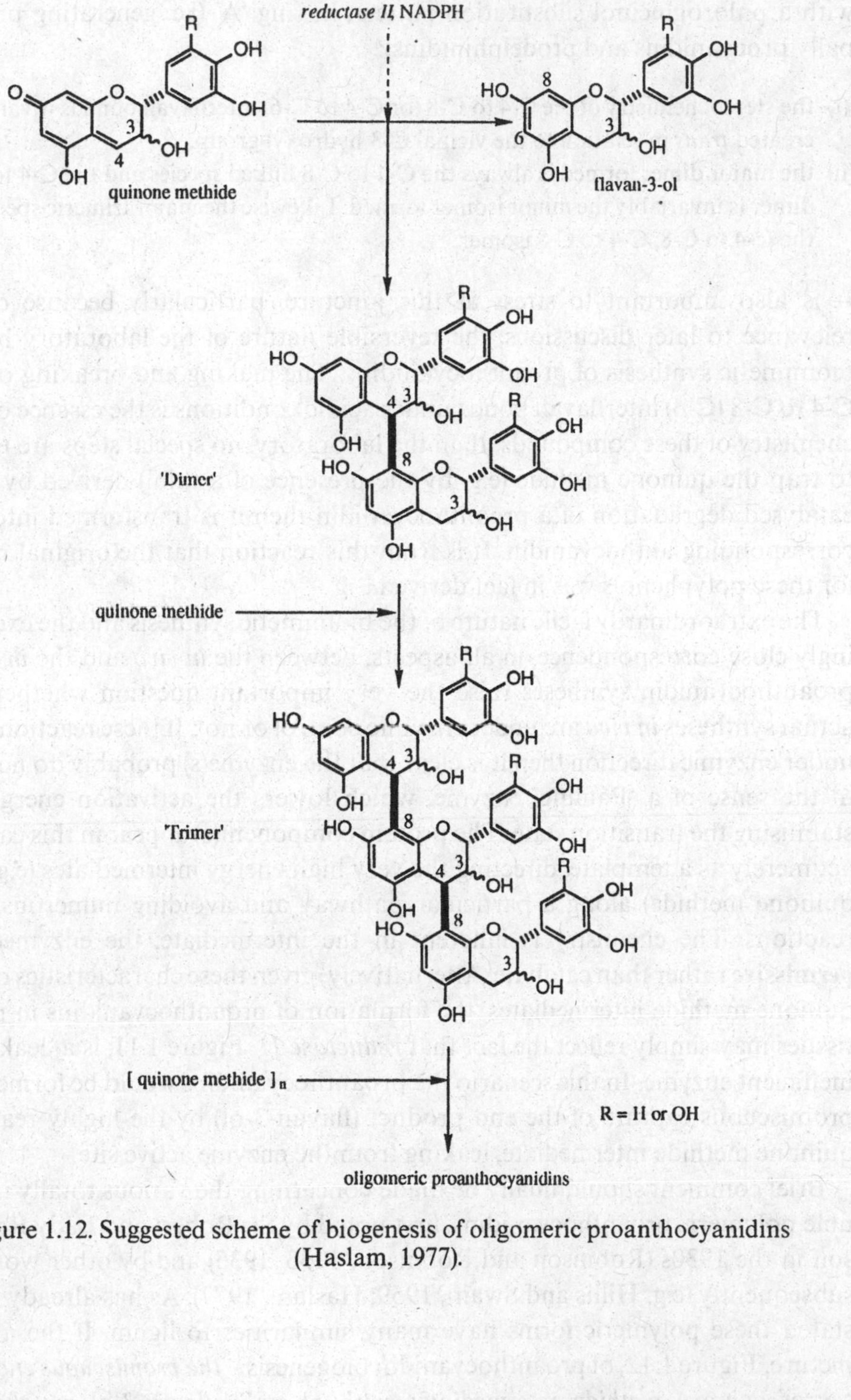

Figure 1.12. Suggested scheme of biogenesis of oligomeric proanthocyanidins (Haslam, 1977).

with a phloroglucinol substitution pattern in ring 'A' (i.e. generating principally procyanidins and prodelphinidins):

(i) the stereochemistry of the C-4 to C-8 (or C-4 to C-6) interflavan bond is invariably created *trans* in relation to the vicinal C-3 hydroxyl group,

(ii) the major dimer formed is always the C-4 to C-8 linked species and the C-4 to C-6 dimer is invariably the minor isomer formed. Likewise the major trimeric species is the C-4 to C-8; C-4 to C-8 isomer.

It is also important to stress at this juncture, particularly because of its relevance to later discussions, the reversible nature of the laboratory based biomimetic synthesis of proanthocyanidins. The making and breaking of the C-4 to C-8 (C-6) interflavan bonds under acidic conditions is the essence of the chemistry of these compounds. If, in the laboratory, no special steps are taken to trap the quinone methide (e.g. by the presence of a thiol) derived by acid catalysed degradation of a proanthocyanidin then it is transformed into the corresponding anthocyanidin. It is from this reaction that the original name for these polyphenols was in fact derived.

The extraordinarily facile nature of the biomimetic synthesis and the exceedingly close correspondence, in all aspects, between the *in vivo* and the *in vitro* proanthocyanidin syntheses raise the very important question whether the actual syntheses *in vivo* are under enzymic control or not. If these reactions are under enzymic direction then it is clear that the enzyme(s) probably do not act in the sense of a 'Pauling' enzyme, which lowers the activation energy by stabilising the transition state. The protein component(s) appear in this case to act merely as a template, directing the very high energy intermediates (e.g. the quinone methide) along a particular pathway and avoiding numerous side reactions. The chemistry is inherent in the intermediate; the enzyme(s) is **permissive** rather than catalytic. Alternatively, given these characteristics of the quinone methide intermediates, the formation of proanthocyanidins in plant tissues may simply reflect the fact that *reductase II*, Figure 1.11, is a 'leaky' or inefficient enzyme. In this scenario the proanthocyanidins would be formed by promiscuous capture of the end product (flavan-3-ol) by the highly reactive quinone methide intermediate, leaking from the enzyme active site.

Brief comment should finally be made concerning the various totally insoluble polymeric proanthocyanidins, first noted by Sir Robert and Lady Robinson in the 1930s (Robinson and Robinson, 1933, 1935) and by other workers subsequently (e.g. Hillis and Swain, 1959; Haslam, 1977). As has already been stated these polymeric forms have many similarities to lignin. If the above picture, Figure 1.12, of proanthocyanidin biogenesis – *the promiscuous encounter of quinone methide intermediates with phenolic flavan-3-ol metabolites*

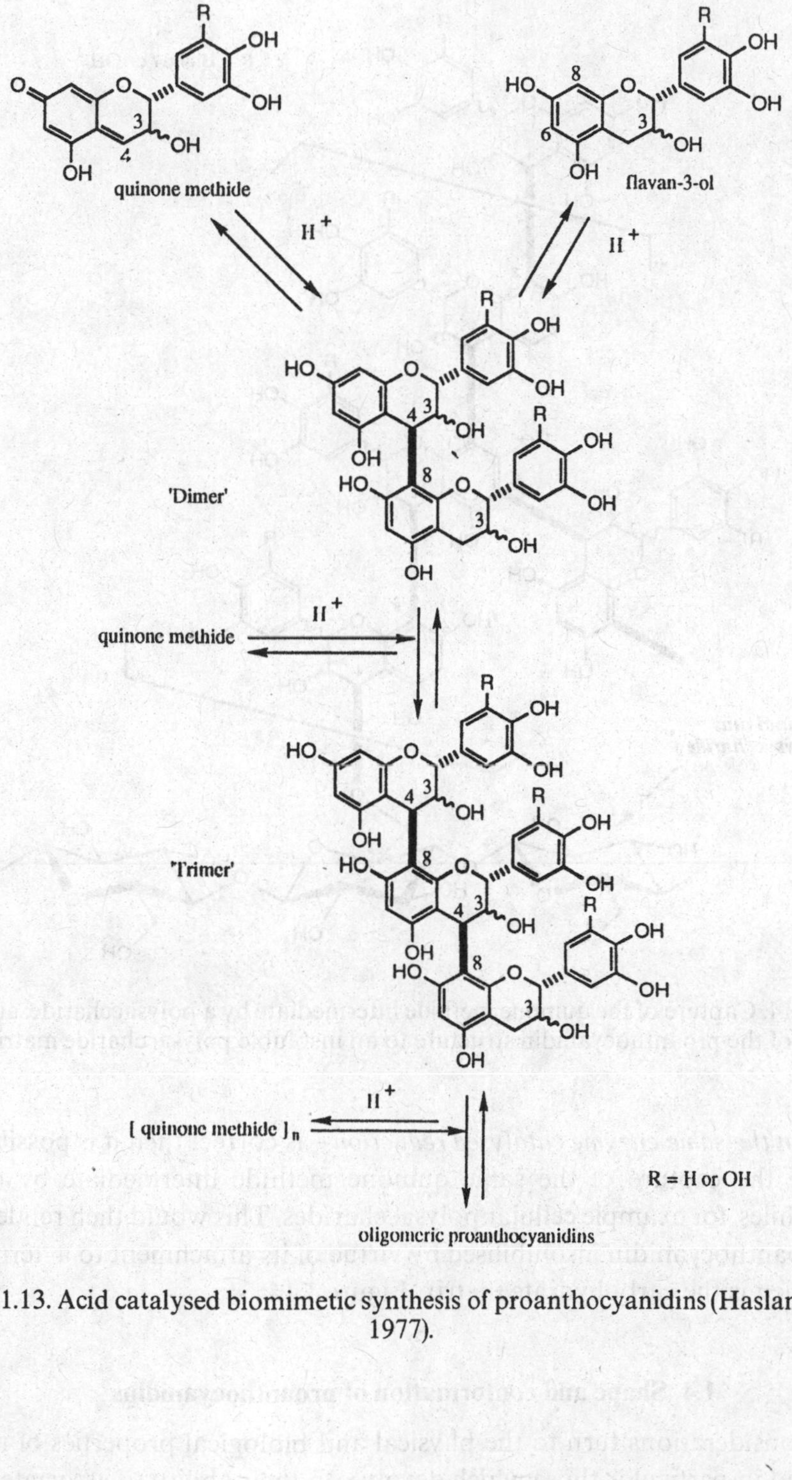

Figure 1.13. Acid catalysed biomimetic synthesis of proanthocyanidins (Haslam, 1974, 1977).

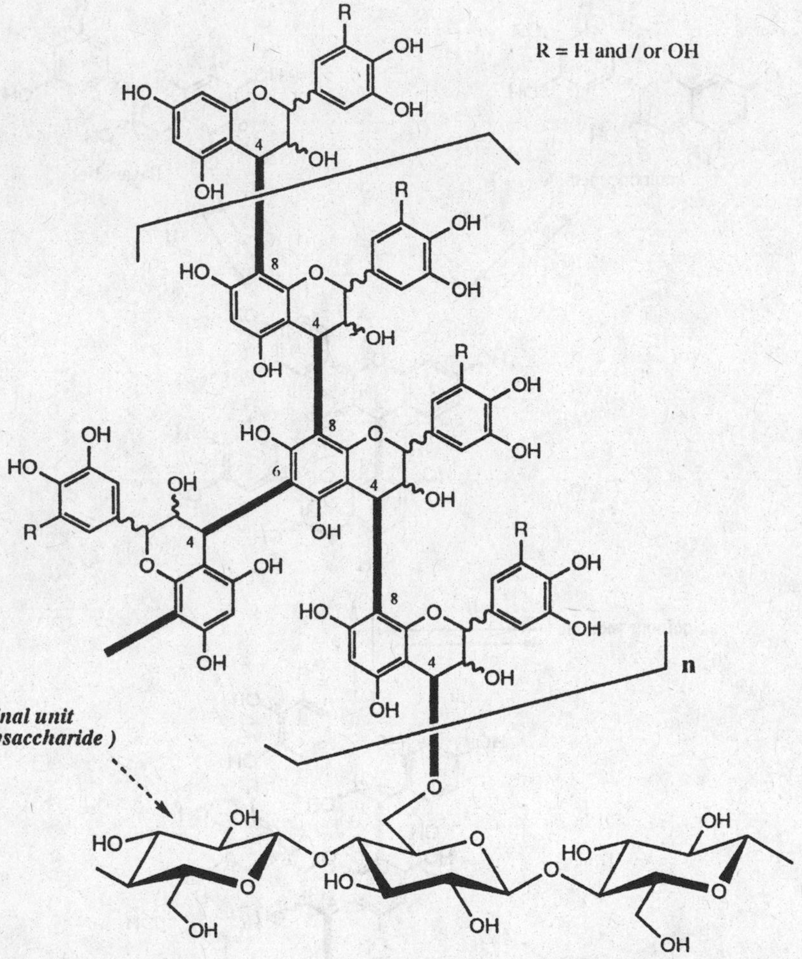

R = H and / or OH

Figure 1.14. Capture of the quinone methide intermediate by a polysaccharide: attachment of the proanthocyanidin structure to an insoluble polysaccharide matrix.

derived in the same enzyme catalysed reduction – is correct then it is possible to envisage the capture of the same quinone methide intermediate by other nucleophiles, for example cellular polysaccharides. This would then render the final proanthocyanidin insolubilised by virtue of its attachment to a 'terminal unit' which is the carbohydrate matrix, Figure 1.14.

1.4 Shape and conformation of proanthocyanidins

When considerations turn to the physical and biological properties of molecules, and in particular those which devolve on their ability to associate with

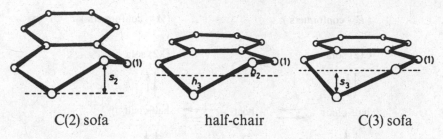

C(2) sofa   half-chair   C(3) sofa

Figure 1.15. Heterocyclic ring 'C' conformations in flavan derivatives (Porter *et al.*, 1986).

other molecules – the property of molecular recognition – attention must be given not only to structure and chemical constitution but also to solvation and to conformational characteristics – the rigidity or flexibility of the shape which these molecules prefer to adopt. Although some progress has been made a great deal remains to be achieved in the area of the proanthocyanidins.

1.4.1 *Flavan-3-ol conformations*

The preferred conformation of the heterocyclic 'C' ring of flavan-3-ols was initially discussed by both Whalley (1962) and by Clark-Lewis, Jackman and Spotswood (1964). The half-chair and the C(2) sofa conformations, Figure 1.15, received particular attention by these groups. The half-chair represents a flattened chair with the oxygen atom, the aromatic ring and C-4 in one plane. True axial and equatorial bonds are found only at C-2 and C-3. The bonds at C-4 are often referred to as pseudo-axial and pseudo-equatorial. The sofa conformation is a modification of the half-chair where C-2 (or C-3) is co-planar with C-4 and the oxygen atom of the flavan ring (Philbin and Wheeler, 1958). Using Drieding models, ^{1}H NMR coupling constants and the Karplus equation, Clark-Lewis concluded that for 2,3-*trans*-flavan-3-ols conformations in which the 2-aryl ring is quasi-equatorial are strongly favoured, but that no clear distinction can be made between the idealised half-chair and C(2) sofa conformations. The data for the 2,3-*cis* compounds suggested that the half-chair and C(2) sofa conformations in which the C-2 aryl group is in a quasi-equatorial position are probably distorted or that the conformer population distribution contained significant proportions of those conformations in which the C-2 aryl group is disposed axially.

Later workers have turned to this problem. Porter and his colleagues (Porter *et al.*, 1986) used empirical modifications of the Karplus equation which, it was suggested, permitted a more accurate correlation of coupling constants (derived from NMR) with heterocyclic ring conformation. They

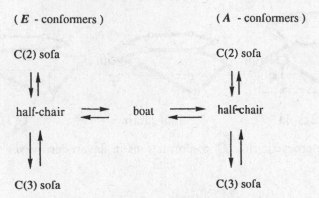

Figure 1.16. Heterocyclic 'C' ring conformer equilibria in flavan-3-ol derivatives (Porter *et al.*, 1986).

concluded that the C-ring flavan conformation could be described in terms of the following equilibria, Figure 1.16, where the *E*- and *A*-conformers are those with the phenolic 'B'-ring at C-2 equatorial or axial respectively.

The boat conformation is the high energy intermediate in the conversion of the *E*- and *A*-conformers; the energy barrier for this conversion is ~ 5–6 kcal mol^{-1} in enthalpy units. An unequal conformational energy for the *E*- and *A*-conformers is manifested in an unequal population of the two states, the one with the lower energy being populated to a greater extent. On the basis of very low temperature ^{1}H NMR spectrometry they showed that both (+)-catechin and (−)-epicatechin undergo conformational exchange in solution and the *E*-: *A*-conformer ratio was 62:38 for the former and 86:14 for the latter.

Crystal structures have been reported for several derivatives of (+)-catechin and (−)-epicatechin and these indicate that the heterocyclic ring may adopt a range of conformations within the framework described by Porter and his collegues. These have been discussed by Mattice and Viswanadhan (1987) and by Haslam (1989). Perhaps the most unusual, and *a priori* unexpected, result is that observed for the penta-acetyl derivative of (+)-catechin (Mattice *et al.*, 1985), which shows both the substituents at C-2 (bisacetoxyphenyl) and C-3 (acetoxy) occupy axial positions on the heterocyclic ring. However only one X-ray crystal structure analysis has been reported for the free phenolic form of a flavan-3-ol (or indeed proanthocyanidin) and this is for (−) epicatechin (Mattice *et al.*, 1984; Spek *et al.*, 1984). No crystal structure is yet available for (+)-catechin. In the structure of (−)-epicatechin, Figure 1.17, the conformation of the heterocyclic ring is described as a half-chair that is distorted towards a C(3) sofa; atoms C-2 and C-3 are respectively 26.3 pm above and 49.5 pm below the mean plane of the 'A' ring. Four of the

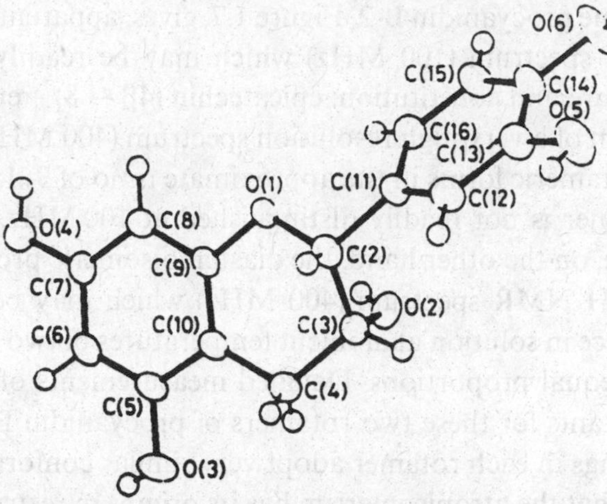

Figure 1.17. Structure of (−)-epicatechin in the crystalline state (Mattice *et al.*, 1984).

five hydroxyl groups participate in intermolecular hydrogen bonds. The conformation of the heterocyclic ring places the dihydroxyphenyl ring 'B' in a pseudo-equatorial position at C-2. Conformational energy calculations performed using the MM2 computer program correctly predict that the heterocyclic ring is a half-chair with distortion towards a C(3) sofa. It is also predicted that (+)-catechin should possess a heterocyclic ring conformation virtually identical to that of (−)-epicatechin. Optimisation by MM2 always favours a heterocyclic ring conformation which places the dihydroxyphenyl ring 'B' in a pseudo-equatorial position at C-2 [as observed in the X-ray analysis of (−)-epicatechin]. Knowledge of the shape and conformation of the various proanthocyanidin dimers and higher oligomers is based largely upon an extrapolation of the data derived for the parent flavan-3-ols, high resolution NMR and MM2 calculations.

1.4.2 *¹H NMR spectroscopy of procyanidin dimers*

Despite the very successful application of NMR methods to facets of the structural and stereochemical properies of the procyanidins, persistent anomalous features of their ¹H NMR spectra (e.g. Thompson *et al.*, 1972) led to speculation by several groups (Jurd and Lundin, 1968; Weinges *et al.*, 1970; du Preez *et al.*, 1971; Thompson *et al.*, 1972) that these natural products exhibited atropisomerism and that, in the case of the dimeric procyanidins, this phenomenon was most probably due to restricted rotation about the C-4 to C-8 interflavan bond. At ambient temperatures the behaviour is solvent dependent and not entirely predictable.

Thus in d_6 acetone procyanidin B-2, Figure 1.7, gives, apparently, a straight-forward ^1H NMR spectrum (100 MHz) which may be readily analysed in terms of the structure and constitution: epicatechin ($4\beta \rightarrow 8$) – epicatechin. In fact close inspection of a very high resolution spectrum (400 MHz) reveals the presence of two rotameric forms in the approximate ratio of 9:1; the presence of the minor rotamer is not readily distinguished at 100 MHz. In the same solvent, d_6 acetone, on the other hand, the diastereoisomeric procyanidin B-3 gives a complex ^1H NMR spectrum (400 MHz) which may be analysed in terms of the presence in solution at ambient temperatures of two atropisomers in approximately equal proportions. Detailed measurements of proton–proton coupling constants for these two rotamers of procyanidin B-3 show that the heterocyclic rings in each rotamer adopt very similar conformations. This strongly suggests that the atropisomerism has its origins in restricted rotation about the interflavan bond and not fluxional changes in the heterocyclic rings.

In the solvent deuterium oxide (D_2O) the properties are reversed. Procyanidin B-3 gives an apparently straightforward first order spectrum ($\sim 95\%$ one rotamer) which can be readily interpreted in terms of the structure: catechin ($4\alpha \rightarrow 8$) – catechin. Conversely, the diastereoisomer procyanidin B-2 exhibits a complex spectrum which is likewise most satisfactorily interpreted in terms of the presence of two major atropisomers (rotamers) where, at the temperature of measurement, the individual proton signals are approaching coalescence.

A comprehensive study of the atropisomerism of procyanidin dimers was carried out (Haslam *et al.*, 1977) and interpreted in terms of two different forms of restricted rotation about the C-4 to C-8 interflavan bond, Figure 1.18, associated with the two distinctive groups of procyanidin dimers – namely those with a C-4α configuration (procyanidins B-3 and B-4) and those with a C-4β configuration (procyanidins B-1 and B-2). Molecular modelling suggested that in the case of procyanidins B-1 and B-2 the restricted rotation originates primarily from the steric interactions between the proton at C-2 and the aromatic π-system of ring 'A' of the 'upper/extension' flavan-3-ol unit and the substituents ortho to the interflavan bond of the 'lower/terminal' flavan-3-ol unit. Conversely the origins of the restricted rotation in procyanidins B-3 and B-4 have several analogies and molecular modelling shows that the oxygen substituents at C-3 and C-5 in the 'upper/extension' flavan-3-ol unit and those in the ortho position to the interflavan bond in the 'lower/terminal' flavan-3-ol unit are primarily responsible for the steric interference.

Typical values determined for ΔG_{rot} = (free energy of activation process) were in the region 15 to 20 kcal mol^{-1}. These results suggest that, although the procyanidin dimers B-1 through to B-8 and their derivatives exhibit atropisomerism, under the conditions of ^1H NMR experimentation, the energy

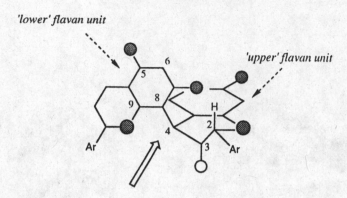

procyanidin B-2

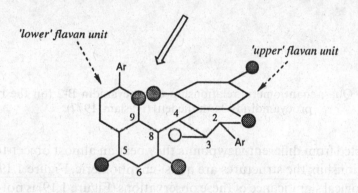

procyanidin B-3

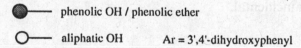

— phenolic OH / phenolic ether

— aliphatic OH   Ar = 3',4'-dihydroxyphenyl

Figure 1.18. Diagrammatic representation of the two forms of restricted rotation about the C-4 to C-8 interflavan bond and the preferred conformations (as determined from molecular models) of procyanidin dimers. Solid arrows indicate viewpoint necessary to visualise quasi-enantiomeric relationship of the procyanidins (Haslam, 1977).

barriers are too small to permit the isolation of the different conformational forms of the procyanidins as separate entities. However, although molecular models show that a number of different conformations about the interflavan bond are accessible, one conformation may well be preferred, Figure 1.18. An interesting feature of the preferred conformations of dimers with the 4α (e.g. procyanidin B-3) and those with the 4β (e.g. procyanidin B-2) configuration is

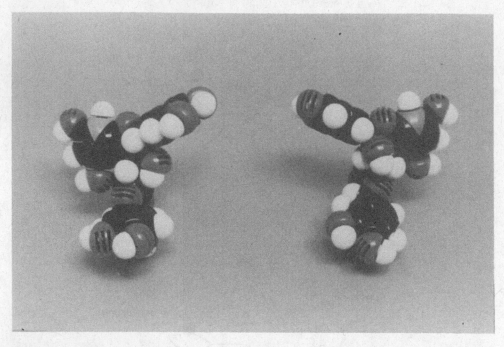

Figure 1.19. Quasi-enantiomeric relationship of procyanidin B-2 (on the right) and procyanidin B-3 (on the left) (Haslam, 1977).

that, inspected from different viewpoints they bear an almost object-to-mirror image relationship; the structures are quasi-enantiomeric, Figure 1.19.

The biological significance of these observations (Figure 1.19) is not yet clear and the relationship described may simply represent a fortuitous assembly of different structural and stereochemical parameters; structural relationships which are merely coincidental.

1.4.3 Procyanidin conformations – MM2 calculations

The molecular mechanics program MM2 has been employed in a conformational analysis of the 16 possible dimers which may be assembled using (+)-catechin and (−)-epicatechin as the two flavan-3-ol monomer units. The 3′,4′-dihydroxyphenyl rings 'B' of both flavan-3-ol monomer units were constrained to the equatorial positions but all possible stereochemical configurations – 4α-8, 4β-8, 4α-6 and 4β-6 – at the interflavan linkage were considered (Mattice and Viswanadhan, 1986, 1987). The results of these calculations bear out some of the qualitative conclusions drawn from the earlier NMR and molecular modelling studies. They show there is a twofold rotation about the interflavan bond between the flavan-3-ol monomer units in the various dimers.

Table 1.2. *Conformational energy differences of procyanidin dimers (Mattice and Viswanadhan, 1986, 1987)*

Procyanidin dimer	ΔE, Energy difference*; kcal mol^{-1}
B-1; [(−)-**EC** − 4β-8 − (+)-**C**]	− 3.6
B-2; [(−)-**EC** − 4β-8 − (−)-**EC**]	− 3.6
B-3; [(+)-**C** − 4α-8 − (+)-**C**]	2.5
B-4; [(+)-**C** − 4α-8 − (−)-**EC**]	1.7
B-5; [(−)-**EC** − 4β-6 − (−)-**EC**]	3.3

(+)-**C** = (+)-catechin; (−)-**EC** = (−)-epicatechin
* The atoms which define the dihedral angle at the interflavan bond are in C-4 to C-8 linked dimers the atoms C(3)-C(4)-C(8)-C(9) and in C-4 to C-6 linked dimers C(3)-C(4)-C(6)-C(5), Figure 1.18. All dimers exhibit twofold minima for this rotation and the mean planes of the two fused ring systems are approximately orthogonal to one another at these minima. The energy difference ΔE is the energy at the minimum near (+)90° minus the energy at the minimum near (−)90°.

The calculated energy differences at the minima are sufficiently small (Table 1.2) such that both minima can be populated to a significant extent in several of the dimers. The two conformational energy minima are quite narrow due to the rapid onset of repulsive steric interactions as the dihedral angle at the interflavan bond rotates away from an equilibrium position. However, it should be borne in mind that MM2 is much more effective in defining the optimal conformation of the heterocyclic ring and the optimal dihedal angle at the interflavan bond than it is in defining the difference in conformational energies ΔE of the two rotational isomers (at the interflavan bond). Figures 1.20 and 1.21 show molecular modelling structural optimisations derived using similar procedures for the natural procyanidin B-2 with the interflavan dihedral angle as +98.2° (Figure 1.20) and −80.8° (Figure 1.21) (Lokman Kahn, Lilley, Williamson and Haslam, unpublished observations).

Heterocyclic rings in the dimers occupy a range of conformations that can best be described as half-chairs with varying degrees of distortion towards C(2) or C(3) sofas. The most frequent distortion is towards the C(2) sofa. Interconversion between most of the heterocyclic ring conformations can be obtained by coordinated motion of C(2) and C(3), over a range of 40 pm, with respect to the mean plane of the fused aromatic ring system. The most 'atypical' heterocyclic ring conformations are usually found in the 'lower' flavan-3-ol unit of 4α-8 linked procyanidin dimers when the interflavan dihedral angle (Φ) is near 90°.

Mattice and Viswanadhan (1987) subsequently employed the MM2 program to analyse the conformations of the same 16 dimers (*vide supra*) but in

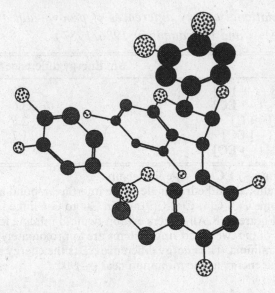

Figure 1.20. Conformational optimisation of procyanidin B-2 with an interflavan dihedral angle of +98.2°. The force field was a modified MM2. Structures were minimised using the Truncated Newton Conjugate Gradient and then the Full Matrix Newton Raphson method (Lokhman Kahn, Lilley, Williamson and Haslam, unpublished observations). All hydrogen atoms are omitted from the stucture; oxygen atoms are depicted as hatched spheres and carbon atoms as grey spheres.

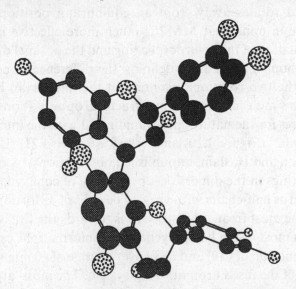

Figure 1.21. Conformational optimisation of procyanidin B-2 with an interflavan dihedral angle of −80.8°. The force field was a modified MM2. Structures were minimised using the Truncated Newton Conjugate Gradient and then the Full Matrix Newton Raphson method (Lohkman Kahn, Lilley, Williamson and Haslam, unpublished observations). All hydrogen atoms are omitted from the structure; oxygen atoms are depicted as hatched spheres and carbon atoms as grey spheres.

which the C(2) – 3′,4′-dihydroxyphenyl substituents (rings 'B') on the hetero-cyclic rings were in axial positions. Similar observations were made on the twofold rotation between monomer units and puckering of the heterocyclic rings. However, the fact that dimers in which the C(2) – 3′,4′-dihydroxyphenyl substituents (rings 'B') occupy equatorial positions are generally more stable than their axial counterparts readily became apparent during seveal optimisations. Thus, for example, the starting structure of procyanidin B-5, [(−)-**EC** – 4β-6 – (−)-**EC**], Figure 1.8, with axial C(2) – 3′,4′-dihydroxyphenyl substituents (rings 'B') and an interflavan dihedral angle < 0° converged to give an optimised structure that placed these substituents in equatorial positions. The reverse flips of the heterocyclic rings were not encountered in earlier studies of the dimers in which the starting structures had the C(2) – 3′,4′-dihydroxyphenyl substituents (rings 'B') in equatorial positions. If the initial starting structures have equatorial substituents at C(2), then so also do the optimised structures.

1.5 Gallic acid metabolism

Spectacular recent advances in the chemistry and biochemistry of plant poly-phenols have recently been made (Haslam, 1989, 1992; Okuda *et al.*, 1989, 1993; Haslam and Cai, 1994) in the field defined by the second structural group of plant polyphenols noted earlier – **the galloyl and hexahydroxydiphenoyl esters and their derivatives** – the most dramatic feature of which has been the increase (by an order of magnitude) of the number of known compounds in this class, which now number well over 750. Much of this work has emerged from two schools in Japan: those of Okuda in Okayama and Nishioka in Fukuoka. Substantial progress has been made possible by the application of new tech-niques of isolation and analysis (Okuda *et al.*, 1989) – mplc and hplc using Sephadex gels, Toyo pearl (TSK HW-40), Diaion HP-20, Mitsubishi MCI gel CHP-20P, reverse phase C-8 and C-18 supports, centrifugal partition and droplet counter current chromatography, etc. High resolution NMR spectro-scopy has provided a mine of structural information and tables of diagnostic (^{1}H and ^{13}C) chemical shift and coupling constant data have been published (Haslam, 1989; Okuda *et al.*, 1989; Yoshida *et al.*, 1984, 1988, Haslam *et al.*, 1982a,b,c). Extensive use has been made of the nuclear Overhuaser effect but particular note should be made of the use of ^{1}H–^{13}C long range 2D NMR spectra which provide specific information concerning the orientation and location of different phenolic acyl groups – such as galloyl, hexahyd-roxydiphenoyl, valoneoyl, dehydrodigalloyl, sanguisorboyl, euphorbinoyl, trilloyl, chebuloyl, elaeocarpinusinoyl, etc. on the polyol (usually D-glucose)

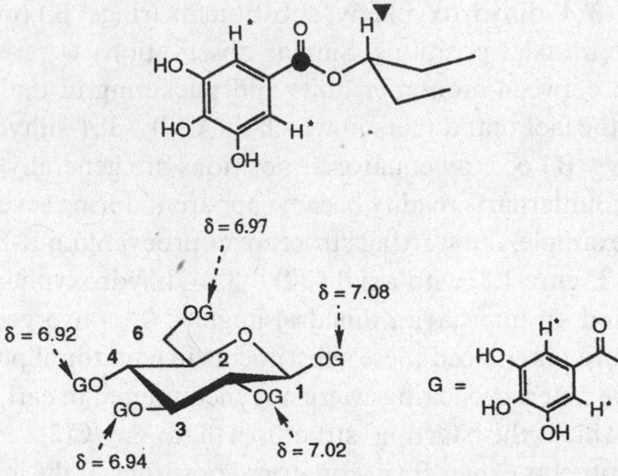

Figure 1.22. Proton chemical shift values for the galloyl ester protons (*) of β-1,2,3,4,6-pentagalloyl-D-glucose (D_2O at 60°C).

core of the polyphenolic ester. This technique is based on the ability to establish connectivity between the aroyl proton(s) of an acyl group (*) and the proton(s) at the position of acylation on the polyol (usually D-glucose) core (▼). This connectivity is established via the three bond couplings of the two groups of proton(s) to the ester carbonyl carbon atom (●, $j_{C,H} = \sim 5$ Hz).

The method has been fully described (Haslam *et al.*, 1990a) and permits, for example, each of the five two-proton singlets associated with each of the five galloyl ester groups of β-1,2,3,4,6-pentagalloyl-D-glucose to be defined, Figure 1.22. This type of information is also of crucial importance for studies of the complexation of polyphenolic esters with other substrates (Haslam *et al.*, 1990b, 1992) e.g. caffeine, cyclodextrins, anthocyanins, peptides and small proteins etc.

1.5.1 *Biosynthesis of galloyl esters*

The distinctive features of gallic acid metabolism in higher plants bear all the hallmarks of secondary metabolism (Bu'Lock, 1961, 1965; Haslam, 1986a, 1995). Thus three prominent characteristics are as follows.

(i) *Structural diversity*. At least 750 metabolites of gallic acid, which fall within the remit of the description polyphenol given above, have now been described. Despite their number the structures of the overwhelming majority of these metabolites may be envisaged as derived by chemical 'embellishment and embroidery' of one key intermediate: β-1,2,3,4,6-penta-*O*-galloyl-D-glucopyranose (Haslam, 1982). In this sense they bear a very close analogy to other groups of secondary metabolites such as the various classes of terpenes and alkaloids where derivation from a common precursor is postulated.

(ii) *Accumulation and storage.* Metabolites containing gallic acid often accumulate in substantial quantities in plant tissues. The apotheosis of this characteristic is the storage (up to 70% of the dry weight) of complex polyphenols of the gallotannin class in Chinese galls (*Rhus semialata*). Similarly the vegetative tissues of the green tea flush (*Camellia sinensis*), may contain up to 25–30% of phenolic flavan-3-ols, prominent amongst which are (−)-epigallocatechin and (−)-epicatechin and their 3-gallate esters (Sanderson, 1972).

(-)-epicatechin-3-O-gallate; R = H
(-)-epigallocatechin-3-O-gallate; R = OH

(iii) *Taxonomic distribution.* Metabolites containing gallic acid are not universally distributed in higher plants. They occur within clearly defined taxonomic limits in both woody and herbaceous dicotyledons (Bate-Smith and Metcalfe, 1957). El-lagitannins are widely distributed in the lower Hamamelidae, Dilleniidae, and Rosidae (the HDR-complex) and have been used as prominent chemotaxonomic markers. It has been suggested that the low degree of diversification in gallate dominated taxa may be as a result of the electron scavenging properties of these metabolites which, in turn, inhibits oxidation, the most important reaction in the synthesis of secondary metabolites (Gottlieb, 1990).

Gallic acid is unique amongst the various naturally occurring hydroxyben-zoic acids (Haslam, 1982, 1986b, 1992) both in respect of its relatively ubiquity in the plant kingdom and of the quantitative significance of its metabolism in many plants. Although some evidence exists to show that gallic acid may arise by oxidative degradation of L-phenylalanine (Zenk, 1964), the weight of experimental data favours the view that it is nevertheless formed primarily via the dehydrogenation of 3-dehydroshikimate (Knowles *et al.*, 1962; Cornthwaite and Haslam, 1965; Dewick and Haslam, 1969), Figure 1.23. Indeed 3-dehyd-roshikimate is readily oxidatively transformed non-enzymically *in vitro* at room temperature in aqueous media at pH around neutrality to gallic acid.

β-Glucogallin (β-1-*O*-galloyl-D-glucose), first isolated from Chinese rhubarb (*Rheum officinale*) in 1903, is, following the extensive studies of Gross and

Shikimate Pathway

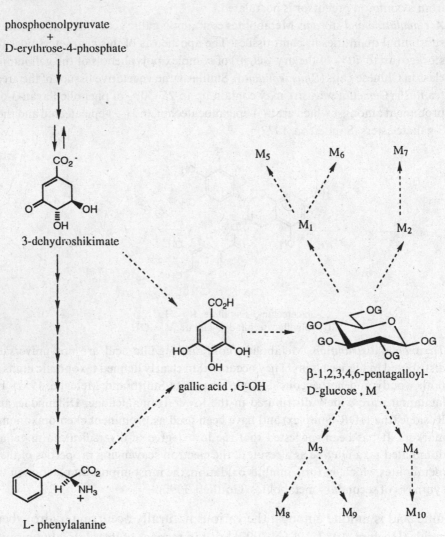

phosphoenolpyruvate
+
D-erythrose-4-phosphate

3-dehydroshikimate

M_5 M_6 M_7

M_1 M_2

gallic acid , G-OH

β-1,2,3,4,6-pentagalloyl-
D-glucose , M

M_3 M_4

M_8 M_9 M_{10}

L- phenylalanine

Figure 1.23. Gallic acid biosyntheis: further metabolism of β-1,2,3,4,6-penta-*O*-gal-
loyl-D-glucose.

co-workers (1982, 1983, 1986a,b, 1988, 1991), considered to be the key intermediate in the biosynthesis of esters of gallic acid. Work with cell-free extracts
from oak leaves and subsequently with the partially purified glucosyl transferase verified that β-glucogallin is generated by the reaction of gallic acid with
UDP-glucose. Thereafter β-glucogallin as substrate undergoes a series of
further galloyl transfer reactions to yield ultimately β-1,2,3,4,6-pentagalloyl-D-
glucose. It is very interesting to note that β-glucogallin acts as the principal

galloyl group donor in these reactions. Thus in the first of these reactions a partially purified enzyme from young oak leaves (EC 2.3.1.90) catalyses the formation of β-1,6-digalloyl-D-glucose from two molecules of β-glucogallin. The sequence continues in analogous fashion, with β-glucogallin as prime galloyl donor, via β-1,2,6-trigalloyl-D-glucose and β-1,2,3,6-tetragalloyl-D-glucose, to give finally β-1,2,3,4,6-pentagalloyl-D-glucose, Figure 1.24. One of the most striking features of this biosynthetic pathway is that the sequence of esterification steps with gallic acid

$$1\text{-OH} > 6\text{-OH} > 2\text{-OH} > 3\text{-OH} > 4\text{-OH}$$

exactly parallels the sequence in the chemically mediated esterification of the hydroxyl groups of D-glucopyranose (Richardson and Williams, 1967). This β-D-glucogallin dependent pathway is however by no means exclusive. Studies with enzyme extracts of sumach (*Rhus typhina*) have shown that in addition to β-D-glucogallin, β-1,6-digalloyl-D-glucose, β-1,2,6-trigalloyl-D-glucose and β-1,2,3,6-tetragalloyl-D-glucose may also act, although with progressively decreasing efficiency, as galloyl group donors (from the 1-position of the galloyl-ester).

Strong circumstantial evidence now exists to support the proposition (Haslam, 1982, 1989, 1992; Haslam and Cai, 1994) that the metabolite β-1,2,3,4,6-pentagalloyl-D-glucose, Figure 1.24, then plays a pivotal role, Figure 1.25, in the formation of the vast majority of gallotannins and ellagitannins which occur in many plants. Its biosynthetic position is analogous to those of norlaudanosoline and strictosidine in the formation of the benzylisoquinoline and the terpene-indole alkaloids respectively. Four distinctive and principal pathways are then presumed to lead from β-1,2,3,4,6-pentagalloyl-D-glucose and to give, by appropriate chemical embellishment, the various classes of metabolites, Figure 1.25 – pathways (a, b, c and d).

1.5.2 Depside metabolites

The ability to metabolise depside derivates of gallic acid (Figure 1.25, pathway (a)), may be used as a guide to inter-relationships in particular plant families (Haslam *et al.*, 1982d) and there is, on present evidence, a close association of this form of metabolism with the Rhoideae tribe in the Anacardiaceae. Many of the products of this form of metabolism were often grouped together in the earlier literature under the generic term 'gallotannin'. The most common and familiar example is Chinese gallotannin or tannic acid (galls, *Rhus semialata*) which possesses the overall composition of a hepta- to octa-galloyl-β-D-glucose and in which, on average, 2 to 3 additional galloyl groups are esterified

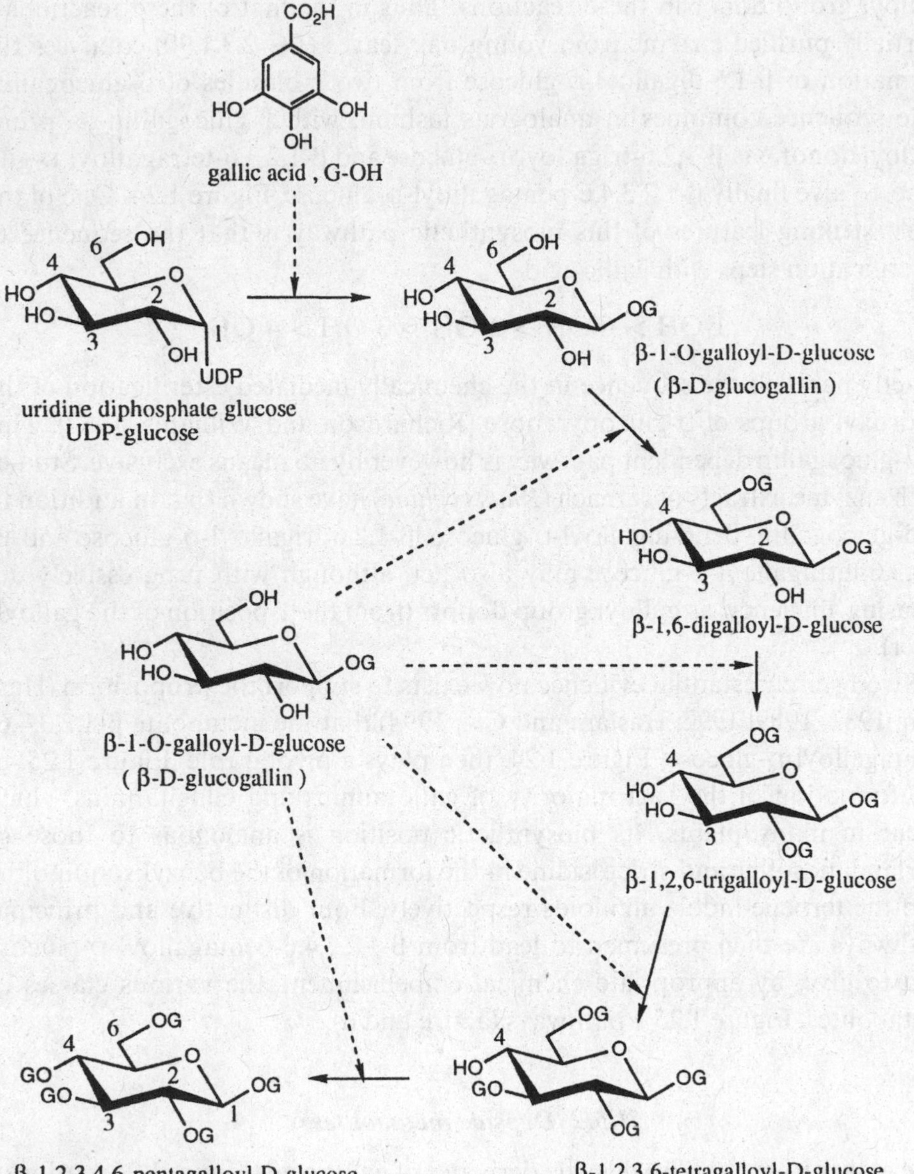

Figure 1.24. Biosynthesis of β-1,2,3,4,6-pentagalloyl-D-glucose; β-D-glucogallin as galloyl group donor.

in depside form to a pre-existing β-1,2,3,4,6-pentagalloyl-D-glucose core. A novel method of analysis (Haslam *et al.*, 1982c; Nishioka *et al.*, 1982) based on the use of hplc and ^{13}C NMR however reveals the heterogeneity of the typical Chinese gallotannin extract. This ranges from β-pentagalloyl-D-glucose itself

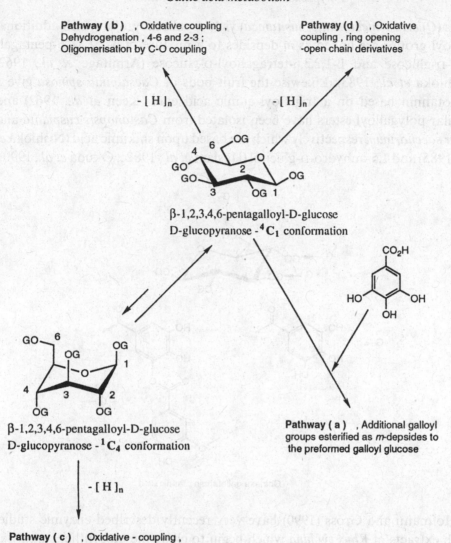

Pathway (b) , Oxidative coupling ,
Dehydrogenation , 4-6 and 2-3 ;
Oligomerisation by C-O coupling

Pathway (d) , Oxidative
coupling , ring opening
-open chain derivatives

$-[H]_n$

$-[H]_n$

β-1,2,3,4,6-pentagalloyl-D-glucose
D-glucopyranose - $^4C_1$ conformation

β-1,2,3,4,6-pentagalloyl-D-glucose
D-glucopyranose - $^1C_4$ conformation

Pathway (a) , Additional galloyl
groups esterified as *m*-depsides to
the preformed galloyl glucose

$-[H]_n$

Pathway (c) , Oxidative - coupling ,
Dehydrogenation , 3-6 , 1-6 ,
and 2-4 ; Dehydrohexahydroxydiphenoyl esters

Figure 1.25. Biogenesis of gallotannins and ellagitannins; the metabolic embellishment of β-1,2,3,4,6-pentagalloyl-D-glucose, principal pathways (Haslam and Cai, 1994).

to compounds with up to 5 or 6 additional galloyl residues linked as *m*-depsides to this core. The proportion of each type determines the final overall composition of the gallotannin extract. Using ^{13}C NMR the position of the additional depside residues has been determined to be predominantly to the galloyl groups at C-2 or C-3, C-4 and C-6. This polygalloyl-D-glucose is the most widely encountered ester of this type found in plants (Haslam *et al.*, 1982d; Haslam, 1989), but others have also been described. The galls of various oaks (*Quercus infectoria, Q. lusitanica*) yield a gallotannin in which additional galloyl groups are linked as *m*-depsides to a mixture of β-1,2,3,4,6-pentagal-loyl-D-glucose and β-1,2,3,6-tetragalloyl-D-glucose (Armitage *et al.*, 1962; Nishioka *et al.*, 1983). Likewise the fruit pods of *Caesalpinia spinosa* give a gallotannin based on a trigalloyl-quinic acid core (Keen *et al.*, 1962) and similar polygalloyl esters have been isolated from *Castanopsis cuspidata* and *Acer saccharinum* respectively which are based upon shikimic acid (Nishioka *et al.*, 1985) and 1,5-anhydro-D-glucitol (Haslam *et al.*, 1982c; Okuda *et al.*, 1990).

Chinese gallotannin , tannic acid

Hofmann and Gross (1990) have very recently described enzymic studies with extracts of *Rhus typhina* which begin to chart the biosynthetic pathway from β-1,2,3,4,6-penta-galloyl-D-glucose to the gallotannins in sumach (mixtures of hexa-, hepta- and octa-galloyl-D-glucose derivatives). Detailed ^{1}H and ^{13}C NMR analysis of the hexagalloyl-D-glucose fraction showed it to contain at least three hexagalloyl esters in which additional galloyl ester groups were linked as *m*-depsides to the galloyl ester groups attached to C-2, 3 and 4 of the D-glucopyranose ring.

1.5.3 *Hexahydroxydiphenoyl esters and related metabolites*

In the 30 year period 1950–80, the Heidelberg school of Otto Schmidt and Walter Mayer made distinguished and seminal contributions to the study of naturally occurring hexahydroxydiphenoyl* esters (Schmidt, 1954, 1955, 1956; Mayer, 1973; Haslam, 1989; Haslam and Cai, 1994). They isolated and identified key metabolites such as corilagin, pedunculagin, chebulinic and chebulagic acids, dehydrodigallic, valoneic and brevifolin carboxylic acids, trilloic acid, terchebin and the dehydrohexahydroxydiphenoyl esters – brevilagin 1 and 2 – vescalin, castalin, vescalagin and castalagin, punicalin and punicalagin, castavaloninic acid, valolaginic acid and isovalolaginic acid. In addition as the work progressed they continued to provide an intellectually satisfying biogenetic rationale for the derivation of this group of naturally occurring polyphenolic esters (*syn.* ellagitannins). This pioneering work has securely underpinned all subsequent developments in this field, particularly the explosive increase in knowledge of the past 15 years which has resulted in the identification of at least 500 discrete compounds in this class. The following discussion does not attempt to be totally comprehensive, but classifies the major groups of metabolites within a structural and biogenetic framework.

Whilst there is, as yet, no formal experimental proof, it is generally assumed, following the hypothesis enunciated by Schmidt and Mayer (1956), that naturally occurring hexahydroxydiphenoyl esters and their derivatives are derived by oxidative coupling of galloyl esters (with the formation of new C–C and C–O bonds) and oxidative and hydrolytic aromatic ring fission. Strong circumstantial evidence now also exists to suggest that the vast majority of metabolites are derived biosynthetically (with subsequent possible modifications *in vivo* by facile hydrolytic reactions), by oxidative transformations of the key precursor β-1,2,3,4,6-pentagalloyl-D-glucose, following a scheme first put forward in 1982 (Haslam, 1982), Figure 1.26.

Amongst the known ellagitannin metabolites, numerous intramolecular 'C–C' linked ester groups have been located in the 'monomers' and similarly various intermolecular 'C–O' linking ester groups have been defined in the formation of the 'oligomeric' structures. The principal members of these two classes of ester group are shown in Figures 1.27 to 1.30.

In the intramolecular formation of a hexahydroxydiphenoyl ester group the generation of a large-membered ring containing two *cis* (*E*) double bonds reduces conformational flexibility and gives the molecule much greater rigidity. Where the bridging occurs 1,6; 3,6 or 2,4 this constrains the D-glucopyranose residue to adopt a thermodynamically unfavourable $^1C_4$ conformation, Figure 1.31.

One specific chirality is imposed upon the hexahydroxydiphenoyl group as

---

* Hexahydroxydiphenyl is used throughout this text as a convenient trivial name for the 6,6'-dicarbonyl-2,2',3,3',4,4'-hexahydroxbiphenyl radical.

β-1,2,3,4,6-pentagalloyl-D-glucose

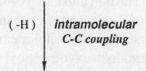

(-H) | ***intramolecular***
      | ***C-C coupling***

hexahydroxydiphenoyl and dehydrohexahydroxydiphenoyl esters
(*'monomers'*)

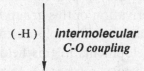

(-H) | ***intermolecular***
      | ***C-O coupling***

[hexahydroxydiphenoyl and dehydrohexahydroxydiphenoyl esters]$_n$
(*'oligomers'*, n = 2 , 3 , 4)

Figure 1.26. Overall patterns of oxidative metabolism of β-1,2,3,4,6-pentagalloyl-D-glucose in higher plants (Haslam, 1982).

it is formed and this chirality is determined by the need of the new ester group to bridge particular positions in the polyol (usually D-glucose) portion of the molecule. The absolute configuration of the twisted biphenyl system in co-rilagin has been determined (Okuda *et al.*, 1982a,b) as (*R*) and the chirality of other hexahydroxydiphenoyl esters may be made by measurements of circular dichroism (CD) and comparison with corilagin. The CD spectra of hexahyd-roxydiphenoyl esters are characterised by a distinctive couplet centred at ∼ 200–210 nm with a positive or negative maximum at 228–238 nm. The Δε values are approximately incremental for the number of hexahyd-roxydiphenoyl ester groups in the molecule (Okuda *et al.*, 1982a,b; Haslam *et al.*, 1982a). It is of interest to note that the same conclusions regarding the chirality of the hexahydroxydiphenoyl ester groups bound to the D-glucopyranose core may be deduced from theoretical considerations (Haslam, 1982; Haslam *et al.*, 1982a). These arguments also strongly suggest that the absolute configurations of the hexahydroxydiphenoyl esters are determined largely if not solely by the inbuilt stereochemistry of the sugar molecule.

The principal mode of transformation (Figure 1.25, pathway (b)) is by oxidative C–C coupling of galloyl ester groups (**4–6** and **2–3**) in the ther-modynamically most stable $^4C_1$ conformation of β-1,2,3,4,6-pentagalloyl-D-glucose followed by oxidative oligomerisation (C–O, coupling). Two further modes of elaboration are also both oxidative in character. Mode (c) in Figure 1.25) occurs via oxidative C–C coupling of galloyl ester groups (**3–6**, **1–6** and

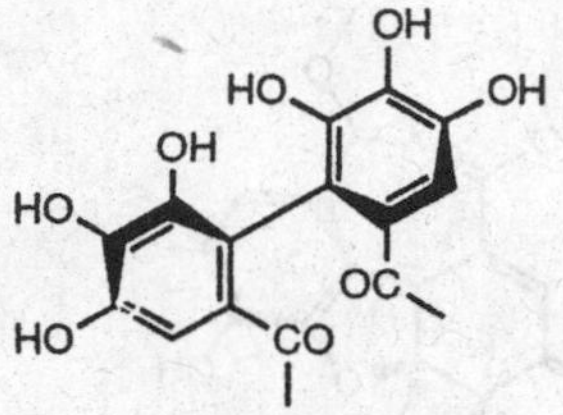

(*R*)-hexahydroxydiphenoyl

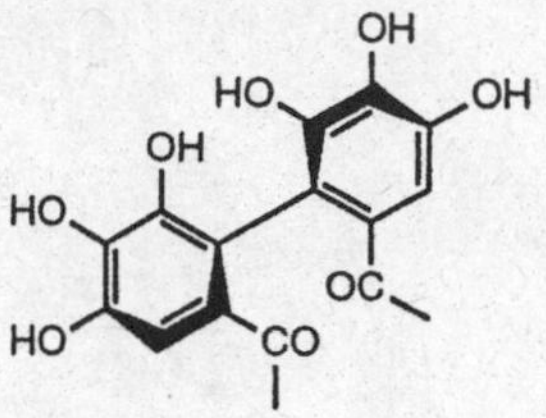

(*S*)-hexahydroxydiphenoyl

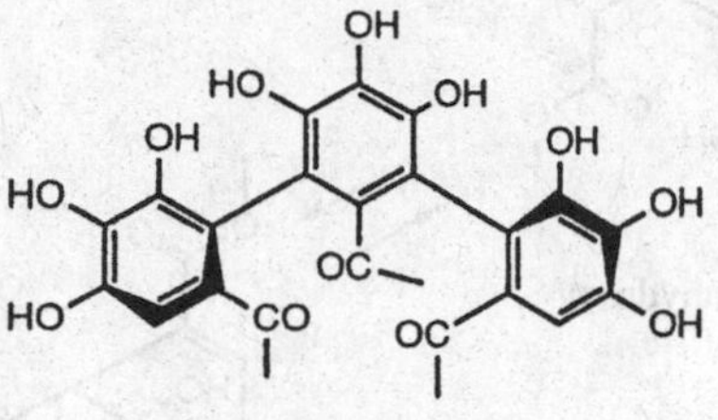

flavogallonyl

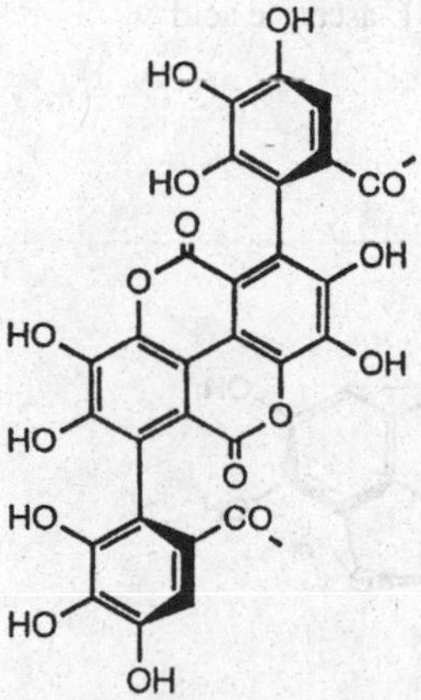

gallagyl

Figure 1.27. Ester groups formed by 'C–C' oxidative coupling. Derivatives of hexahydroxydiphenic acid.

2–4) in the thermodynamically least stable $^1C_4$ conformation of β-1,2,3,4,6-pentagalloyl-D-glucose followed again by oxidative oligomerisation (C–O, coupling). Metabolites in this class are also often characterised by the presence of the dehydrohexahydroxydiphenoyl ester group and derivatives, e.g. chebulinic and chebulagic acids, in which one aromatic nucleus of this functionality has undergone hydrolytic ring fission. In the final pathway, Figure

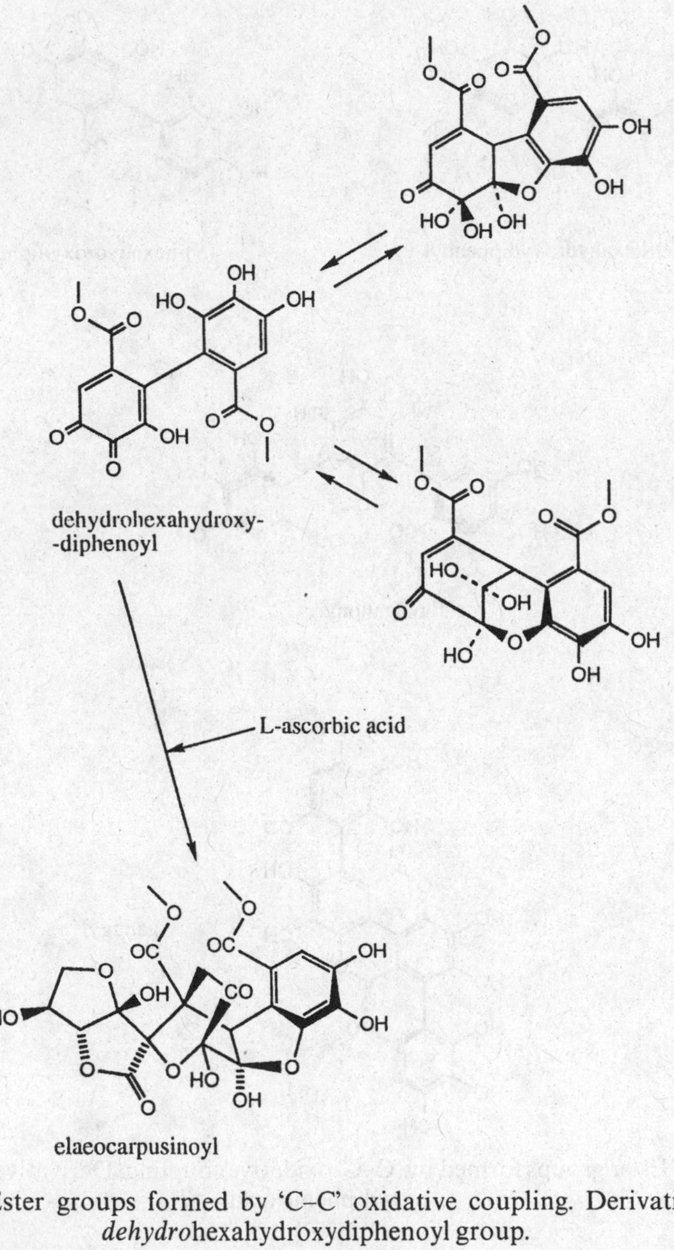

dehydrohexahydroxy-
-diphenoyl

L-ascorbic acid

elaeocarpusinoyl

Figure 1.28. Ester groups formed by 'C–C' oxidative coupling. Derivatives of the
*dehydro*hexahydroxydiphenoyl group.

1.25 (pathway (d)), the oxidative transposition of β-1,2,3,4,6-pentagalloyl-D-glucose takes place by ring opening and the formation of unique open-chain derivatives of D-glucose, such as vescalagin and castalagin and their derivatives.

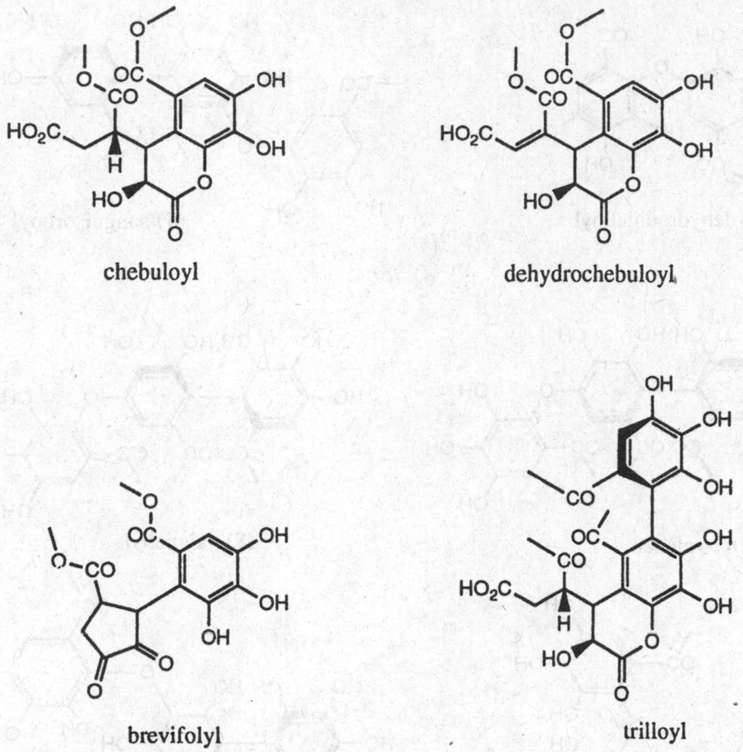

chebuloyl

dehydrochebuloyl

brevifolyl

trilloyl

Figure 1.29. Ester groups formed by 'C–C' oxidative coupling. Derivatives of the hexahydroxydiphenoyl group in which one aromatic ring has undergone hydrolytic cleavage.

1.5.4 Oxidative metabolism – pathway (b)

This pathway represents the most commonly encountered biogenetic route to the ellagitannins. A series of metabolites ('monomers') is first formed by C–C oxidative coupling of vicinal galloyl ester groups 4,6 and 2,3 in β-1,2,3,4,6-pentagalloyl-D-glucose in its thermodynamically preferred $^4C_1$ conformation. Some details of this pattern of metabolism were hinted at in earlier work by Schmidt and Mayer (Mayer, 1973; Haslam, 1989; Haslam and Cai, 1994), Hillis and Seikel (1970), Wilkins and Bohm (1976). Plants whose phenolic metabolism places them within this category furnish, as principal metabolites, one or more of the (S)-hexahydroxydiphenoyl esters shown in Figure 1.32. There are just one or two interesting exceptions to this generalisation, for example, cercidins A and B and cuspinin which contain a 2,3-(R)-hexahydroxy-diphenoyl ester group (for a review of these structures see Okuda *et al.*, 1993; Haslam and Cai, 1994). It is then presumed that, *as a second phase by oxidative*

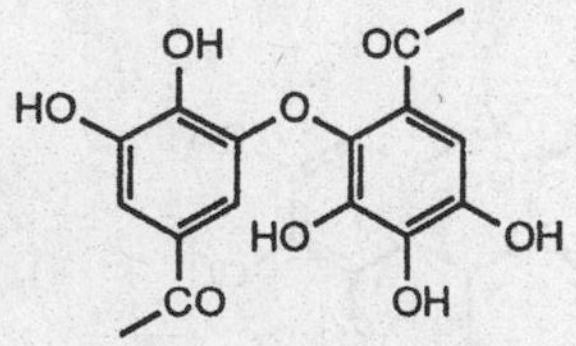

dehydrodigalloyl

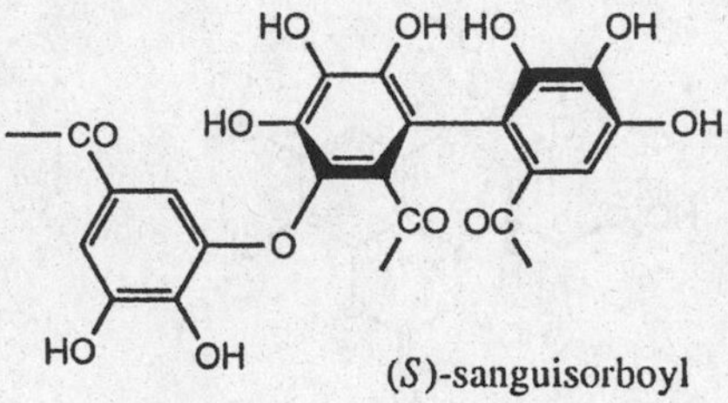

(S)-sanguisorboyl

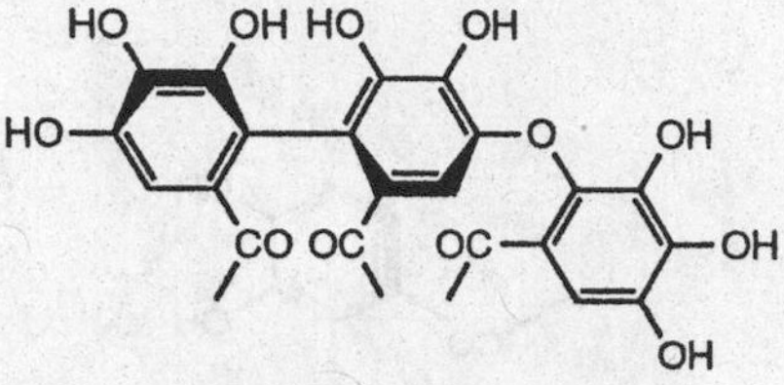

(R)-valoneoyl

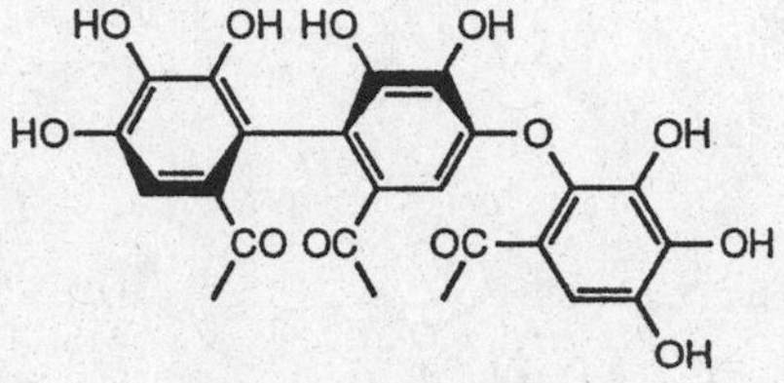

(S)-valoneoyl

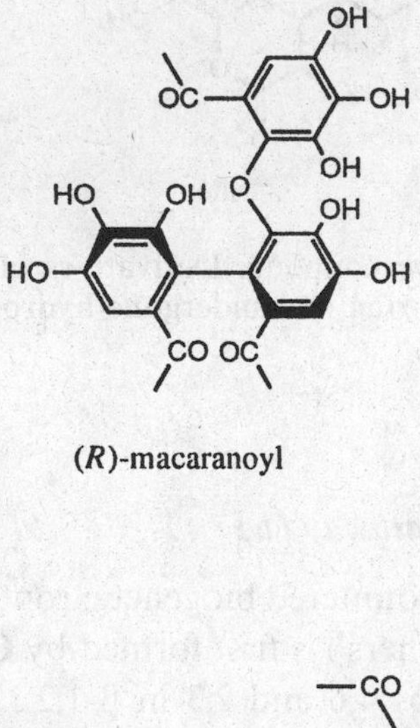

(R)-macaranoyl

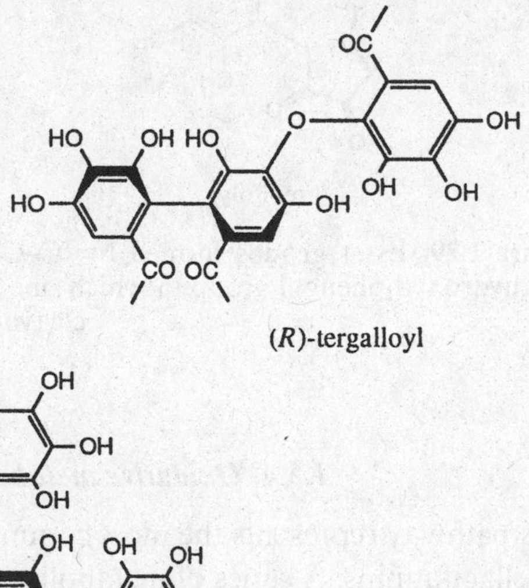

(R)-tergalloyl

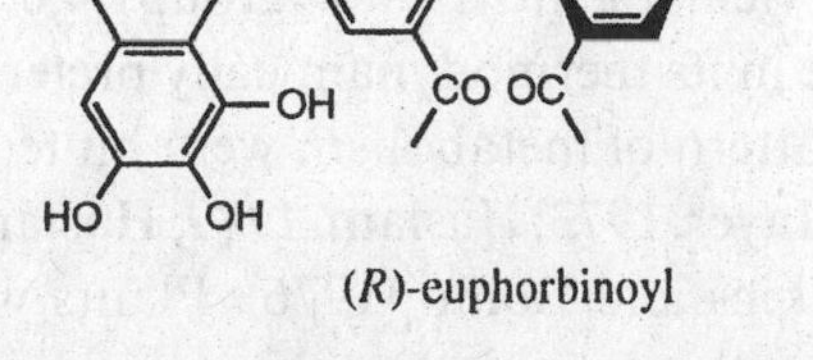

(R)-euphorbinoyl

Figure 1.30. Principal ester groups formed by intermolecular 'C–O' oxidative coupling of galloyl and hexahydroxydiphenoyl esters.

intermolecular C–O coupling, dimeric, trimeric and tetrameric structures are then formed, Figure 1.26 (for a review of these structures see Okuda *et al.*, 1993; Haslam and Cai, 1994).

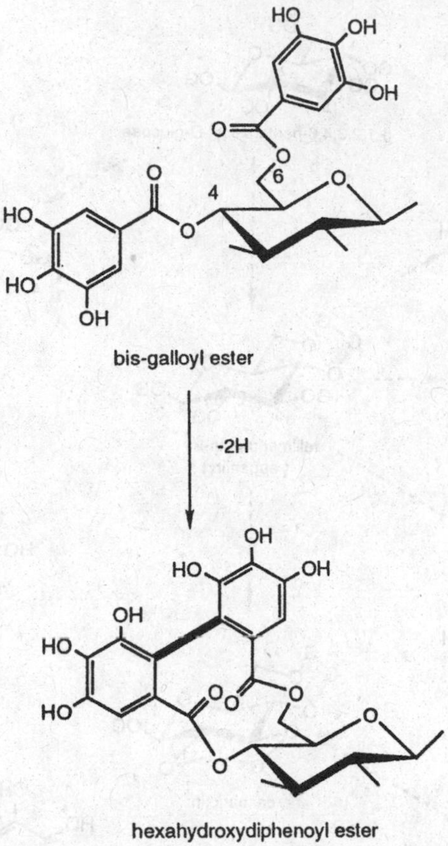

bis-galloyl ester

-2H

hexahydroxydiphenoyl ester

Figure 1.31. Intramolecular generation of a hexahydroxydiphenoyl ester.

The isolation and structure determination of two 'dimers' formed by C–O oxidative coupling of intermediates such as are shown in Figure 1.33 were first reported in 1982 (Haslam *et al.*, 1982a; Okuda *et al.*, 1982c,d). Since that time in the space of just 10 years more than 75 'oligomers' have now been described, (Okuda *et al.*, 1990, 1993; Haslam and Cai, 1994). These are broadly divisible into several sub-types dependent on the mode of C–O coupling between a phenolic hydroxyl group in one 'monomer' and an aromatic ring carbon in another 'monomer'. The three principal modes of C–O coupling, Figure 1.33, are those between two galloyl groups (to give a dehydrodigalloyl linking group), and between one galloyl group and an (*S*)-hexahydroxydiphenoyl ester group to yield either a valoneoyl or, rarely, the positionally isomeric, sanguisorboyl linking ester group. Present evidence suggests that the (*S*)-valoneoyl linking ester group is the one most commonly formed. For some time the most frequently encountered situation was that in which the linkage involved a galloyl ester group at the anomeric centre of at least one of the

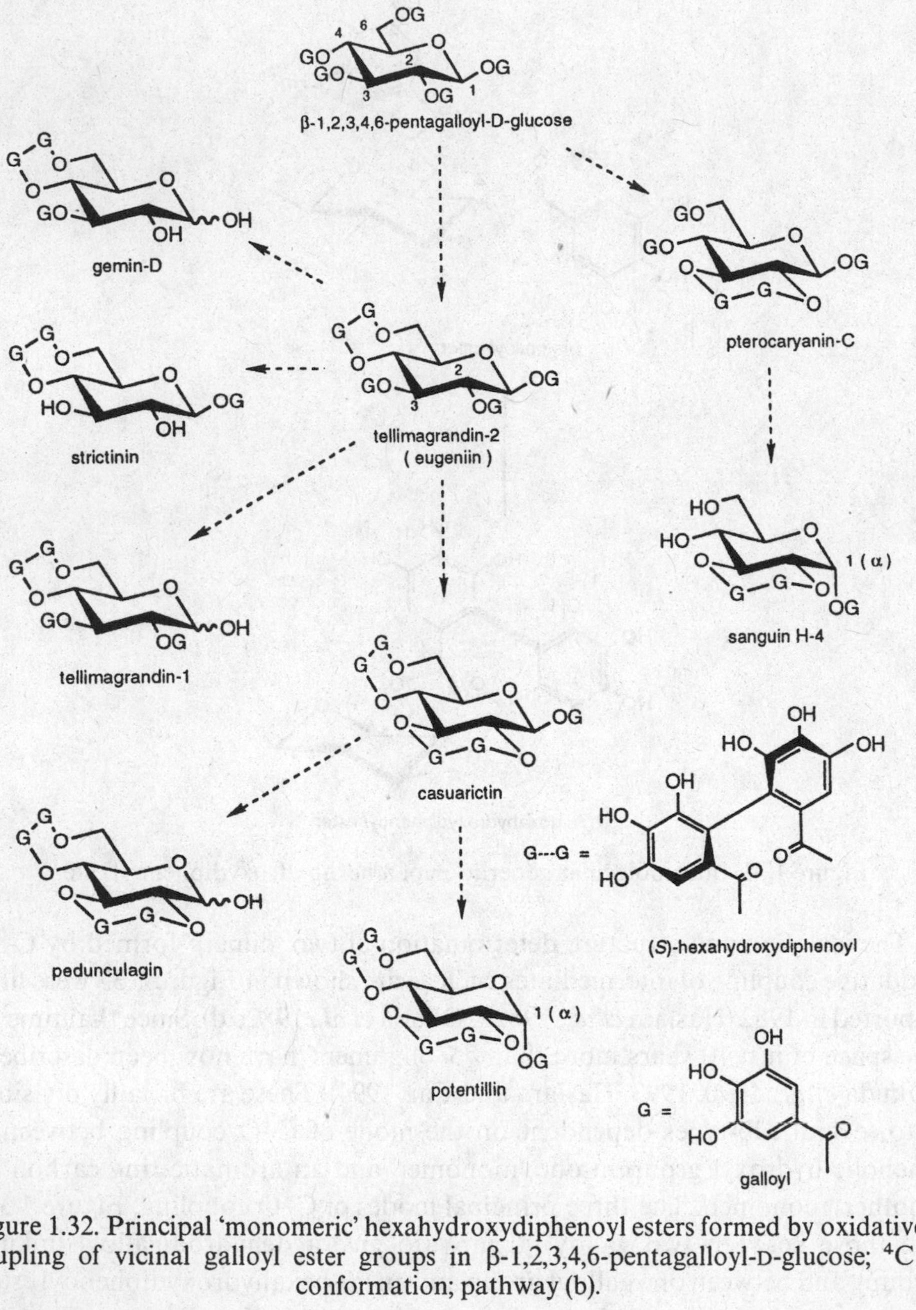

Figure 1.32. Principal 'monomeric' hexahydroxydiphenoyl esters formed by oxidative coupling of vicinal galloyl ester groups in β-1,2,3,4,6-pentagalloyl-D-glucose; $^4C_1$ conformation; pathway (b).

'monomers'. Linkage through galloyl ester groups at other positions on the glucose ring have however now been noted in several instances. Typical illustrations – gemin A, sanguin H-6, and calamanin B [G = galloyl, G–G = (S)-hexahydroxydiphenoyl] – are shown in Figure 1.33. 'Dimers' with

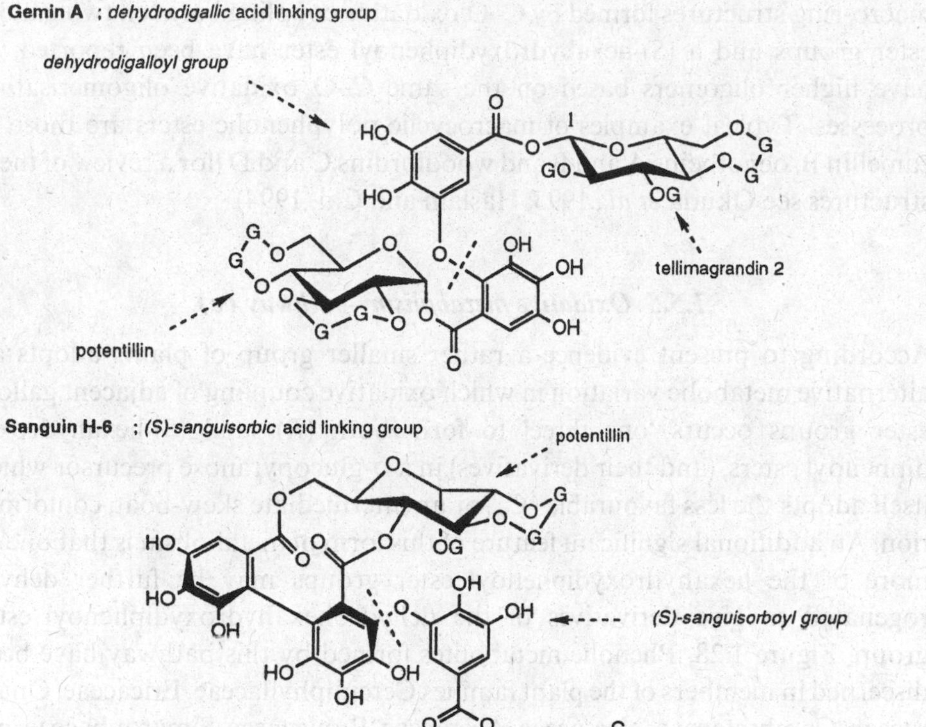

Gemin A ; *dehydrodigallic* acid linking group

dehydrodigalloyl group

tellimagrandin 2

potentillin

Sanguin H-6 ; *(S)-sanguisorbic* acid linking group

potentillin

(S)-sanguisorboyl group

casuarictin

Rugosin-D ; *(S)-valoneic* acid linking group

potentillin

tellimagrandin 2

(S)-valoneoyl group

Figure 1.33. 'Dimeric' ellagitannins: different modes of C–O coupling (- - - -) to give dehydrodigalloyl, sanguisorboyl, and valoneoyl linking ester groups. [G–G = (S)-hexahydroxydiphenoyl, G = galloyl].

macro-ring structures formed by C–O oxidative coupling between two galloyl ester groups and a (S)-hexahydroxydiphenoyl ester have been reported, as have higher oligomers based on the same C–O oxidative oligomerisation processes. Typical examples of macrocyclic polyphenolic esters are those of camellin B, oenotheins A and B and woodfordins C and D (for a review of these structures see Okuda *et al.*, 1993; Haslam and Cai, 1994).

1.5.5 Oxidative metabolism –pathway (c)

According to present evidence a rather smaller group of plants adopts an alternative metabolic variation in which oxidative coupling of adjacent galloyl ester groups occurs 'one–three' to form both (R)- and (S)-hexahydroxy-diphenoyl esters, (and their derivatives) in a D-glucopyranose precursor which itself adopts the less favourable $^1C_4$, or an intermediate skew-boat, conformation. An additional significant feature of this form of metabolism is that one or more of the hexahydroxydiphenoyl ester groups may be further dehydrogenated to give derivatives of the dehydrohexahydroxydiphenoyl ester group, Figure 1.28. Phenolic metabolites formed by this pathway have been discerned in members of the plant families Cercidiphyllaceae, Ericaceae, Onagraceae, Combretaceae, Nyssaceae, Aceraceae, Punicaceae, Simaroubaceae and Geraniaceae (Haslam *et al.*, 1982d). Some of the principal 'monomeric' esters of this class are shown in Figure 1.34.

For its routine metabolic processes Nature appears to prefer the conformationally most stable sugars and amongst the D-aldohexoses glucose, mannose and galactose are widely distributed. Those with higher free energies occur rarely. It is a point of some curiosity therefore that, in this particular form of oxidative metabolism of galloyl esters of D-glucose, the transformations apparently occur with the galloyl D-glucopyranose derivative in an energetically unfavourable chair ($^1C_4$) or related skew-boat conformation. The difference in free energy between the $^1C_4$ and the $^4C_1$ forms of D-glucopyranose has been calculated as $+5.95$ kcal mol^{-1} (24.8 kJ mol^{-1}) and this difference may, in part, explain why oxidative coupling of galloyl ester grops via the $^4C_1$ conformation of the D-glucopyranose precursor (pathway (b), above) is much more widely encountered in the plant kingdom than the alternative (pathway (c)) which is presumed to proceed via a chair ($^1C_4$) or related skew-boat conformation. A key metabolite in this particular sub-class is the beautiful yellow crystalline compound geraniin, first isolated from *Geranium* and *Euphorbia* species by Okuda and by the Sheffield group from *Acer* and *Cercidiphyllum* species (Okuda *et al.*, 1982e; Haslam *et al.*, 1982b). In this, as in other derivatives of the dehydrohexahydroxydiphenoyl ester group, the dehydro-

geraniin

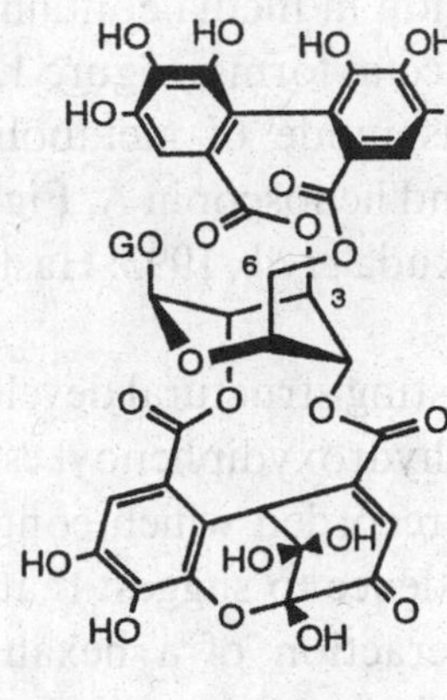

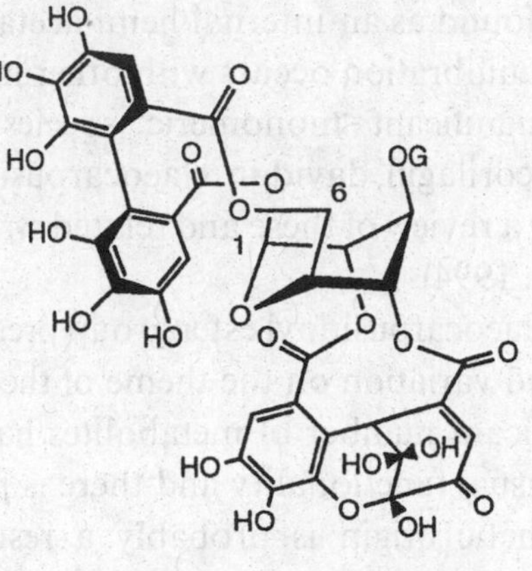

carpinusin

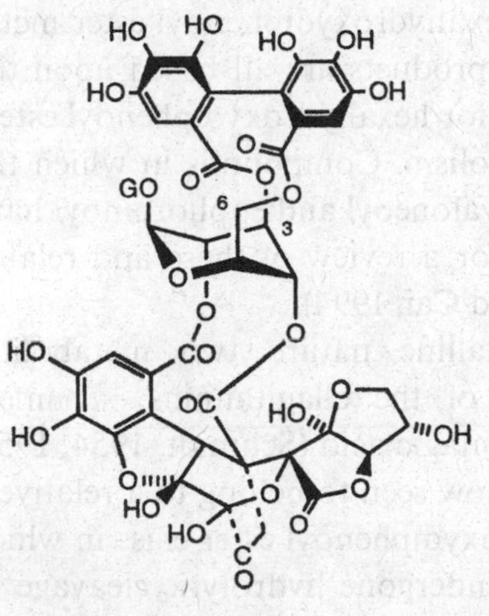

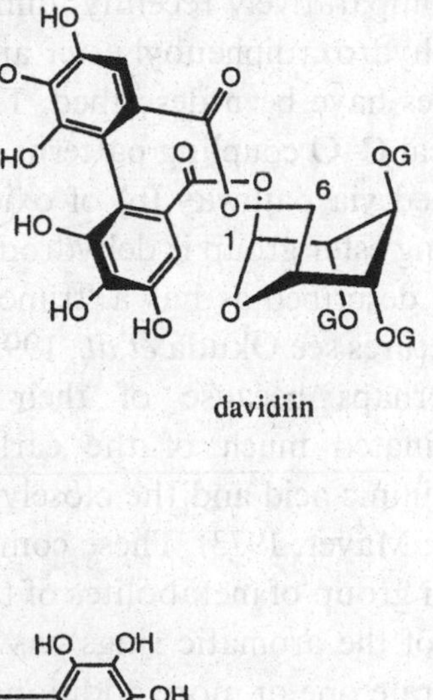

davidiin

elaeocarpusin

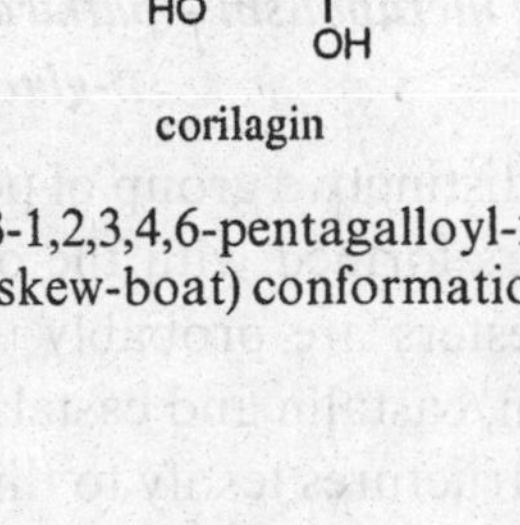

corilagin

Figure 1.34. 'Monomeric' metabolites of β-1,2,3,4,6-pentagalloyl-D-glucose; oxidative C–C coupling via the $^1C_4$ (or related skew-boat) conformation; pathway (c).

ester is found as an internal hemi-acetal. On dissolution in media containing water equilibration occurs with other internal hemi-acetal forms, Figure 1.28. Other significant 'monomeric' species found in this mode of metabolism include corilagin, davidiin, elaeocarpusin, tanarinin and helioscopin A, Figure 1.34 (for a review of these and related structures see Okuda *et al.*, 1993; Haslam and Cai, 1994).

The elaeocarpusinoyl ester group presents an interesting structural development and variation on the theme of the dehydrohexahydroxydiphenoyl ester. A significant number of metabolites have now been recorded which contain this unusual functionality and there is *prima facie* evidence to suggest that its biosynthetic origin is probably a result of the interaction of a hexahydroxydiphenoyl ester group with L-dehydro-ascorbate, during the generation of a dehydrohexahydroxydiphenoyl ester, Figure 1.28 (Nishioka *et al.*, 1987).

Comparatively recently 'dimeric' species involving these various types of hexahydroxydiphenoyl ester and dehydrohexahydroxydiphenoyl ester metabolites have been described. These natural products are all based upon the typical C–O coupling patterns noted earlier for hexahydroxydiphenoyl esters formed via pathway (b) of oxidative metabolism. Compounds in which the linking ester group is dehydrodigalloyl; (*R*)-valoneoyl and euphorbinoyl have been described as has a 'trimeric' species (for a review of these and related structures see Okuda *et al.*, 1993; Haslam and Cai, 1994).

Perhaps because of their ready crystalline nature two metabolites dominated much of the early chemistry of the ellagitannins – namely chebulinic acid and the closely related chebulagic acid (Schmidt, 1954, 1955, 1956; Mayer, 1973). These compounds are now seen to belong to a relatively small group of metabolites of the hexahydroxydiphenoyl ester class in which one of the aromatic rings has apparently undergone hydrolytic cleavage to generate one or more additional carboxylate groups. The significance of this presumed biogenetic process was first highlighted by Schmidt and Mayer (1956), and Haworth (Haworth, 1961; Haworth *et al.*, 1955), Figure 1.35.

1.5.6　Oxidative metabolism – pathway (d) – 'open-chain' derivatives of D-glucose

An intriguing and distinctive group of polyphenolic compounds derived from gallic acid are esters formed with the open-chain form of D-glucose. In this latter sense these esters are probably unique in natural product chemistry. Vescalin, vescalagin, castalin and castalagin are compounds which typify this group and whose structures testify to this uniqueness. They were first isolated from oak and chestnut species and their structures determined by Mayer and

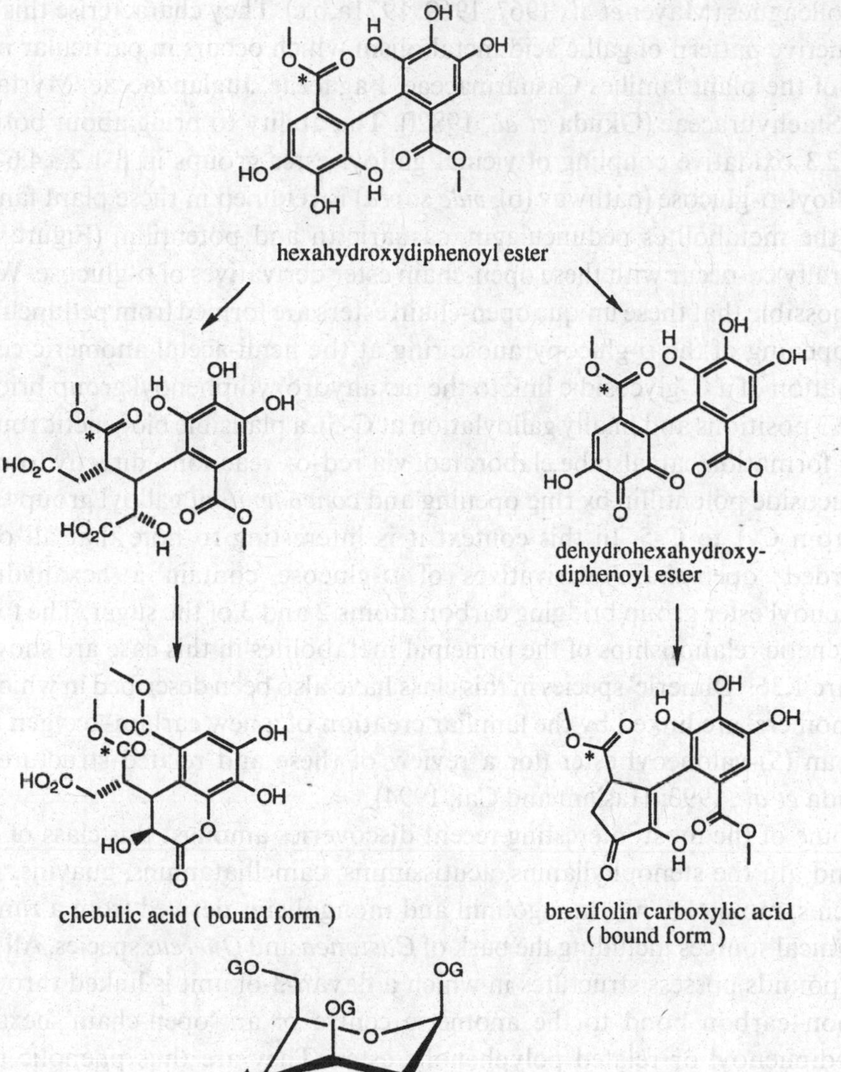

hexahydroxydiphenoyl ester

**dehydrohexahydroxy-
diphenoyl ester**

chebulic acid (bound form)

**brevifolin carboxylic acid
(bound form)**

chebulinic acid , G = galloyl

Figure 1.35. Putative biogenetic pathways to 'ring-opened' hexahydroxydiphenoyl esters – chebulic and brevifolin carboxylic acids. Structure of chebulinic acid.

his colleagues (Mayer *et al.*, 1967, 1969, 1971a,b,c). They characterise this very distinctive pattern of gallic acid metabolism which occurs in particular members of the plant families Casuarinaceae, Fagaceae, Juglandaceae, Myrtaceae and Stachyuraceae (Okuda *et al.*, 1982f). The ability to bring about both 4,6 and 2,3 oxidative coupling of vicinal galloyl ester groups in β-1,2,3,4,6-pentagalloyl-D-glucose (pathway (b), *vide supra*) is retained in these plant families and the metabolites pedunculagin, casuarictin and potentillin (Figure 1.32) generally co-occur with these open-chain ester derivatives of D-glucose. Whilst it is possible that these unique open-chain esters are formed from pedunculagin (by opening of the D-glucopyranose ring at the hemi-acetal anomeric centre, formation of a C-glycosidic link to the hexahydroxydiphenoyl group bridging the 2,3 positions and finally galloylation at C-5), a plausible biogenetic route to their formation can also be elaborated, via red-ox reactions, directly from the α-glucoside potentillin by ring opening and *concommitant* galloyl group transfer from C-1 to C-5. In this context it is interesting to note that all of the recorded 'open-chain' derivatives of D-glucose contain a hexahydroxydiphenoyl ester group bridging carbon atoms 2 and 3 of the sugar. The formal biogenetic relationships of the principal metabolites in this case are shown in Figure 1.36. 'Dimeric' species in this class have also been described in which the 'monomers' are linked by the familiar creation of a new carbon–oxygen bond and an (*S*)-valoneoyl ester (for a review of these and related structures see Okuda *et al.*, 1993; Haslam and Cai, 1994).

Some of the most interesting recent discoveries amongst this class of compound are the stenophyllanins, acutissimins, camelliatannins, guavins, mongolicins, mongolicanin, mongolinin and mongolicins derived from a range of botanical sources including the bark of *Castanea* and *Quercus* species. All these compounds possess structures in which a flavan-3-ol unit is linked through a carbon–carbon bond to the anomeric centre of an 'open-chain' hexahydroxydiphenoyl or related polyphenolic ester. They are thus phenolic metabolites which embrace the structural and chemical features associated with both the hydrolysable and condensed groups of tannins. A typical example of this class of metabolite is acutissimin A [derived presumably from vescalagin/ castalagin and (+)-catechin] (for a review of these and related structures see Okuda *et al.*, 1993; Haslam and Cai, 1994).

The alcoholic hydroxyl group at C-1 of metabolites such as castalin, castalagin, vescalin and vescalagin is benzylic to an electron rich phenolic nucleus and therefore, in principle, subject to possible displacement and the formation of a resonance stabilised carbocation at C-1. This provides a chemical rationale for a possible biogenetic pathway to metabolites such as the stenophyllanins, acutissimins, camelliatannins, guavins, mongolicins, mongolicanin, and

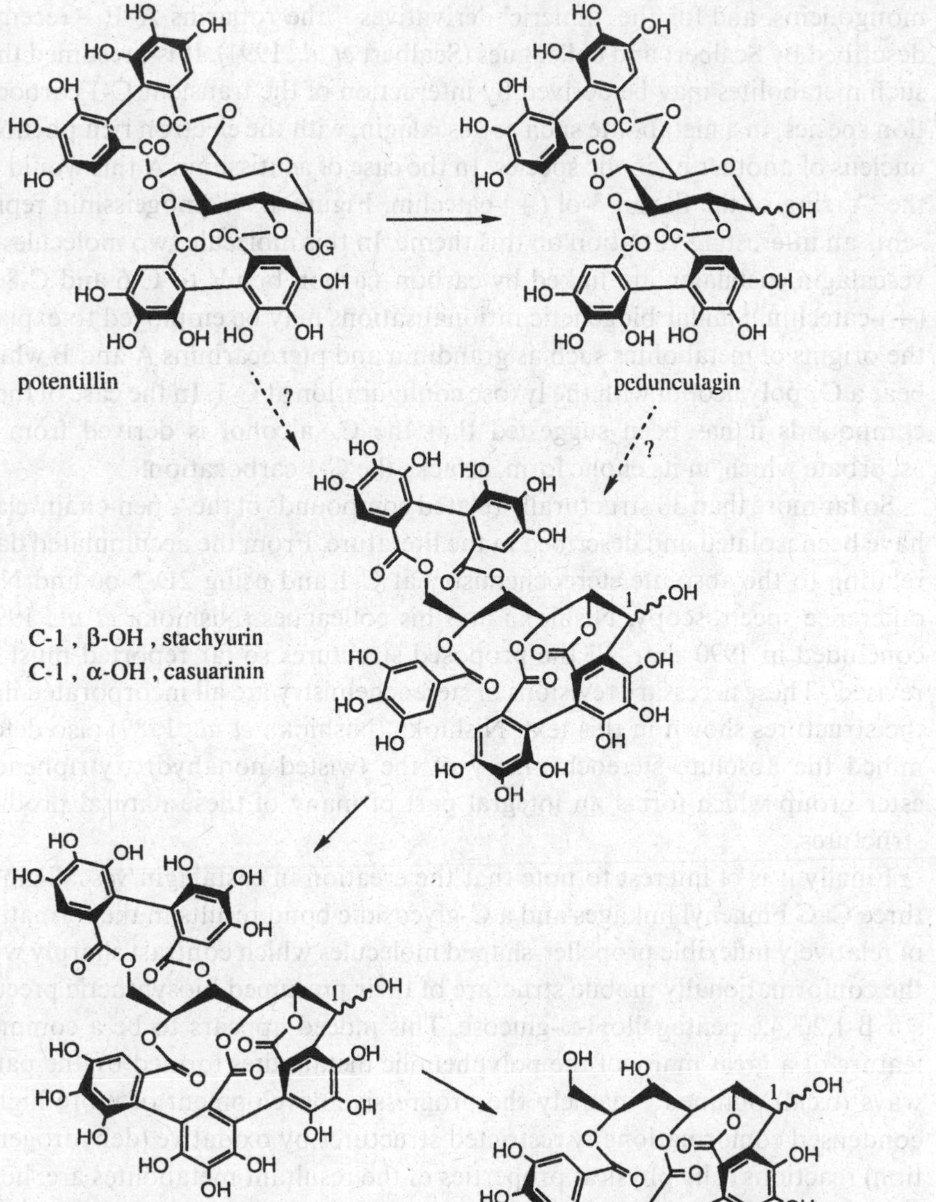

potentillin → pedunculagin

C-1 , β-OH , stachyurin
C-1 , α-OH , casuarinin

C-1 , β-OH , vescalagin
C-1 , α-OH , castalagin

C-1 , β-OH , vescalin
C-1 , α-OH , castalin

Figure 1.36. Oxidative metabolism of β-1,2,3,4,6-pentagalloyl-D-glucose, pathway (d): suggested biogenetic relationships in 'open-chain' polyphenolic esters.

mongolicins, and for the 'dimeric' derivatives – the roburins A–E – recently described by Scalbert and colleagues (Scalbert *et al.*, 1991). It is presumed that such metabolites may be derived by interaction of the transient C-1 carbocation species, in a metabolite such as vescalagin, with the electron rich phenolic nucleus of another phenolic species. In the case of acutissimin A this would be the 'A'-ring of the flavan-3-ol (+)-catechin, Figure 1.37. Anogeissinin represents an interesting variation on this theme. In this molecule two molecules of vescalagin/castalagin are linked by carbon–carbon bonds to C-6 and C-8 of (+)-catechin. Similar biogenetic rationalisations may be employed to explain the origins of metabolites such as grandinin and pterocarinins A and B which bear a C_5 polyalcohol with the lyxose configuration at C-1. In the case of these compounds it has been suggested that the C_5 alcohol is derived from L-ascorbate which, in its enolic form, attacks the C-1 carbocation.

So far more than 30 structurally related compounds of the 'open-chain' class have been isolated and described in the literature. From the accumulated data relating to the absolute stereochemistry at C-1 and using 2D Noe and Noe difference spectroscopy, Nishioka and his colleagues (Nishioka *et al.*, 1990) concluded in 1990 that 'all the proposed structures so far reported must be revised'. These necessary revisions of stereochemistry are all incorporated into the structures shown in this text. Nishioka (Nishioka *et al.*, 1987) also determined the absolute stereochemistry of the twisted nonahydroxytriphenoyl ester group which forms an integral part of many of these natural product structures.

Finally it is of interest to note that the creation in castalagin/vescalagin of three C–C biphenyl linkages and a C-glycosidic bond results in the formation of relatively inflexible propeller-shaped molecules which contrast sharply with the conformationally mobile structure of their presumed biosynthetic precursor β-1,2,3,4,6-pentagalloyl-D-glucose. This indeed appears to be a common feature of a great many of the polyphenolic metabolites formed by the pathways (b,c,d) outlined – namely the progressive development of more highly condensed conformationally restricted structures by oxidative (dehydrogenation) reactions. The physical properties of the resultant metabolites are, however, not readily predicted. Thus vescalagin, castalagin and geraniin are all formally derived from β-1,2,3,4,6-pentagalloyl-D-glucose by the loss of *six* hydrogen atoms. Geraniin (Figure 1.34) is a beautifully crystalline molecule which, despite the presence of 14 hydroxyl groups, is virtually insoluble in water. Vescalagin and castalagin (Figure 1.36) are also both crystalline, possess 15 hydroxyl groups, but in contrast to geraniin and β-1,2,3,4,6-pentagalloyl-D-glucose are highly soluble in water from which they are very difficult to extract with organic solvents. This feature is of very great importance when consider-

vescalagin / castalagin

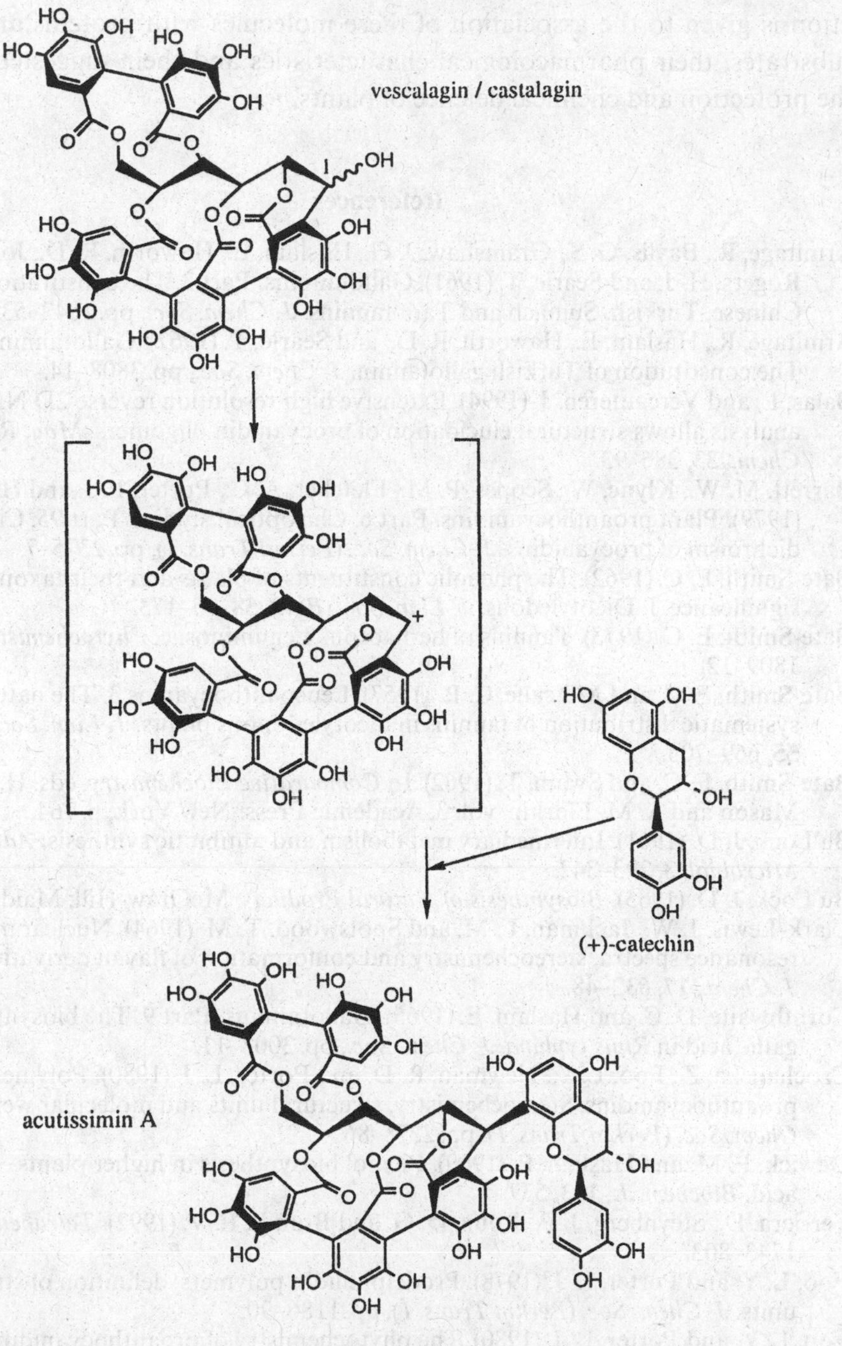

(+)-catechin

acutissimin A

Figure 1.37. Suggested pathway of biogenesis from vescalagin/castalagin of acutissimin A.

ation is given to the association of these molecules with proteins and other substrates, their pharmacological characteristics and their suggested role in the protection and chemical defence of plants.

References

Armitage, R., Baylis, G. S., Gramshaw, J. G., Haslam, E., Haworth, R. D., Jones, K., Rogers, H. J. and Searle, T. (1961). Gallotannins. Part 3. The constitution of Chinese, Turkish, Sumach and Tara tannins. *J. Chem. Soc.*, pp. 1842–53.

Armitage, R., Haslam, E., Haworth, R. D., and Searle, T. (1962). Gallotannins. Part 6. The constitution of Turkish gallotannin. *J. Chem. Soc.*, pp. 3808–14.

Balas, L. and Vercauteren, J. (1994). Extensive high resolution reverse 2D NMR analysis allows structural elucidation of procyanidin oligomers. *Mag. Reson. Chem.*, **33**, 386–93.

Barrett, M. W., Klyne, W., Scopes, P. M., Fletcher, A. C., Porter, L. J. and Haslam, E. (1979). Plant proanthocyanidins. Part 6. Chiroptical studies. Part 95. Circular dichroism of procyanidins. *J. Chem. Soc.* (*Perkin Trans. I*), pp. 2375–7.

Bate-Smith, E. C. (1962). The phenolic constituents of plants and their taxonomic significance. I. Dicotyledons. *J. Linn. Soc.* (*Bot.*), **58**, 95–173.

Bate-Smith, E. C. (1973). Tannins of herbaceous Leguminosae. *Phytochemistry*, **12**, 1809–12.

Bate-Smith, E. C. and Metcalfe, C. R. (1957). Leucoanthocyanins 3. The nature and systematic distribution of tannins in dicotyledonous plants. *J. Linn. Soc.* (*Bot.*), **55**, 669–705.

Bate-Smith, E. C. and Swain, T. (1962). In *Comparative Biochemistry*, eds. H. S. Mason and A. M. Florkin, vol. 3, Academic Press: New York, p. 764.

Bu'Lock, J. D. (1961). Intermediary metabolism and antibiotic synthesis. *Adv. Appl. Microbiol.* **3**, 293–342.

Bu'Lock, J. D. (1965). *Biosynthesis of Natural Products*, McGraw-Hill: Maidenhead.

Clark-Lewis, J. W., Jackman, L. M. and Spotswood, T. M. (1964). Nuclear magnetic resonance spectra, stereochemistry and conformation of flavan derivatives. *Aust. J. Chem.*, **17**, 632–48.

Cornthwaite, D. C. and Haslam, E. (1965). Gallotannins. Part 9. The biosynthesis of gallic acid in *Rhus typhina*. *J. Chem. Soc.*, pp. 3008–11.

Czochanska, Z., Foo, L. Y., Newman, R. D. and Porter, L. J. (1980). Polymeric proanthocyanidins. Stereochemistry, structural units and molecular weight. *J. Chem. Soc.* (*Perkin Trans. I*), pp. 2278–86.

Dewick, P. M. and Haslam, E. (1969). Phenol biosynthesis in higher plants – gallic acid. *Biochem. J.*, **113**, 537–7.

Ferriera, D., Steynberg, J. P., Roux, D. G. and Brandt, E. V. (1992). *Tetrahedron*, **48**, 1743–803.

Foo, L. Y. and Porter, L. J. (1978). Prodelphinidin polymers: definition of structural units. *J. Chem. Soc.* (*Perkin Trans. I*), pp. 1186–90.

Foo, L. Y. and Porter, L. J. (1980). The phytochemistry of proanthocyanidin polymers. *Phytochemistry*, **19**, 1747–54.

Freudenberg, K. and Purrmann, L. (1924). Raumisomere Catechin IV. *Liebig's Annalen*, **437**, 274–85.

Freudenberg, K., Bohme, L. and Purrmann, L. (1922). Raumisomere Catechin II. *Ber. Dscht. Chem. Ges.*, **55**, 1734–47.

Glombitza, K. and Gerstberger, G. (1985). Phlorotannins with dibenzodioxin structural elements from the brown alga *Eisenia arborea*. *Phytochemistry*, **24**, 543–51.

Gottlieb, O. R. (1990). Phytochemicals, differentiation and function. *Phytochemistry*, **29**, 1715–24.

Gross, C. G. (1982). Synthesis of β-glucogallin from UDPglucose and gallic acid by an enzyme from oak leaves. *Fed. Eur. Biochem. Lett.*, **148**, 67–70.

Gross, C. G. (1983). Synthesisof mono- , di- and trigalloyl-D-glucose by β-glucogallin-dependent galloyl transferases from oak leaves. *Z. Naturforsch.*, **38C**, 519–23.

Gross, C. G., Schmidt, S. W. and Denzel, K. (1986a). β-Glucogallin-dependent acyl transferases from oak leaves. I. Partial purification and characterisation. *J. Plant Physiol.*, **126**, 173–9.

Gross, C. G., Schmidt, S. W., Denzel, K. and Schilling, G. (1986b). Enzymatic synthesis of 1,6-digalloyl-D-glucose from β-glucogallin by β-glucogallin: β-glucogallin 6-O-galloyl transferase from oak leaves. *Z. Naturforsch.*, **42C**, 87–92.

Gross, C. G., Denzel, K. and Schilling, G. (1988). Biosynthesis of gallotannins. Enzymatic conversion of 1,6-digalloylglucose to 1,2,6-trigalloylglucose. *Planta*, **176**, 135–7.

Gross, C. G. and Denzel, K. (1991). Biosynthesis of gallotannings. β-Glucogallin dependent galloylation of 1,6-digalloylglucose to 1,2,6-trigalloylglucose. *Z. Naturforsch.*, **46C**, 389–94.

Hahlbrock, K. and Grisebach, H. (1979). Enzymatic controls in the biosynthesis of lignin and flavonoids. *Ann. Rev. Plant Physiol.*, **30**, 105–30.

Hahlbrock, K. and Scheel, D. (1989). Physiology and molecular biology of phenylpropanoid metabolism. *Ann. Rev. Plant Physiol.*, **40**, 347–69.

Harborne, J. B. (1960). The genetic variation of anthocyanin pigment in plant tissues. In *Phenolics in Plants in Health and Disease*, ed. J. B. Pridham, Pergamon Press: Oxford and London, pp. 109–17.

Haslam, E. (1974). Biogenetically patterned synthesis of procyanidins. *J. Chem. Soc. Chem. Commun.*, pp. 594–5.

Haslam, E. (1977). Symmetry and promiscuity in proanthocyanidin biochemistry. *Phytochemistry*, **16**, 1625–40.

Haslam, E. (1982). The metabolism of gallic and hexahydroxydiphenic acid in plants. *Fortschritt. Chem. Org. Naturstoffe*, **41**, 1–46.

Haslam, E. (1986a). Secondary metabolism – fact and fiction. *Nat. Prod. Rep.*, **3**, 217–49.

Haslam, (1986b). Hydroxybenzoic acids and the enigma of gallic acid. In *The Shikimic Acid Pathway*, ed. E. E. Conn, *Recent Adv. Phytochemistry*, **20**, 163–200.

Haslam, E. (1989). *Plant Polyphenols – Vegetable Tannins Revisited*, Cambridge University Press: Cambridge.

Haslam, E. (1992). Gallic acid and its metabolites. In *Plant Polyphenols – Synthesis, Properties and Significance*, ed. R. W. Hemingway and P. E. Laks, Plenum Press: New York, pp. 169–94.

Haslam, E. (1993). *Shikimic Acid: Metabolism and Metabolites*, Chichester-New York: John Wiley.

Haslam, E. (1995). Secondary metabolism – evolution and function; products or processes? *Chemoecology*, **5/6**, 89–95.

Haslam, E. and Cai, Y. (1994). Plant polyphenols (vegetable tannins): gallic acid metablism. *Natural Products Reports* (*Royal Soc. Chem.*), **11**, 41–66.

Haslam, E., Fletcher, A. C., Porter, L. J. and Gupta, R. K. (1977). Plant proanthocyanidins. Part 3. Conformational and configurational studies of natural procyanidins. *J. Chem. Soc.* (*Perkin Trans. I*), pp. 1628–37.

Haslam, E., Haddock, E. and Gupta, R. K. (1982a). The metabolism of gallic acid and hexahydroxydiphenic acid in plants. Part 3. Esters of (*R*) and (*S*) hexahydroxydiphenic acid and dehydrohexahydroxydiphenic acid with D-glucopyranose ($^4C_1$ and related conformations). *J. Chem. Soc.* (*Perkin Trans. I*), pp. 2535–45.

Haslam, E., Haddock, E., Gupta, R. K., Al-Shafi, S. M. K. and Layden, K. (1982b). The metabolism of gallic acid and hexahydroxydiphenic acid in plants. Part 2. Esters of (*S*) hexahydroxydiphenic acid with D-glucopyranose ($^4C_1$). *J. Chem. Soc.* (*Perkin Trans. I*), pp. 2525–35.

Haslam, E., Layden, K., Gupta, R. K., Al-Shafi, S. M. K. and Magnolato, D. (1982c). The metabolism of gallic acid and hexahydroxydiphenic acid in plants. Part 1. Naturally occurring galloyl esters. *J. Chem. Soc.* (*Perkin Trans. I*), pp. 2515–24.

Haslam, E., Haddock, E. A., Gupta, R. K., Al-Shafi, S. M. K., Layden, K. and Magnolato, D. (1982d). The metabolism of gallic and hexahydroxydiphenic acid in plants. Biogenetic and molecular taxonomic considerations. *Phytochemistry*, **21**, 1049–62.

Haslam, E., Spencer, C. M., Cai, Y., Martin R. and Lilley, T. H. (1990a). The metabolism of gallic acid and hexahydroxydiphenic acid in plants. Part 4. Polyphenol interactions. Part 3. Spectroscopic and physical properties of esters of gallic acid and (*S*) hexahydroxydiphenic acid with D-glucopyranose ($^4C_1$). *J. Chem. Soc.* (*Perkin Trans. 2*), pp. 651–60.

Haslam, E., Spencer, C. M., Cai, Y., Martin R. Gaffney, S. H., Magnolato, D. and Lilley, T. H. (1990b). Polyphenol interactions. Part 4. Model studies with caffeine and cyclodextrins. *J. Chem. Soc.* (*Perkin Trans. 2*). pp. 2197–209.

Haslam, E., Spencer, C. M., Cai, Y., Martin R. Gaffney, S. H., Warminski, E., Liao, H., Goulding, P. N., Warminski, E., Liao, H., Goulding, P. N., Luck, G. and Lilley, T. H. (1992). Polyphenol complexation. In *Polyphenols in Food and Health I*, eds. C. T. Ho, C. Y. Lee and M. T. Huang, Washington: American Chemical Society, American Chemical Society Symposia series, **506**, 8–50.

Haworth, R. D. (1961). Pedler Lecture. Some problems in the chemistry of the gallotannins. *Proc. Chem. Soc.*, pp. 401–10.

Haworth, R. D., Pindred, H. K. and Grimshaw, J. (1955). Galloflavin. Part II. *J. Chem. Soc.*, pp. 833–7.

Heller, W. and Forkman, G. (1988). Biosynthesis. In *The Flavonoids – Advances in Research since 1980*, ed. J. B. Harborne, Chapman and Hall: London, pp. 399–425.

Heller, W., Forkman, G., Britsch, L. and Grisebach, H. (1985). Enzymatic Reduction of (+)-dihydroflavonols to *cis*-flavan-3,4-diols with flower extracts from *Matthiola incana* and its role in anthocyanin biosynthesis. *Planta*, **165**, 284–7.

Hemingway, R. W. and Foo, L. Y. (1983). Quinone methide intermediates in procyanidin synthesis. *J. Chem. Soc. Chem. Commun.*, pp. 1035–6.

Hemingway, R. W., Foo, L. Y. and Porter, L. J. (1982). Linkage isomerism in trimeric and polymeric 2,3-*cis*-procyanidins. *J. Chem. Soc.* (*Perkin Trans. I*), pp. 1209–16.

Hillis, W. E. and Swain, T. (1959). The phenolic constituents of *Prunus domestica*. II. The analysis of tissues of the Victorian Plum tree. *J. Sci. Food Agric.*, **10**, 135–44.

Hillis, W. E. and Seikel, M. (1970). Hydrolysable tannins of *Eucalyptus delegatensis* wood. *Phytochemistry*, **9**, 1115–28.

Hofmann, A. S. and Gross, G. C. (1990). Biosynthesis of gallotannins: formation of polygalloylglucoses by enzymatic acylation of β-1,2,3,4,6-pentagalloyl-D-glucose. *Arch. Biochem. Biophys.*, **283**, 530–2.

Hsu, F. L., Nonaka, G. and Nishioka, I. (1985a). Tannins and related compounds, 31. Isolation and characterisation of procyanidins in *Kandelia candel* (L) Druce. *Chem. Pharm. Bull.*, **33**, 3142–52.

Hsu, F. L., Nonaka, G. and Nishioka, I. (1985b). Tannins and related compounds, 33. Isolation and characterisation of procyanidins in *Dioscorea cirrhosa* Lour. *Chem. Pharm. Bull.*, **33**, 3293–8.

Jacques, D., Haslam, E., Greatbanks, D. and Bedford, G. R. (1974). Plant proanthocyanidin A-2 and its derivatives. *J. Chem. Soc. (Perkin Trans. I)*, pp. 2663–9.

Jones, W. T., Broadhurst, R. B. and Lyttleton, J. W. (1976). The condensed tannins of pasture legume species. *Phytochemistry*, **15**, 1407–9.

Jurd, L. and Lundin, R. (1968). Anthocyanidins and related compounds. XII. Tetramethyl-leucocyanidin-phlorogucinol and resorcinol condensation products. *Tetrahedron*, **24**, 2653–61.

Keen, P. C., Haslam, E. and Haworth, R. D. (1962). Gallotannins. Part 7. Tara gallotannin. *J. Chem. Soc.*, pp. 3814–18.

Knowles, P. F., Haslam, E. and Haworth, R. D. (1961). Gallotannins. Part 4. The biosynthesis of gallic acid. *J. Chem. Soc.*, pp. 1854–9.

Kubitzki, K. (1987). Phenylpropanoid metabolism in relation to land plant origin and diversification. *J. Plant Physiol.*, **131**, 17–24.

Kubitzki, K. and Gottlieb, O. R. (1984). Phytochemical aspects of angiosperm origin and evolution. *Acta Botanica Neerlandica*, **33**, 457–68.

McInnes, A. G., Smith, D. G., Walter, J. A. and Ragan M. A. (1985). The high molecular weight polyphloroglucinols of the marine brown alga *Fucus vesiculosus* L.: [1]H and ^{13}C nuclear magnetic resonance spectroscopy. *Canad. J. Chem.*, **63**, 304–13.

Marini-Bettolo, G. B., Delle-Monache, F. and Ferrari, F. (1971). Occurrence of (+)-epicatechin in nature. *Gazz. Chim. Ital.*, **101**, 387–95.

Marini-Bettolo, G. B., Delle-Monache, F., Poce-Tucci, A. and Ferrari, F. (1972). Catechins with the (+)-epi configuration in nature. *Phytochemistry*, **11**, 2333–6.

Mattice, W. L., Fronczek, F. R., Gannuch, G. R., Tobiason, F. L., Brocker, J. L. and Hemingway, R. W. (1984). Dipole moment, solution and solid state structure of (−)-epicatechin, a monomer unit of procyanidin polymers. *J. Chem. Soc. (Perkin Trans. II)*, pp. 1611–16.

Mattice, W. L., Fronczek, F. R., Gannuch, G. R., Tobiason, F. L., Chiari, G., Houglum, K., Shanafelt, A. and Hemingway, R. W. (1985). Preference for occupancy of axial positions by substituents bonded to the heterocyclic ring in penta-acetyl-(+)-catechin in the crystalline state. *J. Chem. Soc. (Perkin Trans. II)*, pp. 1383–6.

Mattice, W. L. and Viswanadhan, V. N. (1986). Conformational analysis of the sixteen C(4)–C(6) and C(4)–C(8) linked dimers of (+)-catechin and (−)-epicatechin. *J. Computational Chem.*, **7**, 711–17.

Mattice, W. L. and Viswanadhan, V. N. (1987). Assessment by molecular mechanics of the preferred conformations of the sixteen C(4)–C(6) and C(4)–C(8) linked dimers of (+)-catechin and (−)-epicatechin with axial or equatorial dihydroxyphenyl substituents at C(2). *J. Chem. Soc. (Perkin Trans. II)*, pp. 739–43.

Mayer, W. (1973). Otto Theodor Schmidt. *Liebig's Annalen*, pp. 1759–76.

Mayer, W., Goll, L. Arndt, E. V. and Mannschrek, A. (1966). Procyanidino-(−)-epicatechin, ein zweiarmig verknuftes, kondensiertes Proanthocyanidin aus *Aesculus hippocastanum. Tetrahedron Letters*, pp. 429–35.

Mayer, W., Einwiller, A. and Jochims, J. C. (1967). Die struktur des Castalins. *Liebig's Annalen*, **707**, 182–9.

Mayer, W., Sietz, H. and Jochims, J. C. (1969). Die struktur des Castalagins. *Liebig's Annalen*, **721**, 186–93.

Mayer, W., Kullman, F. and Schilling, G. (1971a). Die struktur des Vescalins. *Liebig's Annalen*, **747**, 51–9.

Mayer, W., Sietz, H., Schauerte, K., Schilling, G. and Jochims, J. C. (1971b). Die struktur des Vescalagins. *Liebig's Annalen*, **751**, 60–8.

Mayer, W., Bilzer, W. and Schauerte, K. (1971c). Isolierung von Castalagin und Vescalagin aus Valoneagerbstoffen. *Liebig's Annalen*, **754**, 149–51.

Mayer, W., Schilling, G., Weinges, K. and Muller, O. (1973). ^{13}C-NMR Spektroscopische Konstitutionsermittelung der $C_{30}H_{24}O_{12}$Procyanidine. *Liebig's Annalen*, pp. 1471–5.

Mole, S. (1993). The systematic distribution of tannins in the leaves of angiosperms: A tool for ecological studies. *Biochemical Systematics and Ecology*, **21**, 833–46.

Morimoto, S., Nonaka, G. and Nishioka, I. (1983). Tannins and related compounds 13. Isolation and structures of trimeric, tetrameric and pentameric proanthocyanidins from cinnamon. *J. Chem. Soc. (Perkin Trans. 1)*, pp. 2139–45.

Morimoto, S., Nonaka, G. and Nishioka, I. (1985). Tannins and related compounds 35. Proanthocyanidins with a doubly linked unit from the root bark of *Cinnamonium sieboldii* Meisner. *Chem. Pharm. Bull.*, **33**, 4338–45.

Morimoto, S., Nonaka, G. and Nishioka, I. (1986). Tannins and related compounds. Isolation and characterisation of flavan-3-ol glucosides and procyanidin oligomers from cassia bark (*Cinnamomium cassia* Blume). *Chem. Pharm. Bull.*, **34**, 633–42.

Morimoto, S., Nonaka, G. and Nishioka, I. (1988). Tannins and related compounds. 60. Isolation and characterisation of a proanthocyanidin with a doubly linked unit from *Vaccinium Vitis idaea* L. *Chem. Pharm. Bull.*, **36**, 33–8.

Nishioka, I., Nonaka, G., Nishizawa, M. and Yamagishi, T. (1982). Tannins and related compounds. Part 5. Structure of Chinese gallotannin. *J. Chem. Soc. (Perkin Trans. I)*, pp. 2963–8.

Nishioka, I., Nonaka, G., Nishizawa, M. and Yamagishi, T. (1983). Tannins and related compounds. Part 9. Isolation and characterisation of polygalloyl glucoses from Turkish galls (*Quercus infectoria*), *J. Chem. Soc. (Perkin Trans. I)*, pp. 961–5.

Nishioka, I., Nonaka, G. and Ageta, M. (1985). Tannins and related compounds. Part 25. A new class of gallotannins possessing a (−)-shikimic acid core from *Castanopsis cuspidata* var. *sieboldii* Nakai. *Chem. Pharm. Bull.*, **33**, 96–101.

Nishioka, I., Nonaka, G., Ishimura, K., Watanabe, M., Yamauchi, T. and Wan, A. C. S. (1987). Tannins and related compounds. 101. Elucidation of the stereochemistry of the triphenoyl moiety in castalagin and vescalagin and isolation of 1-*O*-galloyl castalagin from *Eugenia grandis. Chem. Pharm. Bull.*, **35**, 217–20.

Nishioka, I., Nonaka, G., Tanaka, T., Sakai, T. and Mihashi, K. (1990). Tannins and related compounds. 97. Structure revision of C-glycosidic ellagitannins, castalagin, vescalagin, casuarinin and stachyurin and related hydrolysable ellagitannins. *Chem. Pharm. Bull.*, **38**, 2151–6.

Okuda, T. Hatano, T., Koga, T., Toh, N., Kuriyama, N. and Yoshida, T. (1982a). Circular dichroism of hydrolysable tannins – 2. Ellagitannins and gallotannins. *Tetrahedron Letters*, **23**, 3937–40.

Okuda, T., Hatano, T., Koga, T., Toh, N., Kuriyama, N. and Yoshida, T. (1982b). Circular dichroism of hydrolysable tannins – 3. Ellagitannins and gallotannins. *Tetrahedron Letters*, **23**, 3941–4.

Okuda, T., Hatano, T. and Ogawa, N. (1982c). Rugosins D, E, F and G. Dimeric and trimeric hydrolysable tannins. *Chem. Pharm. Bull.*, **30**, 4234–7.

Okuda, T., Kuwahara, M., Usman-Memon, M., Shingu, T. and Yoshida, T. (1982d). Agrimoniin and potentillin, an ellagitannin dimer and monomer having an α-glucose core. *J. Chem. Soc. Chem. Commun.*, pp. 163–4.

Okuda, T., Hatano, T. and Yoshida, T. (1982e). Constituents of *Geranium thunbergii*. 12. Hydrated stereostructure and equilibration of geraniin. *J. Chem. Soc. (Perkin Trans. I)*, pp. 9–14.

Okuda, T., Hatano, T., Yazaki, K., Ashida, M. and Yoshida, T. (1982f). Ellagitannins of the Casuarinaceae, Stachyuraceae and Myrtaceae. *Phytochemistry*, **21**, 2871–4.

Okuda, T., Hatano, T. and Yoshida, T. (1989). New methods of analysing tannins. *J. Nat. Prod.*, **52**, 1–31.

Okuda, T., Hatano, T. and Yoshida, T. (1990). Oligomeric hydrolysable tannins, a new class of plant polyphenols. *Heterocycles*, **30**, 1195–218.

Okuda, T., Hatano, T. and Yoshida, T. (1993). Polyphenols of new types and their correlation with plant systematics. *Phytochemistry*, **32**, 507–22.

Philbin, E. M. and Wheeler, T. S. (1958). Conformation of chromans and related compounds. *Proc. Chem. Soc.*, pp. 167–8.

Porter, L. J. (1988). Flavans and proanthocyanidins. In *The Flavonoids: Advances in Research since 1980*, ed. J. B. Harborne, Chapman and Hall: London, pp. 21–62.

Porter, L. J. (1989). Tannins. In *Methods in Plant Biochemistry. Vol. I. Plant Phenolics*, ed. J. B. Harborne, Academic Press: London and New York, pp. 389–419.

Porter, L. J., Wong, R. Y., Benson, M., Chan, B. G., Vishwanadhan, V. N., Gandour, R. D. and Mattice, W. L. (1986). Conformational analysis of flavans. *J. Chem. Research*, pp. 86–7.

Porter, L. J., Newman, R. H., Foo, L. Y., Wong, H. and Hemingway, R. W. (1982). Polymeric proanthocyanidins. ^{13}C NMR studies of procyanidins. *J. Chem. Soc. (Perkin Trans. I)*, pp. 1217–21.

du Preez, I. C., Rowan, A. C., Roux, D. G. and Feeney, J. (1971). Hindered rotation about the sp^2-sp^3 hybridised C-C bond between flavonoid units in condensed tannins. *J. Chem. Soc. Chem. Commun.*, pp. 315–16.

Ragan, M. A. (1985). The high molecular weight polyphloroglucinols of the marine brown alga *Fucus vesiculosus* L: degradative analysis. *Canad. J. Chem.*, **63**, 294–303.

Richardson, A. C. and Williams, J. M. (1967). Selective acylation of pyranosides. I. Benzoylation of methyl-a-pyranosides of mannose, glucose and galactose. *Tetrahedron*, **23**, 1369–78.

Robinson, R. and Robinson, G. M. (1933). XXXI. A survey of Anthocyanins. III. Notes on the distribution of leuco-anthocyanins. *Biochem. J.*, **27**, 206–12.

Robinson, R. and Robinson, G. M. (1935). Leuco-anthocyanins and leuco-anthocyanidins. Part I. The isolation of peltogynol. *J. Chem. Soc.*, pp. 744–52.

Sanderson, G. W. (1972). The chemistry of tea and tea manufacturing. *Rec. Adv. Phytochemistry*, **5**, 247–316.

Scalbert, A., Herve du Penhoat, C. L. M., Michon, V. M. F., Peng, S. and Viriot, C. (1991). Structure elucidation of new dimeric ellagitannins from *Quercus robur* L. – Roburins A–E. *J. Chem. Soc.* (*Perkin Trans. I*), pp. 1653–60.

Schmidt, O. Th. (1954). Ellagengerbstoffe. *Das Leder*, **5**, 129–34.

Schmidt, O. Th. (1955). Naturliche gerbstoffe. In *Moderne Methoden der Pflanzen Analyse*, vol. III, eds. K. Paech and H. Tracey, Springer: Berlin, Gottingen, Heidelberg, pp. 517–48.

Schmidt, O. Th. (1956). Gallotannine und Ellagen-gerbstoffe. *Fortschritt. Chem. Org. Naturstoffe*, **13**, 70–136.

Schmidt, O. Th. and Mayer, W. (1956). Naturliche Gerbstoffe. *Angew. Chem.* **68**, 103–15.

Scott-Moncrieff, R. (1981). The classical period in chemical genetics. *Notes Rec. Royal Soc. London*, **36**, 125–54.

Shen, Z., Haslam, E., Falshaw, C. P. and Begley, M. J. (1986). Procyanidins and polyphenols of *Larix gmelini* bark. *Phytochemistry*, **25**, 2629–35.

Spek, A. L., Kojic-Prodic, B. and Labadie, R. P. (1984). Structure of (−)-epicatechin: (2*R*,3*R*)-2-(3,4-dihydroxyphenyl)-3,4-dihydro-2*H*-1-benzopyran-3,4,5-triol, $C_{15}H_{14}O_6$. *Acta Cryst.*, **C40**, 2068–71.

Stafford, H. A. and Lester, H. H. (1982). Enzymatic and non-enzymatic reduction of (+)-dihydroquercetin to its 3,4-diol. *Plant Physiol.*, **70**, 695–8.

Stafford, H. A. and Lester, H. H. (1984). Flavan-3-ol biosynthesis. The conversion of (+)-dihydroquercetin and flavan-3,4-*cis*-diol (leucocyanidin) to (+)-catechin by reductases extracted from cell suspension cultures of Douglas fir. *Plant Physiol.*, **76**, 184–6.

Stafford, H. A. and Lester, H. H. (1985). Flavan-3-ol biosynthesis. The conversion of (+)-dihydromyricetin to its flavan-3,4-*cis*-diol (leucodelphnidin) and to (+)-gallocatechin by reductases extracted from tissue cultures of *Gingko biloba* and *Pseudotsuga menziesii*. *Plant Physiol.*, **78**, 791–4.

Stafford, H. A., Lester, H. H. and Porter, L. J. (1985). Chemical and enzymic synthesis of monomeric procyanidins (leucocyanidins) from 2*R*,3*R*-dihydroquercetin. *Phytochemistry*, **24**, 333–8.

Thompson, R. S., Jacques, D., Haslam, E. and Tanner, R. J. N. (1972). Plant proanthocyanidins. Part I. Introduction; the isolation, structure and distribution in nature of plant procyanidins. *J. Chem. Soc.* (*Perkin Trans. I*), pp. 1387–99.

Waterman, P. G. and Mole, S. (1994). *Analysis of Plant Phenolic Metabolites*, Blackwell: London, Edinburgh, Boston.

Weinges, K., Bahr, W., Ebert, W., Goritz, K. and Marx, H.-D. (1969). Konstitution, Enstehung und Bedeutung der Flavonoid-Gerbstoffe. *Fortschritte Chem. Org. Naturstoffe*, **27**, 158–260.

Weinges, K., Goritz, K. and Marx, H.-D. (1970). Zur Kenntnis der Proanthocyanidine. XV. Die Rotationsbehinderung an der C (sp²) – (sp³)-Bindung der 4-Aryl substitutienten Polymethoxyflavane. *Chem. Ber.*, **103**, 2336–43.

Weinges, K., Kaltenhauser, W., Marx, H.-D., Nader, E., Nader, F., Perner, J. and Seiler, D. (1968). Procyanidine aus Fruchten. *Liebig's Annalen*, **711**, 184–204.

Whalley, W. B. (1962). The stereochemistry of the flavonoid compounds. In *The Chemistry of Flavonoid Compounds*, ed. T. A. Geissmann, Oxford: Pergamon Press, pp. 441–67.

White, T. (1956). The scope of vegetable tannin chemistry. In *The Chemistry of the Vegetable Tannins*, Soc. Leather Trades' Chemists: Croydon, pp. 7–29.

White, T. (1957). Tannins – their occurrence and significance. *J. Sci. Food Agric.*, **8**, 377–85.

Wilkins, C. K. and Bohm, B. A. (1976). Ellagitannins from *Tellima grandiflora*. *Phytochemistry*, **15**, 211–14.

Yoshida, T., Hatano, T., Memon, M. U., Shingu, T., Okuda, T. and Inoue, K. (1984). Spectral and chromatographic analysis of tannins. *Chem. Pharm. Bull.*, **32**, 1790–8.

Yoshida, T., Hatano, T., Shingu, T. and Okuda, T. (1988). ^{13}C NMR Spectra of hydrolysable tannins. II. Tannins forming anomeric mixtures. *Chem. Pharm. Bull.*, **36**, 2925–33.

Zenk, M. H. (1964). Zur Frage de Biosynthese von Gallusaure. *Z. Naturforsch.*, **19B**, 83–4.

2

Molecular recognition

2.1 Introduction

The infamous battle of the sexes has been likened to a civil war between two entities that have a common ancestry. Each year in nature plants and animals indulge in seemingly chaotic patterns of selection, courtship and finally mating. These have evolved to create new life which in turn is critical to species survival. A crucial and fascinating aspect of reproduction is the strategem, which different species adopt, to choose and attract a partner. The process in nature may involve not only the senses of sight and sound but also those of taste and smell. For man himself, portrayals of the awesome power of sexual desire in the determination of the outcome of that choice continue to fill our literature to the present day. Persse McGarrigle from Limerick is thus the unwitting, if not unwilling, fictional victim in this timeless struggle as it unfolds in the opening pages of *Small World* by David Lodge*.

At that moment the knots of chatting conferees seemed to loosen and part, as if by some magical impulsion, opening up an avenue between Persse and the doorway. There, hesitating on the threshold, was the most beautiful girl he had ever seen in his life. She was tall and graceful, with a full, womanly figure, and a dark creamy, complexion. Black hair fell in shining waves to her shoulders, and black was the colour of her simple woollen dress, scooped out low across her bosom. She took a few paces forward into the room and accepted a glass of sherry from the tray offered to her by a passing waitress. She did not drink at once, but held the stem of the glass up to her face as if it were a flower. Her right hand held the stem of the glass between index finger and thumb. Her left, passed horizontally across her waist, supporting her right elbow. Over the rim of the glass she looked with eyes as dark as peat pools straight into Persse's own, and seemed to smile faintly in greeting. She raised the glass to her lips, which were red and moist, the underlip slightly swollen in appearance, as though it had been stung. She drank, and he saw the muscles in her throat move and slide under the skin as she

* Secker & Warburg Publishers, London, 1984: Penguin Books, London, 1985.

84

swallowed. 'Heavenly God!' Persse breathed, quoting again, this time from *A Portrait of the Artist as a Young Man.*

The philosopher Carl Jung has likened the meeting of two personalities to the contact of two chemical substances and the poet Goethe observed that the attractions and repulsions between man and woman have many similarities with molecules in a reversible chemical process (Rommerts, 1995/1996). Embodied in this first seductive encounter between Persse and Angelica are many subtle allusions and analogies, expressed in human behavioural terms, to the processes of molecular recognition – the attraction (or as is sometimes the case, the repulsion) of one species for another, an attraction which is created primarily by both parties breathing life into probing nascent impulses of recognition and complementarity. An attraction in which, moreover, the context, the milieu plays an equally important, if not dominant, role.

2.2 Molecules of life – DNA

It has been said that DNA (deoxyribonucleic acid) is the blueprint for life and that proteins are the stuff of life. The endeavours which accompanied the early exploration of the structure and the three-dimensional shape of these two groups of natural macromolecules are now part of the folklore of science. The euphoria which greeted these early seminal discoveries gave rise, in the ensuing 40 years, to a mushrooming interest in genetics and molecular biology. Less dramatically, perhaps, they exemplified the crucial importance of weak intra- and inter-molecular interactions as determinants of the shape of natural macromolecules; particular attention was drawn to the roles of **hydrogen bonding** and of '**hydrophobic effects**'. This largely observational and descriptive phase within biology has given rise latterly to a burgeoning interest in the more theoretical aspects of the phenomenon of molecular recognition and the manner in which weak molecular interactions may determine molecular organisation and complexation. Serious attempts continue to be made to seek to quantify (if as yet only approximately) many of these interactions and thus to develop the science of molecular recognition into one capable of making firm predictions.

Although the nucleic acids were first described in 1868 it was not until the 1940s that compelling evidence was obtained to support the view that DNA carries the genetic blueprint for the cell. This went a long way to answering the question: 'what is the nature of the gene?'. Biochemists quickly turned to the corollary of this question, namely: 'how does the DNA replicate itself and transmit genetic information to a daughter cell?'. The answer came in 1953 when the brilliant exposition of Watson and Crick detailed the three-dimensional structure of DNA and from this it was possible, for the first time, to infer a mechanism of replication. Critical to this hypothesis was a particular structural feature of their model for DNA. According to Watson and Crick the molecule of DNA consists of two helically coiled polynuceotide chains, centred around a common axis and with each chain running in opposite directions. The two chains are held together by **hydrogen bonding** between base pairs situated on the inside of the helix. Adenine is always paired with thymine and guanine with cytosine, Figure 2.1. This specificity of base pairing is the most important aspect of the Watson–Crick model for DNA and it immediately suggested a beautifully simple model for its replication (Watson and Crick, 1953).

Now our model for deoxyribonucleic acid is, in effect, a *pair* of templates, each of which is complementary to the other. We imagine that prior to duplication the hydrogen bonds are broken, and the two chains unwind and separate. Each chain then acts as a template for the formation onto itself of a new companion chain, so that eventually we shall have *two* pairs of chains, where we only had one before. Moreover the sequence of the pairs of bases will have been duplicated exactly.

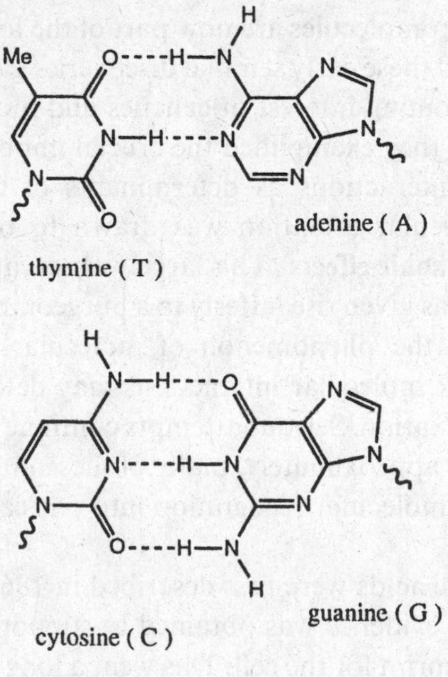

∿ base attachment to C-1 of deoxyribose

Figure 2.1. Base pairing via hydrogen bonding (broken lines) in DNA: adenine (A) – thymine (T) and guanine (G) – cytosine (C) base pairs.

2.3 Molecules of life – proteins

Thousands of different proteins go into the make-up of a living cell; they perform thousands of different functions. How these are organised and executed in physical and temporal terms within the compass of living systems remains a key contemporary problem in biology awaiting resolution. Initial attempts to utilise X-ray crystallography to examine the structures of proteins were discouraging. However in the 1950s progress began to be made, first with amino acids and peptides and then with proteins themselves – myoglobin (Kendrew) and haemoglobin (Perutz). Today the situation is transformed. Since the mid-1970s there has been an annual exponential rise in the number of protein structures solved and now well over 300 three-dimensional structures are recorded – not only of those which play a structural role but also of those whose functions are catalytic or are associated with the transport of molecules and ions. A striking characteristic of the globular proteins which have been examined is that they have well defined three-dimensional shapes (conformations) derived from a specific mode of folding of the linear polypeptide chain(s). It is with these shapes that their biological activity is associated. The conformational stability of almost all naturally occurring globular proteins lies between 5 and 15 kcal mol^{-1}. It thus appears advantageous to living organisms to possess proteins in which the folded, biologically active, conformation is only marginally more stable than an unfolded conformation. The reasons are not clear.

In contrast to the molecule of DNA no spectacular insight or Pauline revelation has yet illuminated the corresponding structural and conformational problems associated with the globular proteins. Whilst, in most cases, it is clear that all the information necessary to define the tertiary fold is enclosed in the particular amino acid sequence the question of how polypeptide chains fold to give discrete, highly organised and tightly packed three-dimensional protein structures is a major unsolved intellectual puzzle in biology and chemistry. However, given that the present evidence suggests that the number of distinctive polypeptide-chain folds is likely to be limited, the prospect of identifying which fold a particular amino acid sequence is probably going to adopt has now become a realistic goal. The complexity of these problems and the length of time over which they have now been addressed has, perforce as a byproduct, born fruit in other ways – most significantly our knowledge of the fundamental underlying mechanisms of molecular recognition has been enhanced.

Pauling and his colleagues (Pauling *et al.*, 1951), based upon their knowledge of the planar peptide bond, bond angles and bond lengths and taking

into account only the steric and chemical constraints imposed by the polypeptide backbone, predicted the occurrence in proteins of two periodically recur-
ring secondary structural motifs – the α-helix and the β-sheet. The α-helix is
rod-like and amino acid side chains extend radially in a helical array from the
coiled polypeptide chain; there is almost no free space inside the helix. This
element of secondary structure is stabilised by interamide hydrogen bonding
between the carbonyl group of one peptide amide bond and the NH group of
another, situated four residues ahead in the sequence. In the same year Pauling
and Corey (1951) enunciated the features of another structural motif – the
β-sheet – in which polypeptide chains are almost fully extended. Adjacent
polypeptide chains may run in parallel or in anti-parallel directions and the
structural arrangement is stabilised by the formation of **interamide hydrogen
bonds** between the carbonyl groups of peptide bonds and the NH groups on
adjacent strands, Figure 2.2. Since these initial predictions, high resolution
X-ray crystallographic and NMR studies of proteins have confirmed these two
'models' as ubiquitous elements of protein structure. There is a decidedly
non-random distribution of amino acids, with certain ones showing very high
probabilities of occurrence in α-helices and others in β-sheets.

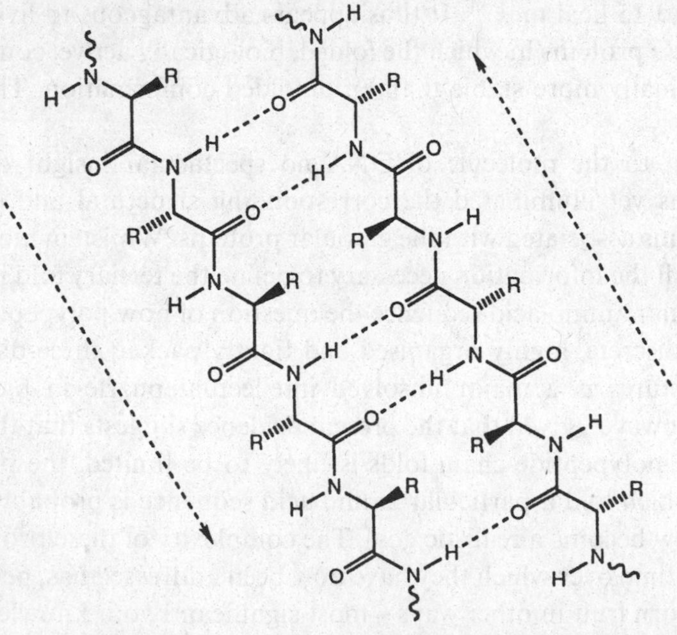

Figure 2.2. Anti-parallel β-pleated sheet. Adjacent strands of the polypeptide chain run
in opposite directions (dotted arrows). Intra-molecular hydrogen bonds (broken lines)
between NH and CO groups on the adjacent strands stabilise the structure. The
polypeptide chains are almost fully extended and the various amino acid side chains
(R) lie above and below the plane of the β-pleated sheet.

In almost all globular proteins the α-helical and the β-pleated sheet segments assemble to give the molecular ensemble its three-dimensional shape. As more protein structures are solved it has become evident that the same folding motifs have been used over and over again for a variety of functional purposes. Thus although protein structures are complex and irregular at the atomic level, their folding patterns are surprisingly few. Typical are the family of α/β barrel enzymes which have a domain with eight β-strands surrounded by seven or eight α-helices. Each inner β-strand is connected to an outer α-helix; the archetypal α/β barrel protein is triose phosphate isomerase. Present arguments suggest that the α/β barrel proteins may all have evolved from a common ancestor (Farber and Petsko, 1990).

It is usual, in attempts to elucidate the principles which govern the overall structure of proteins, to classify the 20 α-amino acid residues in terms of their *hydrophobic* or *hydrophilic* character. This derives from the initial theoretical proposal of Kauzman (1959) that the origin of globular protein stability is the removal from contact with water of hydrophobic residues. Comparison of protein sequences and structures has affirmed this view – namely that the sites which form the interior of proteins are occupied mainly by non-polar residues; those that are highly exposed to the solvent water are nearly always polar or hydrophilic in nature. Similarly the surfaces buried between elements of secondary structure are very hydrophobic in character (Chothia, 1984, 1990). As Perutz plainly and succinctly stated (Perutz, 1992): 'Most water-soluble proteins are waxy inside and soapy outside, because their larger hydrophobic amino acid residues shy away from water and coalesce'. Scientists remain somewhat divided between those who assume '**hydrophobic interactions**' are the result of a positive attraction caused by van der Waals/London dispersion forces on the one hand and those who believe that they result from the strong interactions of water molecules for each other. In this latter view '**hydrophobic interactions**' are thought to be associated with changes in the *structural arrangement* of the water molecules in the vicinity of the solute molecules, *vide infra*. Whatever the precise physico-chemical explanation, these facets of protein structure nevertheless draw particular attention to the role which '**hydrophobic effects**' and the **aqueous environment** play as factors in the development of the three-dimensional structure of natural macromolecules. Along with **hydrogen bonding** these aspects of molecular interactions which facilitate and underly the processes of molecular recognition in aqueous environments are developed in more detail below.

2.4 Non-covalent molecular interactions

To a greater or lesser extent all atoms and molecules interact with one another; at one extreme are the strong interactions characteristic of chemical bonds with stabilisations of some 50 kcal mol^{-1} and more; at the other there are the weak interactions (generally less than 1 kcal mol^{-1}) of two closed shell atoms attributable to dispersion or van der Waals forces. Between these two limits are various non-covalent interactions which are of intermediate strength and are of a quite diverse physical nature and origin. Many molecular recognition processes are a balance between fast and slow, strong highly selective and weak modes of inter- and intramolecular interaction. If one is to achieve an understanding of the features that determine molecular recognition, which parts of molecules are largely instrumental in the recognition process, then it is desirable to have some acquaintance with the physical factors that contribute to non-covalent interactions.

Electrostatic attractive interactions between molecules may be one or more of several types:

(i)   ion–ion,
(ii) ion–dipole,
(iii) ion–induced dipole,
(iv) dipole–dipole,
(v)   dipole–induced dipole, and
(vi) induced dipole–induced dipole.

Their dependence upon the distance of separation (r) and the dielectric constant of the medium (D) is summarised in Table 2.1. Interactions involving ions are thus relatively long ranged, whilst those which only involve permanent or induced dipoles are of significance over a much shorter range. **Ion–ion** interactions are typified by those of a pure electrolyte, such as potassium chloride (K^+Cl^-), between ions of opposite charge. **Ion–dipole** interactions take place when one of the species has a permanent dipole, i.e. molecules that carry no net charge but have a permanent charge separation due to the nature of the electronic distribution within the molecule itself. The molecule of water illustrates this facet of molecular structure. Oxygen is more strongly electronegative than hydrogen and the 'centre of gravity' of the electron density in each of the covalent bonds may be envisaged as displaced towards oxygen. This imbalance results in a partial negative charge on oxygen and correspondingly partial positive charges on each hydrogen. The vector addition of the electrical forces derived from the charge separation leads to the creation of an electric dipole. The resultant dipole moment (μ) is expressed as the product of

Table 2.1. *Dependence of electrostatic interactions upon distance (r) and effective dielectric constant of the medium (D)*

Interaction	Energy of interaction
ion–ion	$- K_1 \cdot q_1 \cdot q_2 / r \cdot D$
ion–dipole	$- K_2 \cdot q_1 \cdot \mu_2 / r^2 \cdot D$
ion–induced dipole	$- K_3 \cdot q_1{}^2 \cdot \alpha_2 / r^4 \cdot D^2$
dipole–dipole*	$- K_4 \cdot \mu_1{}^2 \cdot \mu_2{}^2 / r^6 \cdot D$
dipole–induced dipole*	$- K_5 \cdot \mu_1{}^2 \cdot \alpha_2 / r^6 \cdot D^2$
induced dipole–induced dipole	$- K_6 \cdot \alpha_1 \cdot \alpha_2 / r^6 \cdot D^2$

$K_1, K_2, \ldots K_6$ = constants; q_1, q_2 = *ionic charges*; μ_1, μ_2 = dipole moments; α_1, α_2 = polarisabilities; r = distance between the interacting species: D = effective dielectric constant of the medium; * orientationally averaged.

the charge (q) and the distance of separation (d). The existence of a dipole moment in the water molecule also requires that the molecule is bent.

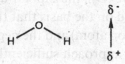

Although molecules such as carbon dioxide have no *net* dipole, a *transient* dipole may be induced when they are exposed to an electrical field. Electronic interactions of this type are described as **ion–induced dipole** and they are dependent upon the ability to induce a dipole in the neutral molecule, that is upon its polarisiability, α.

When molecules are in very close proximity, attractive forces – van der Waals interactions – occur. Van der Waals forces are all cohesive forces and vary with respect to distance (Table 2.1, $1/r^6$) and become dominant where close fitting and proximity of molecular species can be satisfied. Van der Waals attraction thus depends upon the approximate 'area' of contact between two molecules – the greater this area, then generally the greater the attractive force. The van der Waals radius of an atom or group is a measure of the effective size of the group. As two molecules approach each other the van der Waals attractive force increases to a maximum and then becomes repulsive. The van der Waals radius is defined as one-half the distance between two equivalent atoms at the point of the energy minimum. It is an equilibrium distance and some typical values are shown in Table 2.2.

Dipole–dipole interactions occur when molecules with permanent dipoles approach each other. In this type of interaction the forces which act to align the dipoles will be countered by forces whose effects are to randomise the dipoles.

Table 2.2. *Some typical van der Waals radii (in ångströms)*

H	N	O	F
1.2	1.5	1.4	1.35
–CH$_2$–	P	S	Cl
2.0	1.9	1.85	1.8
–CH$_3$			Br
2.0			1.95

Thus **dipole–dipole** interactions are sensitive to both temperature and distance. The magnitudes of the individual dipole moments are therefore not the only factors which influence the strength of the dipolar attraction. Similarly how close the negative and positive termini of the dipoles can approach is also important and here considerations of the van der Waals radii are relevant. Thus in the case of the alkyl halides (although they possess a significant dipole, e.g. methyl chloride – 1.94 debye (D)), the dipolar attraction is comparatively small. This has been rationalised on the basis that the positive terminus of the dipole (associated with the carbon atom) and the negative terminus (associated with the halogen atom) cannot approach sufficiently closely to bring about a strong interaction, Figure 2.3.

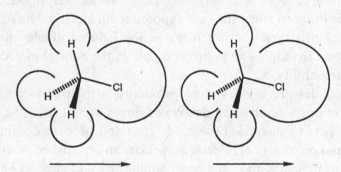

Figure 2.3. Influence of van der Waals radii upon dipole–dipole interactions.

In a similar fashion to that noted earlier for ion–induced dipole interactions, molecules with a permanent dipole may induce, in a hitherto formally neutral molecule, an induced dipole. Thus are born **permanent dipole–induced dipole interactions**. This type of interaction depends upon the polarisability of the neutral molecule. Brief note should finally be made of **induced dipole–induced dipole** effects. These forces of attraction result from intermolecular electron correlation and are often referred to as London or dispersion forces. They are sensitive to distance and are significant for molecules that are close to each other – but not too close. In such a state the motion of the electrons in the different molecules is correlated to produce, at all times, a weak net interaction; as molecules get too close to each other and there is an appreciable overlap of electron clouds then repulsive forces dominate. Van der Waals forces of this type depend upon the approximate area of contact between the two interacting molecules – the greater this area, the greater the attractive force. In the case of hydrocarbons the energy of attraction is approximately 1–1.5 kcal mol^{-1} and in a homologous series each additional methylene group provides an additional incremental area of contact that enhances the total intermolecular attractive force.

Much of the chemistry in biological systems takes place in a bulk environment which is largely aqueous in character. Water is a unique compound whose physical properties set it very clearly apart. It has one of the largest recorded dipole moments (78.5 D at 298 K) and is therefore one of the most polar of solvents. The dielectric constant (Table 2.1, *D*) represents the ability of the solvent to reduce the effective charge on an ion or dipole and hence reduce the strength of electrostatic interactions. Water thus interacts strongly with charged or dipolar groups on solute molecules, causing the charge to be dissipated through the bulk solvent. In the interior regions of natural macromolecules such as proteins which are 'hidden' from water the effective dielectric constant is reduced and hence electrostatic interactions, such as the deployment of interamide hydrogen bonds, Figure 2.2, are enhanced in their strength. Likewise in enzyme molecules the catalytic activity is often associated with the binding of the substrate molecule to the active site – a cleft or region on the surface of the protein molecule where access of the solvent water is restricted.

2.5 Hydrogen bonds and hydrogen bonding

The realisation of the importance of hydrogen bonds in the structure of biological macromolecules came well before the determination, at atomic resolution, of protein and nucleic acid structures, *vide supra*. Processes of hydrogen bonding are also now recognised as equally important functional aspects of intermolecular recognition in living systems. Intermolecular recognition, which is often a prelude to information transfer, is characterised by its specificity and its speed (often close to the time scale of $\sim 10^9$ s^{-1}). Specificity is achieved, not by the deployment of one or two strong and selective interactions, but by the simultaneous formation of several weak interactions between regions of molecules which are complementary in their steric and chemical make-up. Weak intermolecular interactions, such as those of a typical hydrogen bond (~ 3 kcal mol^{-1}) permit fast association *and* dissociation so that in a very short space of time a great many modes of interaction may be sampled and explored prior to the establishment of the final geometry of the complex. Moreover if the hydrogen bond donors and acceptors are arranged in particular spatial relationships then the hydrogen bonds not only determine specificity but they have the added bonus of being additive and cooperative in their overall effects.

Although first mention of the hydrogen bond in the chemical literature occurred around 1912–13 it was in 1920 that Latimer and Rodebush first recognised the significance of hydrogen bonding in the determination of the structure and properties of molecules. Thus in the context of the water molecule they speculated in a percipient manner:

...a free pair of electrons on another water molecule might be able to exert sufficient force on a hydrogen held by a pair of electrons on another water molecule to bind the two molecules together...

Thus was born the concept of the hydrogen bond. During the intervening three-quarters of a century ideas of what constitutes a hydrogen bond have mirrored changing knowledge and attitudes in this field. In so far as biological chemistry is concerned, most important have been the visionary contributions of Pauling (1940) who noted that certain functional groups (such as NH and C=O) exhibit a much higher propensity to participate in hydrogen bonding than others. An all embracing, yet simple, definition of a hydrogen bond has nevertheless, according to some authors, proved elusive (Aakeroy and Seddon, 1993). In the ensuing discussion therefore the definition of Pimentel and McClellan (1960) is taken as a starting point from which to develop ideas of most significance to biological systems and thus ultimately to questions related to the complexation of plant polyphenols.

Pimentel and McClellan stated:

A hydrogen bond exists between a functional group A–H and an atom or groups of atoms B in the same or a different molecule when:
     (a) – there is evidence of bond formation (association or chelation),
     (b) – there is evidence that this new bond linking A–H and B specifically involves the hydrogen atom already bonded to A.

The hydrogen bond is thus an attraction between a *proton donor* (*A–H*) and a *proton acceptor* (*B*) which, as implied by Pauling, in biological molecules is principally electrostatic in its origins.

However it should be noted that the nature of the donor (*A*) and acceptor (*B*) atoms does not enter into the definition of Pimentel and McClellan which is catholic in its breadth and embraces the hydrogen bond acceptor capacities of alkenes and aromatic 'π' electron clouds and likewise the hydrogen bonding potential, as donors, of such groups as the methine hydrogen atom (H–C) and the phosphine hydrogen (H–P). An example, typical of the former category, is seen in the crystal structure of 2-methoxy-1,4-benzoquinone (Keegstra *et al.*, 1994) in which intermolecular dipole –dipole and –C–H · · · O– interactions are each thought to play a decisive role in the establishment of the crystal structure. Thus the crystal consists of planar sheets parallel to the *ab* plane, Figure 2.4; in the *c* direction the quinone rings pack efficiently (intersheet distance 3.3 Å) with a 'π–π' stacking motif. The occurrence of –C–H · · · O– hydrogen bond formation was inferred from the observed H · · · O– distances and their directionality. They were designated by the authors as either linear or bifurcated with the quinone carbonyl and methoxy oxygen atoms acting as acceptors, Figure 2.4.

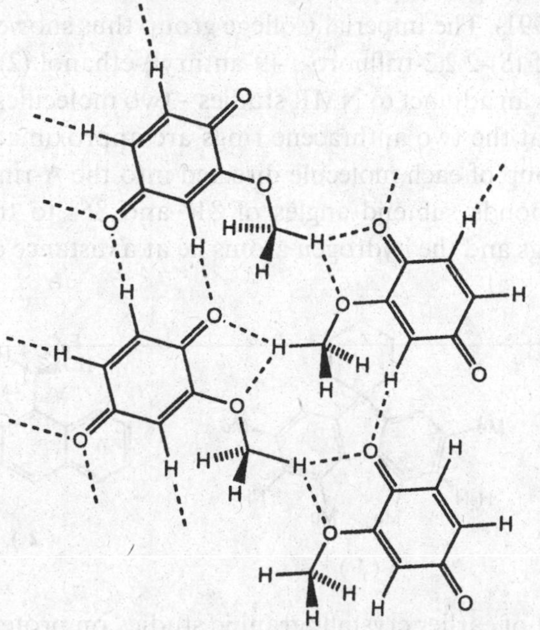

Figure 2.4. Sheet structure of 2-methoxy-1,4-benzoquinone parallel to the *ab* plane of the crystal. Linear and bifurcated –C–H · · · O– hydrogen bonds (Keegstra *et al.*, 1994).

Etter (1990) has proposed methods to predict packing arrangements in crystal structures based upon hydrogen bonding patterns. The underlying principle is that the strongest hydrogen bond acceptor should be associated with the strongest hydrogen bond donor. This procedure is continued until all the hydrogen bonding functionalities are suitably matched. In cases in which a particular molecule has an excess of hydrogen bond acceptors (*B*) the generally superfluous hydrogen bond acceptors 'find' the next most acidic hydrogen atoms and form C–H $\cdots$ B hydrogen bonds. Another common structural motif in such situations, which can relieve a deficiency of hydrogen bond donors, is the bifurcated three-centre hydrogen bond in which one hydrogen bond donor is shared between two hydrogen bond acceptors. Both of these structural features are exhibited in the crystal structure of 2-methoxy-1,4-benzoquinone, Figure 2.4.

In the case of molecules, or molecular assemblies, that are deficient in hydrogen bond acceptors (**B**) Hunter and his colleagues (Hunter *et al.*, 1992) have shown that the π electrons of an aromatic ring may then represent adequate hydrogen bond acceptors. The crystal structure of the bis-amine (**1**) thus displayed various intermolecular interactions including the formation of –NH $\cdots$ π hydrogen bonds. Analogous cases of –OH $\cdots$ π hydrogen bonds have also been reported by Hardy and MacNichol (1976) and Williams (Williams *et al.*, 1991). The Imperial College group thus showed that in the unit cell of crystals of (*S*)-2,2,2-trifluoro-1-(9-anthryl)-ethanol (**2**) – a chiral resolving agent used as an adjunct to NMR studies – two molecules are arranged in such a manner that the two anthracene rings are approximately parallel, with the hydroxyl group of each molecule directed into the A-ring π face of the other. The –O–H bonds subtend angles of 81° and 86° to the planes of the two aromatic rings and the hydrogen atoms lie at a distance of $\sim$ 2.2 Å from these planes.

(1)

(2)

As a result of earlier crystallographic studies on proteins Levitt and Perutz (1988) embarked on an investigation designed to discover whether the energy of conventional electrostatic interactions between amino- and benzene-groups is sufficiently strong to be able to speak of a hydrogen bond. Simple energy calculations showed that there is in fact a significant interaction between a hydrogen bond donor, such as –NH$_2$, –NH and –NH–C(=NH)–NH$_2$, and the centre of a benzene ring, which acts as a hydrogen bond acceptor. According to Levitt and Perutz this aromatic hydrogen bond arises from the small partial charges centred on the ring carbon (δ^-) and hydrogen atoms (δ^+). This interaction is about half as strong as a typical hydrogen bond, but they suggested that it may be expected to play a significant role in molecular associations.

Important as such slightly unusual examples are the overwhelming burden of evidence is that, in biological systems, the most significant hydrogen bonds which are likely to be formed are those in which the donor (*A–H*) hydrogen atom is bound to an electronegative atom *A*, such as oxygen, nitrogen or sulphur. Likewise the acceptor atoms are usually oxygen or nitrogen. Some examples, typical of those encountered in biology, are shown in Figures 2.1, 2.2 and 2.5.

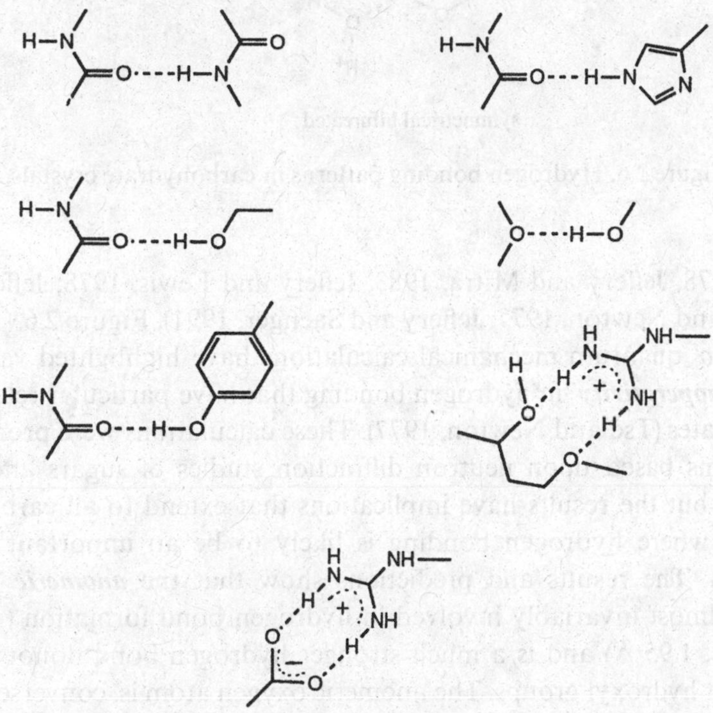

Figure 2.5. Patterns of hydrogen bonding in biological systems.

2.5.1 Carbohydrates

Carbohydrate crystal structure analysis has proved to be a rich source of information relating to –O–H···O– hydrogen bonds. The majority of these studies have taken place with simple pyranose sugars and methyl pyranosides. There is with these subjects little variation in the types of hydrogen bonds encountered; the donors are thus –C–O–H groups and the acceptors are –C–O–H groups and ring and glycosidic oxygen atoms –O– (Jeffery and

hydroxyl - hydroxyl   hydroxyl - acetal

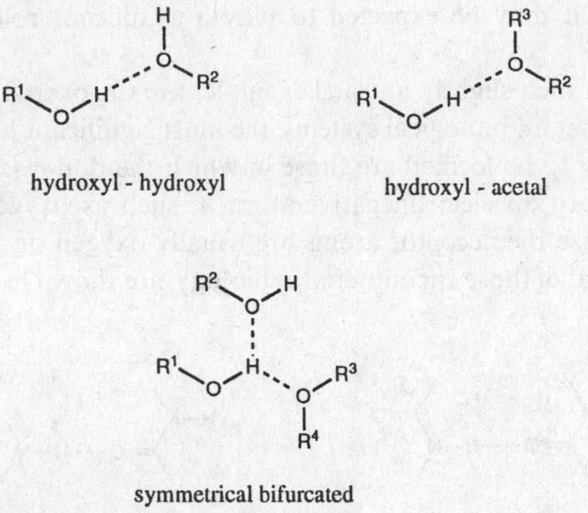

symmetrical bifurcated

Figure 2.6. Hydrogen bonding patterns in carbohydrate crystals.

Takagi, 1978; Jeffery and Mitra, 1983; Jeffery and Lewis, 1978; Jeffery *et al.*, 1977; Tse and Newton, 1977; Jeffery and Saenger, 1991), Figure 2.6.

'*Ab initio*' quantum mechanical calculations have highlighted various aspects of '***cooperativity***' in hydrogen bonding that have particular relevance to carbohydrates (Tse and Newton, 1977). These calculations were prompted by observations based upon neutron diffraction studies of sugars and methyl glycosides but the results have implications that extend to all carbohydrate molecules where hydrogen bonding is likely to be an important cohesive interaction. The results and predictions show that the ***anomeric*** hydroxyl group is almost invariably involved in hydrogen bond formation (–H$\cdots$O– distances < 1.95 Å) and is a much stronger hydrogen bond donor than the other sugar hydroxyl groups. The anomeric oxygen atom is, conversely, rarely a hydrogen bond acceptor, and, when it is, the bond is generally weak. Secondly, and perhaps more significantly, the donor–acceptor relationships between hydroxyl groups which give rise to the formation of *chains of hydrogen bonds* far exceed in number and frequency those where the hydroxyl group acts simply as a donor. The bond lengths are shorter (1.80 vs $\sim$ 1.87 Å) and with this is the implication that the bonds are stronger. Hydroxyl groups for which the oxygen atom also accepts a hydrogen bond form stronger hydrogen bonds than those where the hydroxyl group is only a hydrogen bond donor. This is, presumably, a consequence of the additional decrease in electron density around the proton, when the hydroxyl group is a hydrogen bond acceptor.

Thus in carbohydrate molecules the arrangement of molecules will tend to

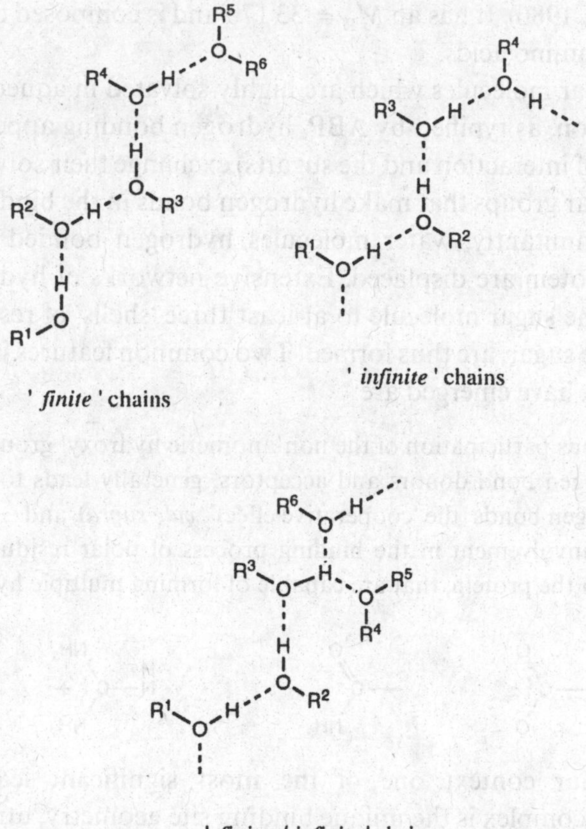

' *finite* ' chains

' *infinite* ' chains

' *finite* / *infinite* ' chains

Figure 2.7. Hydrogen bonding chains in carbohydrate crystals.

be such that the hydroxyl groups are aligned, through hydrogen bonding, into finite and infinite chains. Ring and glycosidic oxygen atoms and anomeric hydroxyl groups (because of their weak acceptor properties) function in such cooperative arrays as chain terminators. Some typical examples of patterns of chains of hydrogen bonds which have been observed in carbohydrate crystals are shown in Figure 2.7.

There has been considerable advancement over the past three decades in our understanding of protein–sugar interactions at the molecular level, beginning with the difference Fourier analysis of the structure of lysozyme and its complexation with sugars. Other carbohydrate–protein complexes have been similarly analysed, and the complex of L-arabinose with the L-arabinose binding protein (ABP) illustrates many of the essential features of sugar–protein interactions. The L-arabinose binding protein (ABP) is a periplasmic protein that serves as an initial component of a high affinity transport

system (Quiocho, 1986). It has an $M_R = 33\,170$ and is composed of a polypeptide chain of 306 amino acids.

Sugars are polar molecules which are highly solvated in aqueous media. In
complex formation, as typified by ABP, hydrogen bonding appears to be the
dominant form of interaction and the sugar(s) exchange their solvation shell of
water for the polar groups that make hydrogen bonds in the binding site of the
protein. Concommitantly, water molecules hydrogen bonded to the polar
groups of the protein are displaced. Extensive networks of hydrogen bonds,
radiating from the sugar molecule to at least three 'shells' of residues around
the vicinity of the sugar, are thus formed. Two common features in many of the
complexes which have emerged are:

(a) the simultaneous participation of the non-anomeric hydroxyl groups of the sugar
    as both hydrogen bond donors and acceptors, generally leads to stronger than
    average hydrogen bonds (the 'cooperative effect', *vide supra*), and
(b) the extensive involvement in the binding process of polar residues with planar
    side-chains, on the protein, that are capable of forming multiple hydrogen bonds,
    viz:

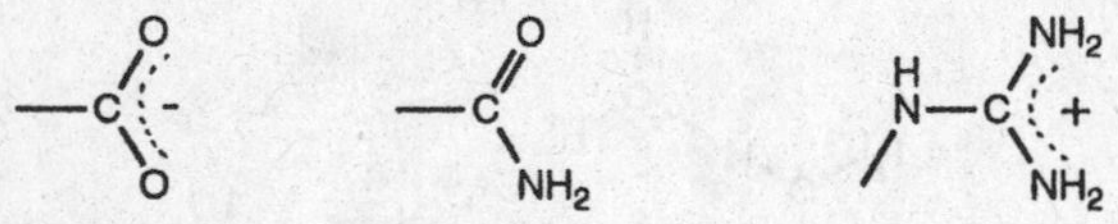

In this particular context one of the most significant features of the
ABP–arabinose complex is the unique binding site geometry, utilising the side
chain carboxylate group of aspartic acid (60), which accommodates both the α-
and the β-anomeric forms of the L-arabinose substrate, Figure 2.8, with little
change of overall geometry.

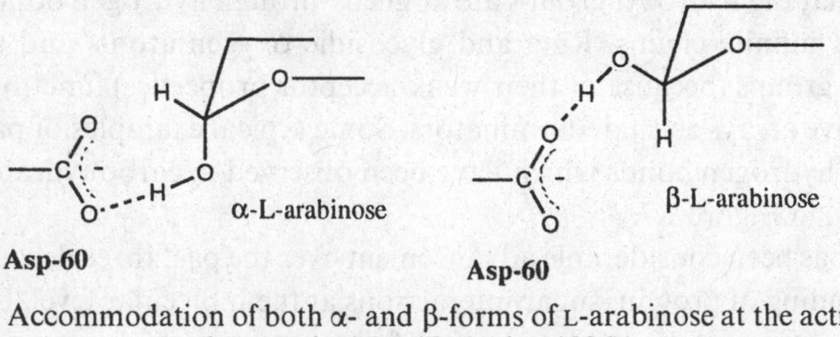

Figure 2.8. Accommodation of both α- and β-forms of L-arabinose at the active site of
the protein ABP (Quiocho, 1986).

Many van der Waals contacts, less readily described than hydrogen bonds,
are also undoubtedly formed in the sugar–protein complexes. In the L-arabinose molecule, for example, the relative disposition of the equatorial C-
3–OH and the axial C-4–OH on one side of the pyranose ring creates a cluster

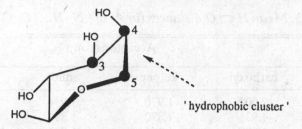

Figure 2.9. 'Hydrophobic cluster' of carbon atoms in β-L-arabinose.

of non-polar atoms (C-3, C-4 and C-5), Figure 2.9, and this hydrophobic face is partially stacked with the aromatic residue tryptophan (16) in the ABP–L-arabinose complex. However, the extent of such contributions to the stability of the sugar–protein complex is difficult to evaluate.

2.5.2 Geometric considerations

Over the past two decades, generating structural information on the nature of the hydrogen bond and hydrogen bonding from X-ray, and neutron spectroscopy has been an extremely fruitful field of endeavour. However, these structural parameters refer to the hydrogen bond in the crystalline state and interpretations placed thereon are therefore subject to the proviso that crystal field effects may well be the dominant factor in defining the crystal geometry (Taylor and Kennard, 1984). Similarly, although the problems are in principle no different from the use of molecular structural data derived from crystal structure analysis to describe the shape and conformation of globular proteins in solution, the perturbations which arise from the crystal field environment must necessarily be borne in mind when extrapolating the properties of the hydrogen bond as observed in the crystalline state to those in solution. Finally of course hydrogen bonding in biological molecules and in the formation of intermolecular complexes in aqueous media is a complicated phenomenon because water itself is a strong competitor for hydrogen bonding sites (Kauzman, 1959). With these important caveats in mind some of the salient features of the patterns of hydrogen bonding encountered in organic molecules are discussed and illustrated below.

The hydrogen bond is largely an electrostatic phenomenon (due to the Coulombic interaction between $A-H^{\delta+}$ and $B^{\delta-}$) and as a consequence the length of the hydrogen bond may be correlated with the nature of the donor (A) and acceptor (B) atoms. Kennard and her colleagues (1984a,b) examined over one thousand $-N-H \cdots O=C-$ intermolecular bonds. For the acceptor atom (B) mean $H \cdots O$ distances tended to increase in the series: $-CO_2^- < -CONH_2 <$

Table 2.3. *Mean H ... O distances (in Å) of –N–H ... O=C bonds*

Donor	Acceptor group			
	carboxyl	ketone	amide	carboxylate
–N–H	2.002	1.970	1.934	1.928
$RNH_3{}^+$	1.936	1.872	1.891	1.841
$R_2NH_2{}^+$	1.887	1.966	1.793	1.796
R_3NH^+	—	1.938	1.845	1.722

Data taken from Taylor and Kennard (1984).

R^1–CO–R^2 < CO_2H; for the donor group (A–H) increased H $\cdots$ O distances followed the pattern: R_3NH^+ < $R_2NH_2{}^+$ < $RNH_3{}^+$ < R_2NH; Table 2.3. Hydrogen bond distances were also correlated with the number of hydrogen bonds in which the acceptor group is involved. In general, bonds in which the –C=O group accepts just one hydrogen bond are shorter than those in which the –C=O group accepts two. This may well be due to steric effects, i.e. repulsions between two (or more) donor species.

single acceptor   dual acceptor

Two centre hydrogen bonds (A–H $\cdots$ B) are by far the most commonly observed in all classes of hydrogen bond but there is strong evidence that three centre/bifurcated hydrogen bonds also occur and more frequently than had been previously anticipated; most commonly in crystal structures in which there is an excess of acceptors over donors.

three centre / bifurcated

Hydrogen bonds observed in crystal structures are rarely linear. Kennard and her colleagues (1984a,b) in their survey of –N–H $\cdots$ O=C– intermolecular hydrogen bonds thus showed a mean value of 161.2° for the –N–H $\cdots$ O– bond angle and for the corresponding intramolecular hydrogen bonds a mean value of 132.5°. Similar distributions have been demonstrated for intermolecular $-O-H \cdots O-$ (mean value 163.1°) and $-C-H \cdots O-$ (mean value 152.7°) hydrogen bonds. The $-N-H \cdots O-$ bond angles observed with intermolecular $-N-H \cdots O=C-$ hydrogen bonds show that short bonds tend to be more linear than long ones. A suggested rationale for this observation is that as the $-H \cdots O-$ distance is decreased the nitrogen/oxygen non-bonded interaction becomes more unfavourable. Thus short bonds tend to straighten out so as to increase the nitrogen/oxygen separation.

Some uncertainty surrounds the question of whether hydrogen bonds tend to align along lone-pair directions of the acceptor atom. X-ray data analysis suggested that the answer to this question was *no* (Taylor and Kennard, 1984). However, determination of the equilibrium angular geometry of dimers of the form $(A-H \cdots B)$ using rotational spectroscopy (albeit in the gas phase using hydrogen fluoride as the proton donor – conditions infrequently encountered in living systems!), shows that the $A-H \cdots O$ bonds lie, on average, in approximately the same directions as the oxygen lone pairs (Legon and Millen, 1987, 1993). Several classes of substrate (B) were identified:

(i) those in which each acceptor atom (oxygen or sulphur) may be considered as having two equivalent non-bonding electron pairs that do not lie in the molecular plane (e.g. where B is water, oxetane, oxirane, 2,5-dihydrofuran or hydrogen sulphide),

(ii) those (e.g. where B is methanal) in which the electron acceptor atom (oxygen) carries two non-bonding electron pairs trigonally disposed and lying within the molecular plane, and

(iii) those (such as where B is ethene and ethyne) in which π electrons may be presumed to fulfil the role of lone pair electrons.

The observed gas phase angular geometries of dimers of the type $(B \cdots H-X)$ can then be predicted using the following general guidelines:

(a) the axis of the HX molecule coincides with the supposed axis of a non-bonding electron pair as conventionally envisaged, or

(b) *if B has no non-bonding electron pairs but has π-bonding electron pairs*, the axis of the H–X molecule intersects the internuclear axis of the atoms forming the π-bond and is perpendicular to the plane of symmetry of the π-bond,

(c) rule (a) is definitive when B has both non-bonding and π-bonding electron pairs.

Thus water, oxetane, oxirane, 2,5-dihydrofuran and hydrogen sulphide form complexes with H–F which are pyramidal at oxygen (or sulphur); the methanal – H–F complex is trigonal at oxygen; in the ethene – H–F complex the H–F molecule lies along the perpendicular C_2 axis of the ethene molecule.

[water - H-F]

[methanal - H-F]

[ethene - H-F]

Legon and Millen were, however, careful to add the following rider to their conclusions.

Thus, essentially we are relating the experimental angular geometries of hydrogen-bonded dimers to the simple nonbonding (or π-bonding) pair model of the acceptor molecule B that now has a familiar place in all modern textbooks of Chemistry. We must emphasize that we are *not* attempting to prove that nonbonding pairs exist as 'rabbits ears' or 'water wings'. Indeed, we shall make it clear that conventional diagrams representing nonbonding pairs in this way are, at the very least, gross exaggerations. Nevertheless, we find in all cases examined that the potential energy of a nonperturbing electrophilic probe in the vicinity is a minimum along just those directions normally associated with the axes of non-bonding (or π-bonding) pairs, as conventionally envisaged by chemists.

2.5.3 *Proton donor and proton acceptor scales*

A hydrogen bond is weaker than a conventional covalent bond. Its strength is variable; for neutral molecules in aqueous media it normally lies in the range 0–2.5 kcal mol^{-1} (Williams *et al.*, 1991). The effect of molecular structure on the strength of the hydrogen bond and the determination of the relative hydro-bonding abilities of different organic functional groups have been the subjects of intense interest and concern. Most of these investigations have been based upon measurements of solution association constants under equilibrium conditions. Thus for example extensive studies of the effects of structural variation upon base strengths toward a common reference acid, as measured by the formation constant K_f for the 1:1 hydrogen-bonded complex, have been carried out. Gurka and Taft (1969) reported work in which *p*-fluorophenol was employed as the reference acid and measurements were made by fluorine NMR spectroscopy. The extent of the proton transfer in the hydrogen-bonded complexes of *p*-fluorophenol is comparatively small and therefore, not surprisingly, it was established, for a variety of common organic functional groups, that the aqueous pK_A value of the base (proton acceptor) and the $\log K_f$ (in CCl_4 at 25°C) were unrelated. Nevertheless Taft and his colleagues (Taft *et al.*, 1969) assigned a parameter pK_{HB} ($\log K_f$ for the formation of the hydrogen-bonded complex of *p*-fluorophenol in CCl_4 at 25°C), to a base to represent its relative strength as an acceptor in hydrogen-bonded complex formation with any suitable –OH reference acid. Values for 117 bases representing 16 different functional groups were presented.

Although comparison of pK_A and pK_{HB} scales of base strength showed them to have little in common it was established for primary amines and for carbonyl derivatives that plots of pK_A vs pK_{HB} gave parallel straight lines which therefore satisfactorily describe the relationship between these descriptors for these two classes of functional group. In this context it is of interest to note that the amides acetamide and *N,N*-dimethylformamide are, on this reckoning, members of the carbonyl and not amine group of proton acceptors. In turn this offers support to the contention that these substrates preferentially undergo *O*-protonation as opposed to *N*-protonation in the hydrogen-bonded complex equilibria, Figure 2.10.

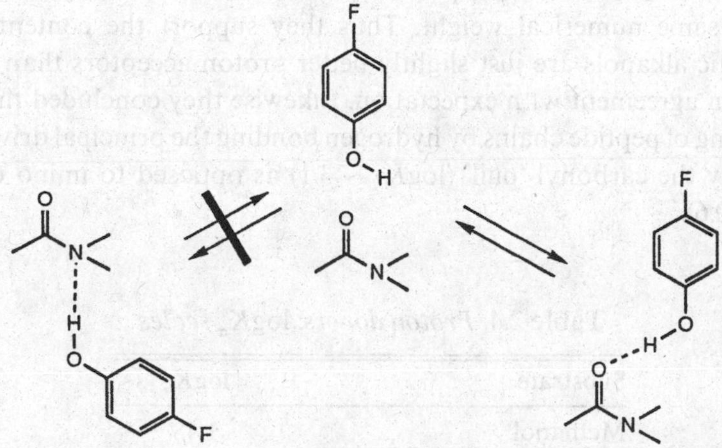

Figure 2.10. *O*-Protonation as opposed to *N*-protonation of amides with *p*-fluorophenol in CCl_4.

These data were ultimately incorporated into the β-scale of hydrogen bond acceptor (HBA) basicities and a corresponding α-scale of hydrogen bond donor (HBD) acidities was constructed (Kamlet and Taft, 1976a,b, 1979). Some care should be exercised in the use of the α- and β-scales particularly in the case of alcohols – amphiprotic self-associating solvents, able to act simultaneously as hydrogen bond donors (HBD) and hydrogen bond acceptors (HBA)

(HBA). Thus alcohols appear to be stronger HBAs than bases such as pyridine, triethylamine, and dialkylethers. Kamlet and Taft rationalised these observations on the basis that in the neat solvent one was more likely to be dealing with self-associated oligomers/polymers than the simple alcohol monomer. An alcohol molecule, associated through its hydroxyl hydrogen with another alcohol molecule, should have increased electron density on the oxygen of that alcohol group. Effectively this should lead to alcohol dimers and oligomers being significantly stronger HBA bases than the corresponding monomeric species. Similar anomalies in the rank order of amphiprotic solvent HBD values have been rationalised in terms of the differing structures of the associating solvent donors.

More recently, hydrogen-bonding equilibrium constants have been measured in 1,1,1-trichloroethane for a large selection of proton donors against a common acceptor (N-methylpyrrolidone) and for proton acceptors against a common donor (p-nitrophenol) (Abraham *et al.*, 1989). Together these were employed to create alternative log K_α (proton donor) and logK_β (proton acceptor) scales, Tables 2.4 and 2.5.

Abraham and his colleagues laid no claim to the universality of their particular scales. However they pointed out that their scales may, fortuitously, carry the same numerical weight. Thus they support the contention that amphiprotic alkanols are just slightly better proton acceptors than they are donors – in agreement with expectation. Likewise they concluded that in the cross linking of peptide chains by hydrogen bonding the principal driving force is probably the carbonyl 'pull' (log$K_\beta \sim 3.1$) as opposed to imino donation (log$K_\alpha \sim 0.6$).

Table 2.4. *Proton donors*: logK_α *scales*

Substrate	logK_α
Methanol	1.48
Ethanol	1.21
t-Butanol	0.78
2,2,2-Trifluoroethanol	2.00
Phenol	2.14
2-Methylphenol	1.75
4-Methoxyphenol	2.18
4-Nitrophenol	3.12
Acetic acid	2.04
Trifluoroacetamide	1.52
N-Methylpivalamide	0.70

Data taken from Abraham *et al.*, 1989.

Table 2.5. *Proton acceptors*: $\log K_\beta$ *scales*

Substrate	$\log K_\beta$
Ethanol	1.41
Tetrahydrofuran	1.69
Anisole	0.30
Dioxan	1.28
Propan-2-one	1.61
Cyclohexanone	1.70
N,N-Dimethylformamide	2.81
N-Methylpyrrolidone	3.12
Ethyl acetate	1.43
Pyridine	2.52

Data taken from Abraham *et al.*, 1989.

2.5.4 *Hydrogen bonding and the construction of molecular aggregates*

It has recently been recognised that consideration of patterns of hydrogen bonding is an indispensable tool in the design of molecular aggregates and supramolecular aggregates. Etter and her colleagues (Etter, 1990, 1991) have exploited the vast source of information present in the Cambridge Crystallographic Data Base to draw up a set of empirical 'hydrogen bonding rules' to assist in the determination of the preferred modes of hydrogen bonding in molecular aggregates. Different structural groups thus display clear preferences for the formation of specific hydrogen bond patterns in their crystal structures, despite the influence of other lattice forces. Etter thus identified the pattern preferences displayed by a range of organic functional groups and molecular types – amides, imides, carboxylic acids, 2-aminopyridines, etc. Three very useful rules of thumb were thus enunciated:

(i) all acidic hydrogens available in a molecule will be used in hydrogen bonding in the crystal structure of the molecule, and

(ii) all good hydrogen bond acceptors will be used in hydrogen bonding when hydrogen bond donors are available, and

(iii) the best hydrogen bond donor and the best hydrogen bond acceptor will preferentially form hydrogen bonds to one another.

The last of these rules is beautifully exemplified by the co-crystallisation of three different molecules, with the one with the hydrogen bond acceptor capacity selecting one of two molecules, with hydrogen bond donor capacities, in the crystallisation. Thus 4-phenyl pyridine and ethyl isonicotinate were both shown to be capable of completely separating the acid of lower pK_A from one

Table 2.6. *Selectivity in the co-crystallisation of pyridine derivatives and carboxylic acids*

ΔpK_A

(a) O_2N—⬡—CO_2H | 2.35 | Me_2N—⬡—CO_2H
(O_2N)

(b) O_2N—⬡—CO_2H | 0.74 | O_2N—⬡—CO_2H
(O_2N)

(c) O_2N—⬡—CO_2H | 0.41 | ⬡—CO_2H (Cl)

Data from Etter (1990, 1991).
The acid in each pair (a, b, c) on the left hand side of the table was in the co-crystal.
The acid on the right hand side was found uncomplexed.

with a higher pK_A by co-crystallisation. Neutral hydrogen bonded co-crystals form with a single hydrogen bond between the nitrogen of the pyridine and the hydroxyl group of the carboxylic acid component. The implications of these observations are that the strength of the hydrogen bond in solution, to the extent that this is reflected in the pK_A value, has determined which species nucleates crystal growth, Table 2.6.

Diacetamide provides some intriguing examples of selectivity in co-crystallisation with phenolic substrates. It crystallises in two crystal forms: the *cis–trans* conformation forms diacetamide dimers leaving two of the four carbonyl groups unbonded (Jeffery *et al.*, 1988). Diacetamide co-crystallises with phenols preserving this stable dimer pattern whilst utilising the residual two carbonyl groups to hydrogen bond with the phenolic hydroxyl group.

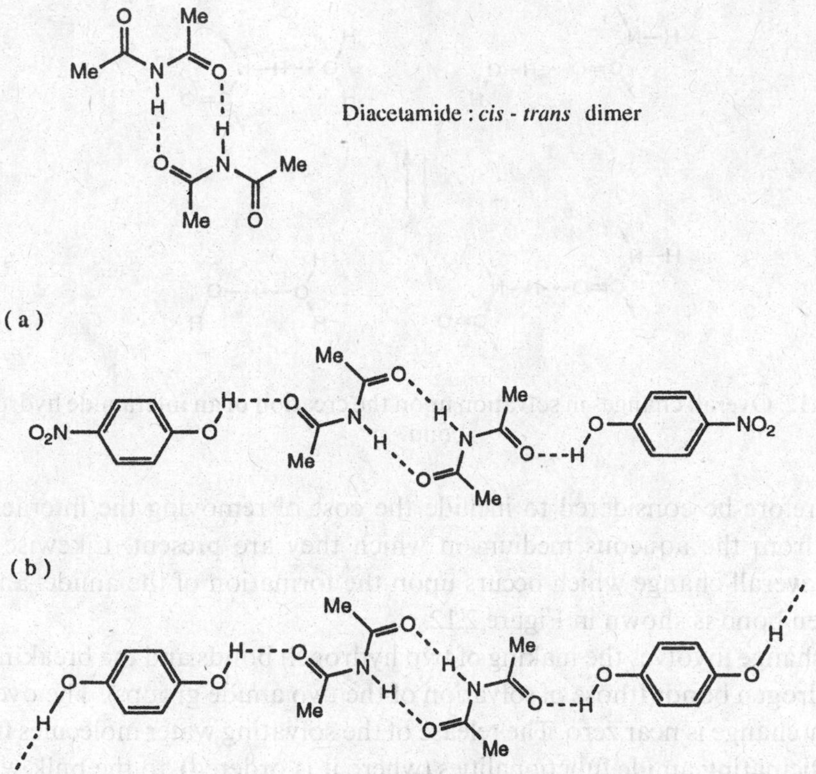

Diacetamide : *cis - trans* dimer

(a)

(b)

Figure 2.11. Diacetamide *cis–trans* dimer. Co-crystallisation with (a) *p*-nitrophenol and (b) quinol.

Thus diacetamide can be converted from a dimer into a complex with four molecules by co-crystallising with *p*-nitrophenol, and into an infinite hydrogen bonded chain by co-crystallising with quinol (hydroquinone), Figure 2.11. However, in the co-crystallisation of diacetamide with *p*-hydroxybenzoic acid the *cis–trans* dimer is not preserved but a heterodimer is formed instead.

2.6 Solvation

Molecular recognition is ultimately a manifestation of the balance which is achieved between the tendencies which molecular species or chemical groups possess to interact with each other and their propensities to be solvated by the environment to which they are exposed. Qualitatively speaking, 'good' solvation generally inhibits strong intermolecular association. In biological systems it is thus the compromise between solute–solute interactions and solvation in a predominantly aqueous environment which determines the strength of inter-molecular association. Thus the non-covalent binding of substrates to enzymes

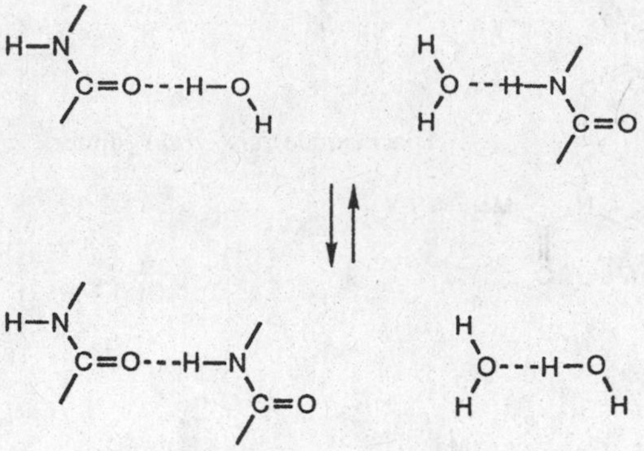

Figure 2.12. Overall changes in solvation upon the creation of an interamide hydrogen bond.

can therefore be considered to include the cost of removing the interacting species from the aqueous medium in which they are present. Likewise the typical overall change which occurs upon the formation of the amide–amide hydrogen bond is shown in Figure 2.12.

The change involves the making of two hydrogen bonds and the breaking of two hydrogen bonds (those of solvation of the two amide groups). The overall *enthalpy* change is near zero. The release of the solvating water molecules from the participating amide functionalities (where it is ordered), to the bulk water medium underlies the favourable *entropy* change which accompanies interamide bond formation. Understanding many of the questions which surround molecular recognition and related phenomena thus necessitates some knowledge (or at least an educated guess) of the nature and degree of solvation of solute species. Our *quantitative* appreciation of these solvent effects at the molecular level is, however, still poor; attempts to fill this serious lacuna in our knowledge have met with limited and varying degrees of success (Williams, 1991).

Most experimentalists find it useful to refer to certain compounds or chemical groupings as being relatively '*hydrophilic*' or '*hydrophobic*'. The affinity of a compound for an aqueous medium can, in principle, be determined by measuring the dimensionless equilibrium constant for its transfer from the dilute vapour phase, in which each molecule exists in virtual isolation, to an infinitely dilute aqueous solution in which solute–solute interactions can be ignored and in which each solute molecule may be envisaged as surrounded by water. In Figure 2.13 water-to-vapour distribution coefficients of some representative uncharged organic compounds are arranged on a scale that spans ten orders of magnitude (from Wolfenden, 1983). Some of the trends evident in this scale may be rationalised by simply enumerating the groups in each solute which

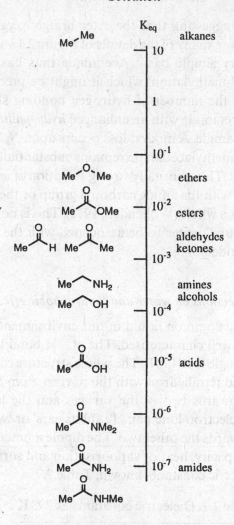

principal solvating groups (*)

Figure 2.13. Vapour/water distribution coefficients of some organic compounds (from Wolfenden, 1983): $K_{eq} = \text{mol}\,l^{-1}$ (vapour)/$\text{mol}\,l^{-1}$ (aqueous).

might be expected to form strong hydrogen bonds to the solvent water. Thus on this basis ethyl acetate, acetic acid, acetamide and methyl guanidine (with respectively one, two, three and four such groups) lie approximately in the order predicted (Figure 2.13). Esters and ketones are similar to each other in hydrophilic character, suggesting that the ester bridge oxygen confers little affinity for water. However, many of the details of Figure 2.13 are not so readily rationalised on this very simple basis. Acetamide thus has an exceptional affinity for water but *N*-methylation, which it might be predicted using this rationale would reduce the number of hydrogen bonding sites available to water, gives *N*-methylacetamide with an enhanced *hydrophilic* character compared to the parent acetamide. A modest loss occurs upon *N,N*-dimethylation but the product *N,N*-dimethylacetamide remains substantially more strongly solvated than acetic acid. This unusually strong solvation of amides is believed to be associated mainly with the amide carbonyl group or the dipolar character of the amide group as a whole (Wolfenden, 1978). This is borne out by X-ray studies which show 'bound' water to be associated with the carbonyl rather than the NH of the peptide bonds.

2.6.1 Reflections on water and 'hydrophobic effects'

Water is by far the most common liquid in our environment. The individual water molecule itself is well characterised. The –O–H bond length is 0.957 Å and the H–O–H bond angle is 104° 31′. The whole structure can be represented by a somewhat distorted tetrahedron with the oxygen atom at the centre; the protons are directed towards two of the vertices and the lobes of electron density – the so-called electron lone pairs ('rabbits ears' or 'water wings', *vide supra*) – are directed towards the other two. The dipole moment of water is 1.83 D, it has a high heat capacity, heat of vapourisation and surface tension, and one of the highest dielectric constants known, Table 2.7.

Table 2.7. Dielectric constants at 298 K

Substance	Dielectric constant
Water	78.5
Methanol	32.6
Ethanol	24.0
Benzene	2.2
Carbon tetrachloride	2.2

Because water was probably indispensable for the genesis of life and because of its unique properties, in a pure form and as a solvent, the structure of water (and ices) has been the subject of intense work since the last century. The literature is voluminous (Franks, 1972–80; Frank, 1970; Eisenberg and Kauzman, 1979; Stillinger, 1980; Rice and Sceats, 1981; Henn and Kauzman, 1989; Symons, 1981). Agreement on a comprehensive theory of the molecular structure of bulk water has nevertheless proved elusive. That the structure and properties of bulk water derive from extensive hydrogen bonding interactions between individual water molecules is accepted, it is the nature and arrangement of these interactions which is still the subject of debate. From the point of view of intermolecular hydrogen bonding the individual water molecule is well placed in having both double donor (protons) and double acceptor (lone pairs) hydrogen bond functionality based about one central oxygen atom.

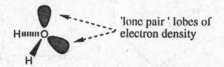

This dual functionality is most clearly seen in the crystal structure of ordinary hexagonal ice, Figure 2.14, in which each water molecule has four nearest neighbours to which it is hydrogen bonded. These four hydrogen bonds are spatially arranged with local tetrahedral symmetry; that is the oxygen atoms of the neighbouring water molecules occupy the vertices of a regular tetrahedron surrounding the oxygen atom of the central molecule. At 0 K the distance between nearest neighbour oxygen atoms is 2.74 Å. Starting at any oxygen atom it is possible to sketch a closed pathway of contiguous hydrogen bonds by moving always to nearest neighbour oxygen atoms. Polygonal paths returning to the original oxygen atom with an *even* number of steps equal to or exceeding six are possible.

This propensity of water to enter into extensive hydrogen bonding networks extends to the various clathrate hydrates which it forms. As in ice the hydrogen bonded water molecules are four coordinate in a distorted tetrahedral arrangement. They are 'ice-like' but differ in having large internal cages (Jeffery, 1969). Owing to the steric requirements imposed by the inclusion of other molecules as guests within these voids their hydrogen bond networks are organised differently from that of hexagonal ice. The most commonly encountered polyhedral void is that formed by twelve regular pentagonal faces – the pentagonal dodecahedron $[H_2O]_{20}$ in which each oxygen atom is hydrogen bonded to *four* neighbours with coordinate angles that differ from the tetrahedral by only a few degrees. On the other hand in the 'un-hydrogen-bonded' *t*-butylamine hydrate $(Me)_3CNH_2 \cdot [9.75] \, H_2O$ the guest molecule occupies a heptakaidecahedron cage which has square, pentagonal, hexagonal and heptagonal faces. The role of the guest molecules in these clathrates is to fill the voids and thereby stabilise these 'ice-like' structures. The stoicheiometry of the hydrates is determined by the physical dimensions of the guests in relation to those of the voids.

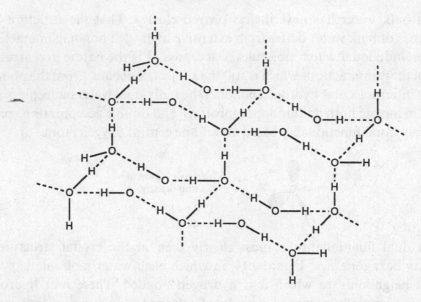

Figure 2.14. Part-structure of hexagonal ice. Hydrogen bonds are shown as broken lines; covalent bonds as full lines.

The melting of ice to produce ordinary liquid water clearly entails fundamental changes in the way in which water molecules are arranged relative to one another. Essentially two models have been developed for the molecular structure of liquid water (Eisenberg and Kauzman, 1969). Frank's *'flickering cluster'* theory invokes the concept of clusters of three or four coordinate water molecules with a mean lifetime of about a nanosecond. Thus Frank and Wen (1957) and Nemethy and Scheraga (1962a, 1964) visualised bulk water as composed of two or more 'states' in equilibrium. One state – the *cluster* – is composed of more or less ordered hydrogen bonded molecules; the other consists of an array of non-hydrogen bonded, disordered molecules. The *clusters*, of varying shapes and sizes, are short lived and are envisaged to continually exchange water molecules with the surrounding unstructured phase. This dynamic picture of the structure of the bulk water medium is based, in part, on the postulate that the formation of hydrogen bonds in water is a cooperative phenomenon such that when one bond forms then so do many others simultaneously. Conversely when one bond breaks then typically a whole group or *cluster* will dissipate.

A second type of model, of which a number of variations have been proposed, is the uniform continuum model, originally suggested by Bernal and Fowler (1933). Essential to these models (Pople, 1951; Rice and Sceats, 1981) is the assumption that *all* of the hydrogen bonds in liquid water are intact but that they may be stretched, bent and distorted to an extent that varies with temperature and pressure; the existence of 'broken' hydrogen bonds is not specifically recognised. The random network model of Rice and Sceats (described delightfully by the authors as a simple, modestly accurate description but with deficiencies) is based upon two assumptions – that the dominant structural feature of the liquid form is a locally tetrahedral continuous hydrogen bonded network and that molecular motion in that network occurs on two time scales. In the immediate vicinity of a given water molecule the arrangement of the quasi-equilibrium positions resembles that in ice, but that further away from this centre the arrangement and connectivity may well be significantly different. In particular it is likely that the $O-H\cdots O$ hydrogen bonds will be bent and stretched to varying degrees. The thermodynamic properties of water in this model are separated into two contributions: (i) a lattice vibrational contribution arising from oscillations of the water molecules about the quasi-equilibrium positions and (ii) a configurational contribution defined by the arrangement of quasi-equilibrium positions and the deformation of hydrogen bonds that they require (Henn and Kauzman, 1989).

Water is a notoriously poor solvent for apolar compounds, such as hydrocarbons and the noble gases, at moderate temperatures and pressures. This reluctance to dissolve in water has been popularly attributed to the hydrophobicity of these substances – their *fear of water*. In a sense the term may be misleading and whilst the phenomenon is readily observed it has proved to be difficult to satisfactorily explain. For an apolar compound to dissolve in water it must intrude into a liquid that is characterised by an extended network of hydrogen bonds and has a high cohesive energy (*vide supra*). Any valid explanation of '*hydrophobic effects*' must obviously be based upon a reasonably valid model for the structure of liquid water itself. Many rationalisations nevertheless have focussed upon the large losses of entropy which accompany the dissolution of non-polar solutes, such as a noble gas or a hydrocarbon, in water. Miller and Hildebrand (1968), using the Lennard–Jones and Pople model of water, suggested a mechanism for the loss of entropy consequent upon the dissolution of inert gases (ranging from Ne to $n\text{-}C_4H_{10}$) in water that did not postulate the formation of rigid structures of any sort. However for others the large losses of entropy which accompany the dissolution of non-polar solutes have, quite naturally, been interpreted as 'structure-making' phenomena in that medium.

Frank and Evans (1945) first sought to rationalise the unusual thermodynamic properties of non-polar solutes in water by postulating a particular ordering of water molecules around the solute. They described the process as '*iceberg*' formation:

when a rare-gas or a non-polar molecule dissolves in water at room temperature it modifies the water structure in the direction of greater '*crystallinity*' – the water, so to speak, builds a microscopic iceberg around it.

Although many may regard the explanation as inadequate (Blokzijl and Engberts, 1993) the concept has nevertheless provided a useful picture upon which to base qualitative explanations for a number of phenomena which may be placed under the umbrella description of '*hydrophobic effects*'. If a non-polar solute molecule is to be placed in water, the random hydrogen bonded network of the liquid medium must re-organise to accommodate the non-polar solute, but at the same time cause as little disruption to the network as possible. The reported crystal structures of the clathrate hydrates (*vide supra*) suggest that such re-organisations are indeed possible to facilitate the creation of solvent cages to accommodate molecules with large non-polar groupings (e.g. *t*-butylamine). Likewise the interaction of non-polar groups or regions in biopolymers is thought to be a major determinant in selecting the native conformation of the macromolecule. The generally accepted qualitative explanation for these effects and interactions is that, when non-polar molecules are juxtaposed in aqueous media, their *joint* solvent cage entails less entropy reduction than when these solutes are constrained to remain far apart. In slightly more colourful language Perutz (1992) described, with seductive simplicity, the key role which 'hydrophobic forces' play in protein folding:

He (Kauzman) suggested that the water molecules' anarchic distaste for the orderly regimentation imposed upon them by the hydrophobic side chains of the protein forces these side chains to shy away from water and to congregate in the centre of the protein.

Without describing the detailed structural features of the typical '*iceberg*' Nemethy and Scheraga (1962b) have developed a quantitative description of the perturbed water shell which surrounds a non-polar substrate, such as a hydrocarbon, in terms of modified energy levels of the solvent in the vicinity of the solute. More recently, Gill *et al.* (1985) have re-examined the '*iceberg*' concept in the context of the large heat capacity difference between gaseous and dissolved non-polar solutes in water. They concluded that the '*iceberg*' is a monomolecular layer of water in which each molecule behaves independently. This point of view thus contrasts with the implied cooperative behaviour in '*iceberg*' formation initially suggested by the model of Frank and Evans (1945).

Direct experimental evidence for the idea of the '*iceberg*' principle of Frank and Evans was first realised in the X-ray analysis of the small hydrophobic protein crambin (46 amino acids, $M_R = 4720$) which is found in the embryonic tissue of seeds from *Crambe abyssinica*, commonly known as Abyssinian cabbage, (Hendrickson and Teeter, 1981; Teeter, 1984; Teeter and Whitlow,

1987). The molecular conformation of crambin has the shape of the Greek letter gamma (γ) and its surface has an amphipathic character. Six charged groups and other hydrophilic side chains are segregated from what is otherwise a largely hydrophobic surface – unlike other water soluble proteins – and includes fully exposed leucine, isoleucine, valine and proline side chains. Klotz (1958) originally proposed that hydrophobic amino acid residues might organise surface water into five-membered ring arrays linked analogously to those in the water clathrate cage hydrate structures. However, if such cages were formed and not anchored in some way to the protein, they might become disordered and contribute little to the X-ray experiment. In the crystal structure of crambin four pentagonal rings of water molecules (designated A, C, D and E) lie along a hydrophobic surface of the protein at an intermolecular contact and a fifth (ring B) extends into the solvent. The average O–O distance for the sixteen water molecules which constitute rings A–E is ~ 2.80 Å. Rings A, C and E, which share an apex, clearly create a 'cap' which covers the $C_{\delta 2}$ methyl group of leucine-18. Critically, in addition to these hydrophobic contacts the rings are extensively hydrogen bonded to the protein structure. The general presumption underlying these observations is that, although the crystals of crambin diffract to at least 0.88 Å, and it may well be a fortuitous example, ordered structures formed by water of hydration around hydrophobic surfaces may be more common than had been assumed.

There is ample experimental evidence that relatively non-polar molecules have a favourable net free-energy of interaction with each other in aqueous media and that these '*hydrophobic effects*' (for want of a better definition) are probably the most important single factor, the most significant driving force for non-covalent intermolecular interactions in aqueous media. Given this importance it is perhaps surprising that there are as yet no clear definitive criteria which exist to explain these effects. On the one hand it may be argued that these interactions are the result of a positive attraction caused by van der Waals–London dispersion forces between the non-polar solute species. On the other hand there is the view, elaborated above, that it is the much greater affinity of water molecules for each other, than for many solute species, which is primarily responsible for the greater part of this free-energy of solute–solute interaction. As the example of the protein crambin shows, there is evidence that water molecules in the vicinity of the surfaces of non-polar solutes are more 'structured' than in the bulk medium and it therefore follows that the extent of this structural organisation is reduced when solute species interact with one another. Serious attempts to distinguish the relative contributions to the overall free energy of interaction of non-polar solutes from these two mechanistic possibilities are, as yet, uncommon.

Diederich and Smithrud (1990) have recently made some critical observations on this question and the role of solvents generally in a model system – namely the complex formed between a macrocyclic cyclophane 'host' and the aromatic hydrocarbon pyrene as 'guest'. The stability of this complex was studied in water and seventeen other solvents which covered the entire polarity range. Complexation strength decreased progressively from that with water as solvent ($- \Delta G^{\ominus} = 9.4$ kcal mol^{-1} at $T = 303$ K) to that shown in apolar solvents such as carbon bisulphide ($- \Delta G^{\ominus} = 1.3$ kcal mol^{-1} at $T = 303$ K). Whilst the most stable complex formed in water, strong binding also occurred in formamide and in some alcohols. 2,2,2-Trifluoroethanol, ($- \Delta G^{\ominus} = 7.8$ kcal mol^{-1} at $T = 303$ K) came nearest to water in its ability to promote complexation. The authors concluded that the large differences in binding strength derive, almost exclusively, from solvation effects.

Binding is strongest in solvents characterised by low molecular polarisabilities and by high cohesive interactions. Solvent molecules with high cohesive interactions interact more favourably with bulk solvent molecules than with the complementary apolar surfaces of free host and guest, and therefore, energy is gained upon the release of surface-solvating molecules to the bulk during the complexation step. Upon complexation, the less favourable dispersion interactions between solvent molecules of low polarisability and highly polarisable hydrocarbon surfaces are replaced with more favourable interactions between the complementary surfaces of host and guest. **No special concepts are necessary to explain the ability of water to promote apolar complexation. The solvation properties of water can be rationalised on the basis of its physical properties. Water has the highest cohesive interactions and possesses by far the lowest molecular polarisability.** *Both effects taken together* **make it the best solvent for apolar complexation.**

2.7 π–π Interactions

The association of aromatic surfaces with one another is believed to play a significant role not only in 'host–guest' chemistry where it appears to control a range of molecular recognition and self-assembly phenomena (Stoddart *et al.*, 1991) but also in the folding and complexation behaviour of biopolymers (Burley and Petsko, 1985, 1986). Likewise its relevance to the structure and properties of nucleic acids is very well documented (Saenger, 1984). Non-covalent interactions between delocalised π-systems, including those between aromatic molecules, are frequently described as *π–π interactions*, although such an all-embracing nomenclature may ultimately become to be regarded as something of a misnomer. Recent work has amply illustrated that *π–π interactions* have quite strong geometrical requirements such that in both crystal structures and the gas and liquid phase they are directional to a much greater degree than had previously been anticipated.

2.7.1 'Face to face' *π–π* interactions

Many early observations of *π–π interactions* were made with charge transfer complexes which are often created when molecules of low ionisation potential interact with molecules of high electron density (Prout and Wright, 1968; Wright, 1987). The formation of these complexes is recognised by the almost invariable appearance of new intense broad spectral bands in the ultra violet/ visible spectrum. A classical example is provided by the complex formed from benzene and tetracyanoethylene. Each of the components is colourless but they yield a bright orange complex when solutions are mixed.

Quinhydrone, formed from an equimolar mixture of hydroquinone (quinol) and *p*-benzoquinone, forms dark green crystals, m.p. 171°C.

quinhydrone

These, and analogous, experimental observations were interpreted on a theoretical basis by Mulliken (1952) and gave rise to Mulliken's 'Overlap and Orientation' principle which indicates that maximum charge transfer occurs when the relative orientations of the two molecules provides maximum over-lap of the filled donor orbital (HOMO) and the vacant acceptor orbital (LUMO). In resonance terms the complexes are characterised as a hybrid of two resonance structures (D = donor; A = acceptor):

The charge transfer structure makes only a small contribution to the total electronic structure but provides a part of the bonding that keeps the complex together. Typical charge transfer donors are aromatic nuclei with electron donating groups such as: –OH, –OMe, –Me and –NMe$_2$; common acceptors are quinones or aromatic nuclei bearing several cyano or nitro groups.

It should be noted, however, that Hunter and Sanders (1990) have pointed out that the widespread use of the electron donor–electron acceptor concept can be misleading. They argue that it is the properties of the atoms at the points of intermolecular contact rather than the overall redox properties of the molecules which determine how the two π-systems may interact. According to this view the charge transfer transitions observed for such complexes are a **consequence** and **not a cause** of the π–π *interaction*; electron donor–electron acceptor complexes are simply a special case of π–π *interaction*.

All crystals of 1:1 charge transfer complexes are formed from approximately plane to plane stacks of alternate donor and acceptor molecules. Three characteristic patterns of packing the stacks together in the crystals that are readily identifiable are depicted in Figure 2.15a, as are sketches of the relative vertical orientation (overlap) of particular donor and acceptor molecules in the stacks, Figure 2.15b. In the complex of *N,N,N',N'*-tetramethyl-*p*-phenylene diamine with chloranil the aromatic rings are *exactly* superposed in the eclipsed position (a slight offset is in fact shown in Figure 2.15(b)–(v)); in the majority of other cases involving simple benzene derivatives one aromatic ring is usually displaced relative to the other and the overlap of the aromatic nucleii is small. Where benzoquinone derivatives act as acceptors in charge transfer complexes the highly polar carbonyl group usually, but not invariably, lies over the centre of the aromatic ring of the donor molecule (Figure 2.15(b) – (vi)). In addition to the charge transfer complexes many large planar molecules assemble as crystal structures composed of molecules packed 'face to face'; structures containing stacks of molecules of just one type are frequently termed *homosoric* (Greek *sorus*, pile or stack). The crystal structure of hexamethylbenzene (Brockway and Robertson, 1939) typifies the characteristic stacking patterns of this type commonly encountered in homomolecular crystals, Figure 2.16. Successive layers of molecules are displaced laterally. As well as achieving effective molecular packing these stacking arrangements maximise dispersion forces between planar molecules having delocalised π-electron systems.

In an analysis of 110 crystal structures of peptides and related materials containing non-polar side chains, including phenylalanine, Walkinshaw and his associates (1985) observed specific modes of interaction between aromatic groups. In this extensive analysis any atoms of a phenyl group within 4 Å of a carbon atom or 3.5 Å of a hydrogen atom of another phenyl group were regarded as an interacting pair. A plot of the interplanar angle between two interacting phenyl groups showed a significant number where the two planes were parallel (interplanar dihedral angle as $\sim 0°$), Figure 2.17. It is significant that in all the members of this particular group there were no examples of exact plate-like stacking of phenyl groups, i.e. $d(xy) = 0$, Figure 2.17. The stacking patterns observed were such that $d(z)$ was in the range 3.1–3.4 Å and $d(xy)$ lay

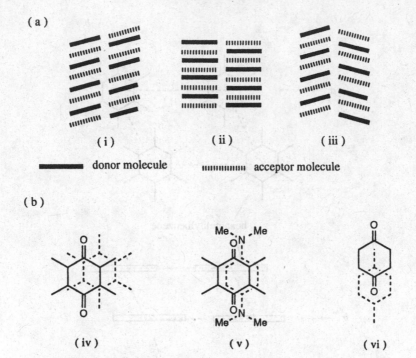

(a)

(i)   (ii)   (iii)

━━━━━ donor molecule   ıııııııııııı acceptor molecule

(b)

(iv)   (v)   (vi)

Donor molecules in 'lower' position (broken lines) in each diagram :
(iv) - hexamethylbenzene ; (v) - tetramethyl- *p*-phenylenediamine ;
(vi) - quinol
Acceptor molecules in 'upper' position (full lines) : (iv) - chloranil ;
(v) - chloranil ; (vi) -*p*-benzoquinone .

Complex (donor / acceptor)	Interplanar perpendicular distance in Å	Angle of inclination between planes of donor and acceptor molecules
(iv) - hexamethylbenzene / chloranil	3.50	2.1°
(v) - tetramethyl-*p*-phenylene diamine / chloranil	3.26	
(vi) - quinol / *p*-benzoquinone (quinhydrone)	3.16	

Figure 2.15. The geometry of some simple charge transfer complexes: (a) typical donor–acceptor stacks in π–π complexes; (b) patterns of overlap in charge transfer complexes (data taken from Prout and Wright, 1968).

in the range 3.6–4.3 Å. This conformational alignment, the authors concluded, 'allows a mutual interaction of one phenyl group with the π-electron cloud of the other ring.'

The elucidation of the three-dimensional structure of DNA by Watson and

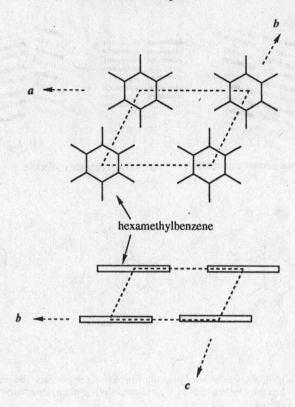

Figure 2.16. Homomolecular crystals: stacking arrangement of molecules of hexa-methyl benzene (data taken from Wright, 1987).

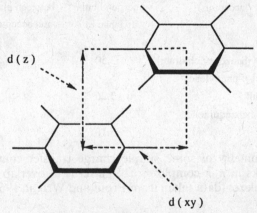

Figure 2.17. 'Face to face' stacking of the aromatic nuclei of L-phenylalanine deriva-tives, data from Walkinshaw *et al.*, 1985, d(xy) lies in the range 3.1–3.4 Å; d(z) lies in the range 3.6–4.3 Å.

Crick not only answered numerous vital questions central to biology but it also opened up many new avenues of research – from genetics to quantum chemistry. In the context of the present discussion, knowledge of the DNA double helix prompted wide ranging studies on the nature and complementarity of base stacking in the DNA molecule. The classical **B** form of DNA consists of a right-handed double helix composed of two polynucleotide strands twisted around each other. Successive base pairs in a strand stack, '*like rolled coins*' (Saenger, 1984), on top of each other. They are separated by 3.4 Å along the helix axis and are related by a rotation of 36°; hence the structure repeats after ten residues in each strand. To accord with Chargaff's and Gulland's data, Watson and Crick proposed specific hydrogen bonding between base pairs: adenine (A) with thymine (T) and guanine (G) with cytosine (C), *vide supra*. The forces which stabilise the DNA double helix are related to two kinds of interaction between bases:

(i) those in the plane of a particular base pair, due to hydrogen bonding, Figure 2.1, and

(ii) those perpendicular to the planes of the bases in a particular strand ('base stacking').

Diverse factors have, in the intervening years, been implicated in the drive to achieve vertical base stacking in the DNA double helix, viz: hydrophobic effects, π–π, dipole–dipole and dipole–induced dipole interactions (Saenger, 1984).

Initial scrutiny focussed primarily on the bases (and analogues) themselves where stacking arrangements are dominant amongst crystal structures examined (Bugg *et al.*, 1971). The relative geometry of the bases in the stacks is, however, quite variable and it has been suggested that in many cases these orientations cannot be rationalised on the basis of permanent molecular dipole–dipole interactions and may owe more to the influence of dipole–induced dipole interactions. Observations of solution thermodynamic and spectroscopic studies support the view that extensive association of bases and of nucleosides occurs in aqueous media, predominantly by vertical stacking of the bases. The degree of association proceeds in several cases well beyond the dimer stage and the stacking process broadly follows isodesmic behaviour, i.e. each step is independent and displays the same kinetic and thermodynamic parameters, suggesting that addition of one base to another, or to a pre-existing stack of bases, is additive and *not* cooperative. NMR studies (Tso, 1974; Tso *et al.*, 1967, 1969) of purine nucleosides point to the conclusion that it is the six- , as opposed to the five- , membered ring of the bases which participates preferentially in the stacking arrangements. Data

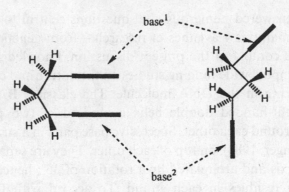

Figure 2.18. Nucleotide base pairing in molecules based upon the 'trimethylene spacer theme': open conformation on the right, folded or closed base stacking conformation on the left (Leonard, 1979).

obtained for both purine and pyrimidine nucleosides in aqueous media show that the stacking interactions are generally weak and follow the trend:

$$purine–purine > purine–pyrimidine > pyrimidine–pyrimidine$$

Leonard (1979) has reviewed results from a range of experiments based upon the 'trimethylene-bridge theme' designed to investigate the propensity of the nucleic acid bases (and their analogues) to interact with each other in solution. Compounds of the general form **Base^1–(CH$_2$)$_3$–Base^2**, where **Base^1** and **Base^2** are 9-substituted adenine or guanine or 1-substituted cytosine, thymine or uracil groups, were synthesised and studied in solution, under conditions which precluded the formation of *inter*molecular complexes by base pairs, by a variety of spectroscopic methods (UV, fluorescence, luminescence). The evidence clearly suggested that, for many of these dinucleoside analogues in aqueous solution, the heterocyclic rings were in close proximity in stacked orientations, Figure 2.18.

Similar observations were made on the heterodimeric molecules, such as (**3**), in which 9-amino-6-chloro-2-methoxyacridine is linked by polymethylene chains of variable length (3, 4 or 5 methylene units) to the bases adenine, thymine and guanine. Detailed spectroscopic analysis showed that all the molecules existed mainly as 'folded' conformations in the temperature range 0–90°C, with the acridine and the nucleotide base stacked one on top of the other (Constant *et al.*, 1988). In the case of (**3**) this exists totally as 'folded' conformations in water (Bolte *et al.*, 1982), underlining the very specific contribution π–π interactions make in the determination of conformation. Interestingly in this context, the crystal structure of 1,3-bis-(8-theophylline) propane (**4**) shows that each molecule adopts a 'folded' conformation (**5**, structures

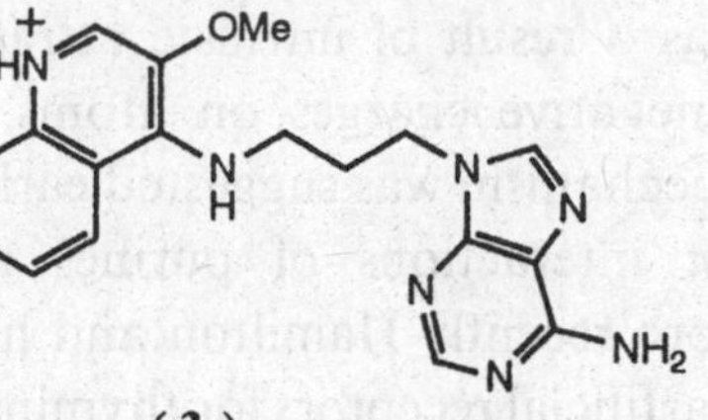

(3)

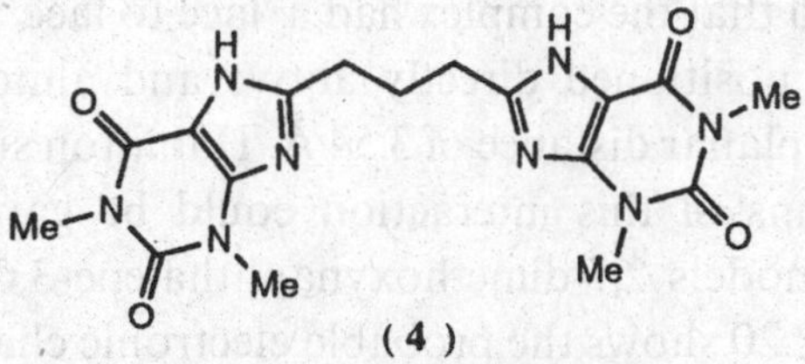

(4)

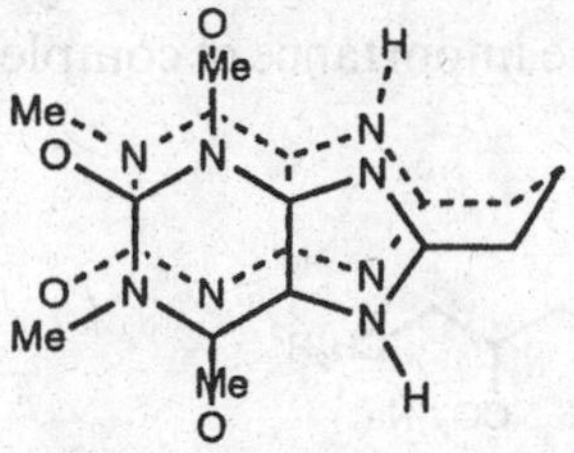

(5)

Figure 2.19. Conformational stacking of heterocyclic rings in synthetic molecules with trimethylene spacers.

partially offset) and each theophylline group stacks with its partner (Rosen and Hybl, 1971), Figure 2.19. However it should be noted that the xanthine rings are 'reversed' and that each –C=O group is eclipsed by an –N–Me group of its partner.

Newcomb and Gellman (1994) have employed the trimethylene spacer idea as a tool with which to investigate the forces that drive aromatic–aromatic interactions. Adenine–adenine stacking in aqueous media was studied by comparing the bis-adenine derivative (6) to the mono-adenine compound (7). Similar comparisons were made with the bis-naphthyl derivative (8) and the corresponding mono-derivative (9) and the unsymmetrical adenine–naphthyl compound (10). Intramolecular π–π stacking was observed for both (6) and (10) but **not** for the bis-naphthyl compound (8). Newcomb and Gellman concluded from this evidence that the favourable stacking between the two heterocyclic rings in (6) and the naphthyl and adenine groups in (10) is not a result of hydrophobic effects or dispersion forces but is consistent with the view that

π–π stacking occurs as a result of intrinsic attractive interactions between partial positive and negative charges on atoms in the two neighbouring aromatic rings. This mechanism was suggested earlier (Tso, 1974; Kopple and Marr, 1967) for π–π interactions of purines and pyrimidines and for diketopiperazines. More recently Hamilton and his colleagues (Hamilton *et al.*, 1988) in a study of artificial receptors for thymine showed that the naphthyl diester (**11**) forms a strong complex with 1-butylthymine (**12**). ^{1}H NMR and X-ray analysis showed that the complex had a 'face to face' geometry in which the naphthyl ring is positioned directly above and almost parallel to the pyrimidine at an interplanar distance of 3.54 Å. Hamilton suggested that some insight into the origins of this interaction could be gained from MNDO calculations on the models 2,7-dimethoxynaphthalene-3,6-dicarboxylic acid and thymine. Figure 2.20 shows the probable electronic charge distribution in the complex and indicates a close alignment of *five* pairs of oppositely charged atoms, thus confirming the importance of complementary electrostatic interactions in parallel stacking.

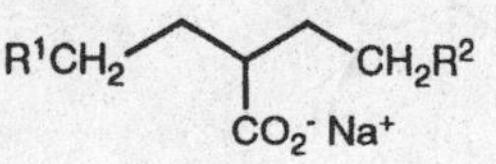

(**6**) - $R^1 = R^2 =$

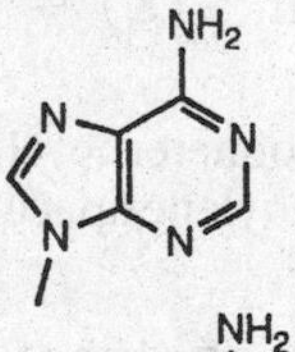

(**7**) - $R^1 = H$; $R^2 =$

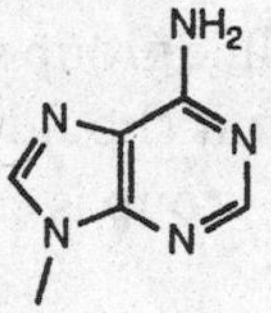

(**8**) - $R^1 = R^2 =$

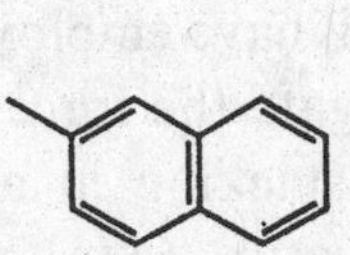

(**9**) - $R^1 = H$; $R^2 =$

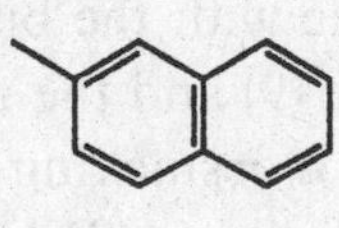

(**10**) - $R^1 =$ 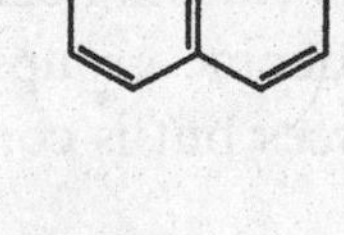   ; $R^2 =$

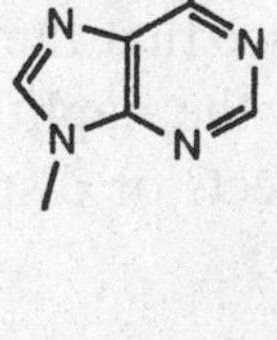

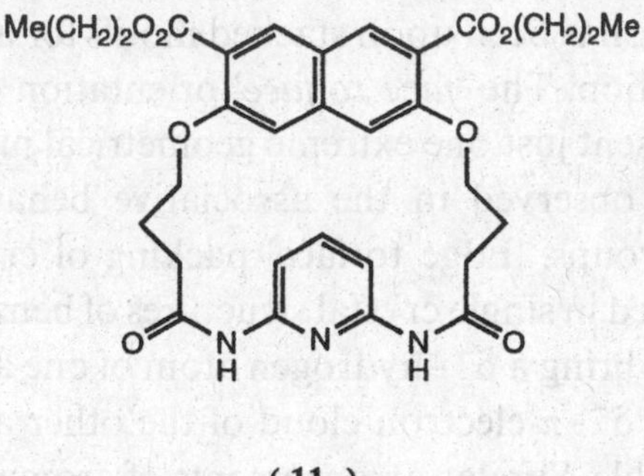

(11)

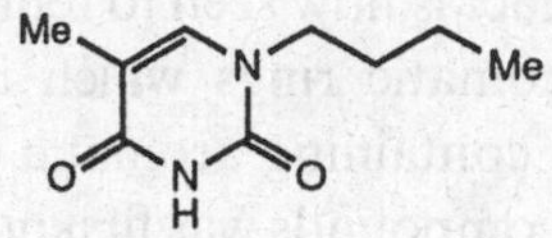

(12)

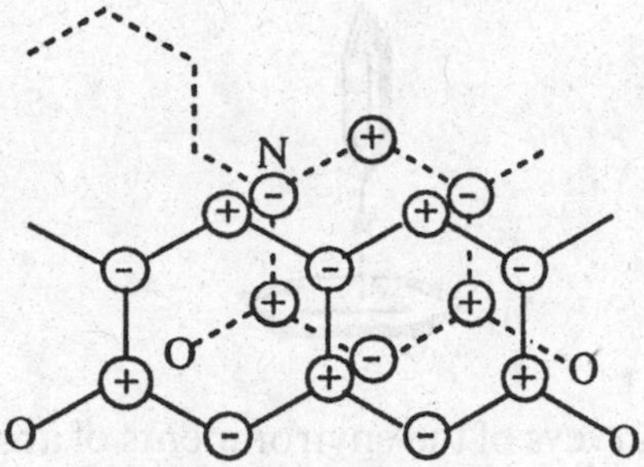

Suggested stacking pattern - position of partially
charged atoms : naphthyl derivative(**11** , above) and
N-butylthymine (**12** , below).

Figure 2.20. Aromatic–aromatic interaction between synthetic receptor (**11**) and *N*-butylthymine (**12**). Top view of the complex with electronic charge distributions (sign only) virtually superimposed (Hamilton *et al.*, 1988).

The significance of atomic charge patterns in π–π stacking arrangements has also been noted in a quite different context. Although the structures of DNA have been studied in some detail it is not yet well understood why DNA is stabilised in a particular helical conformation. Jernigan and his associates (1988) have suggested that the origins of the **B**-form double helix can be attributed in large part to the atomic charge pattern in base pairs and the resultant electronic interactions which favour specific helical stacking of the bases.

2.7.2 'Edge to face' π–π interactions

Possibly because of the early work on charge transfer complexes and the very powerful impact which the discovery of the double helical structure of DNA has had on subsequent developments in this and related areas of science the focus, until comparatively recently, has been upon stacked motifs for aromatic rings in intermolecular complexation. The *'face to face'* orientation (parallel stacks) in fact is now seen to represent just one extreme geometrical juxtaposition of aromatic rings which is observed in the associative behaviour of molecules containing aromatic groups. 'Edge to face' packing of crystalline aromatic compounds was first noted in single crystal structures of benzene and its derivatives. Such arrangements bring a δ^+–hydrogen atom of one aromatic group into close contact with the δ^-–π electron cloud of the other aromatic ring and are energetically favourable. Similar arrangements of aromatic rings

have been detected in surveys of the environments of aromatic amino acid side chains in peptides and proteins (Walkinshaw *et al.*, 1985; Burley and Petsko, 1985). Theoretical studies on the interaction of two benzene rings have suggested an energy minimum when the two rings are perpendicular, but more recent calculations (Jorgensen and Severance, 1990) favour a slightly tilted '*T*' structure. These authors also carried out gas phase optimisations for the complexes of benzene with naphthalene and anthracene. A distorted '*T*' structure still has the lowest interaction for the benzene naphthalene pair but upon increasing the system size to anthracene, then the preference emerged for a slightly tipped stacked structure. The stacked structure has a benzene hydrogen atom near the centre for each of the three aromatic rings in anthracene, Figure 2.21. The binding for all these systems is dominated by the van der Waals ($1/r^6$) attraction. Jorgensen and Severance indicated that this work suggested:

(i) stacked structures become more favourable with increasing size of the arenes, and
(ii) the preferred geometry for stacked structures is offset with hydrogen atoms roughly over ring centres.

The stacked versus '*T*' shaped issue is an important one in the design of organic host molecules for aromatic guest substrates. Hunter (1991) exemplified the construction of a synthetic host molecule which would complex a guest (*p*-benzoquinone) using 'edge to face' π–π interaction or the '*T*' structure as a specific mode of association, Figure 2.22. Thus the macrocyclic tetra-amide uses these 'edge to face' interactions (H_a of the benzoquinone to ring A, H_b to

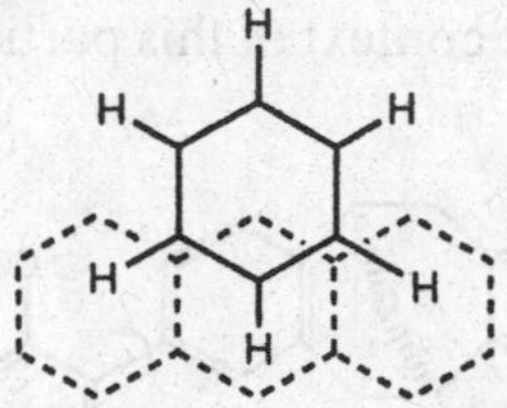

Figure 2.21. Optimised 'stacked' geometry in structure of benzene (upper): anthracene (lower) complex (Jorgensen and Severance, 1990).

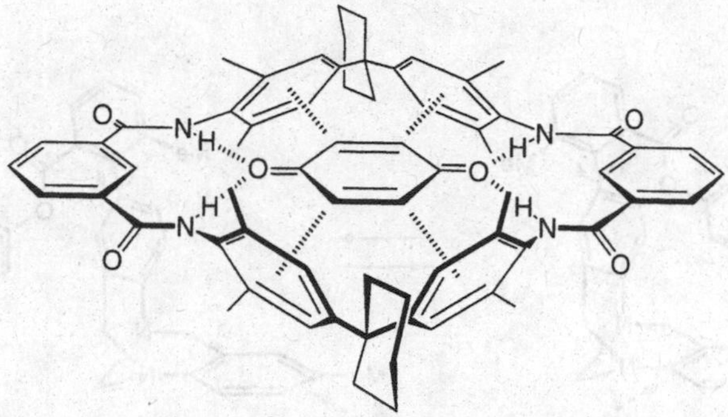

Figure 2.22. Complexation of *p*-benzoquinone by a synthetic macrocyclic tetra-amide host (Hunter 1991).

ring B, etc.) and also maximises the hydrogen bonding potential of the two quinone carbonyl groups (to the four amide hydrogens) to achieve successful complexation.

Conformational isomerism has been exploited by Wilcox and his colleagues (1994) as a probe of 'edge to face' aromatic–aromatic interactions. Aryl esters of the general type (**13**, R = substituted phenyl group) were synthesised. Each ester has two principal conformations that interconvert by slow rotation (**13a** and **13b**); for the folded conformation (**13a**) the substituted phenyl ring lies over the aromatic ring 'A' and is oriented in a tilted '*T*' position relative to that ring. In the unfolded conformation (**13b**) this interaction with ring 'A' is absent. X-ray crystallographic analysis confirmed this 'edge to face' interaction for the *p*-nitrophenyl ester (Y = NO$_2$), with the relevant centroid–centroid distance of 4.95 Å. The folded conformer (**13a**) was also readily identified in solution by ^{1}H NMR spectroscopy because the proton *meta* to the C–O bond in the phenyl group was shifted upfield as a result of the magnetic anisotropy of the ring 'A'. The phenyl ester (**13**, Y = H) is characteristic of the general trend, namely that the folded state is preferred by $\Delta G = -0.24$ kcal mol^{-1}. However, the observation that the cyclohexyl and *t*-butyl esters favour the folded state more than the phenyl is one which, in the context of this particular rationalisation, should be borne in mind!

(13)

ring ' A '

(13a)   ring ' A '   (13b)

The above discussion amply illustrates the wide variations and patterns observed in aromatic π–π interactions, and very recently Hunter and Sanders (1990) have advanced a model whose purpose is to rationalise this diverse behaviour and with it provide an explanation of the nature of aromatic π–π interactions. The model satisfactorily explains the continuum of attractive geometries that extend from the 'edge to face', apparent in the crystal structure of many aromatic molecules, to the offset geometry of 'face to face' stacked π-systems such as are observed in many charge transfer complexes, Figure 2.15. The key feature of the model is that it considers the σ-framework and the π-electrons separately. The treatment concludes that net favourable π–π inter-actions are *not* due to an attractive interaction between the two π-systems but that these occur when the attractive interactions between the π-electrons and the atoms of the σ-framework outweigh unfavourable π-electron repulsion. Some general rules to predict the geometry of aromatic π–π interactions were derived by Hunter and Sanders, and these are particularly successful for non-polarised π-systems. These rules thus indicate that for non-polarised π-systems π–π *repulsions* dominate in 'face to face' orientations of the two π-systems but that π–σ attraction dominates in 'edge to face' or '*T*' shaped and 'off-set' π-stacked geometry, Figure 2.23.

Experimental evidence for the general validity of these rules comes from

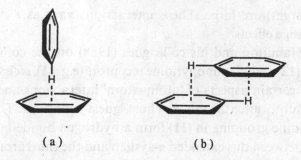

(a)   (b)

Figure 2.23. Favourable intermolecular geometries in non-polarised π-systems: π–σ attraction (Hunter and Sanders, 1990). (a) '*T*' shaped; (b) 'offset' π-stacked.

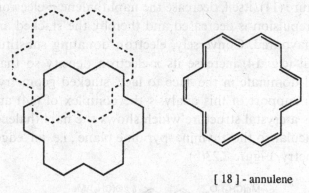

[18] - annulene

Figure 2.24. The geometry of π-stacking interactions in the crystal structure of [18]-annulene.

observations on the crystal structures of simple aromatic compounds. Two types of geometry are frequently observed: 'edge on' or '*T*' shaped, which give rise to the typical herring bone pattern; and 'off-set' stacking of the aromatic systems. The crystal structure of [18]–annulene illustrates this point. In one dimension the π-systems are parallel, stacked but off-set so that the π-system of one molecule lies over the π-cavity at the centre of its nearest neighbour, Figure 2.24. In the other dimensions the molecules are aligned so as to give perpendicular, 'edge on', '*T*' shaped, interactions with their neighbours.

The presence of strongly polarising atoms, such as oxygen and nitrogen, has a major effect upon the electrostatic interactions between π-systems. The 'face to face' stacked geometry is always favoured by van der Waals interactions but is generally disfavoured by π–π repulsion. Hunter (1993, 1994) has described these charge–charge interactions as atom–atom, and according to Hunter and Sanders (1990) they dominate the interactions between highly polarised aromatics, such as the DNA bases, where the partial atomic charges (the sum of the π and σ charges) are large. These interactions vary as r^{-1} and so are relatively long range effects.

The work of Hamilton and his colleagues (1988) on the complexation of N-butylthymine (**12**) and diaminopyridine receptors (e.g. **11**), *vide supra*, Figure 2.20, exemplifies certain aspects of 'atom–atom' interactions and the way in which they may subtly alter the mode of host–guest behaviour. The guest base (**12**) and the pyridine grouping in (**11**) form a hydrogen bonded complex and the interactions between this extended π-system and the third aromatic grouping – the naphthalene unit in (**11**) – were investigated by Hamilton. The naphthalene unit in (**11**) is constrained by molecular design to lie over the π-systems of the two bases. Electron withdrawing groups in the naphthalene, such as in structure (**11**), itself, decrease the naphthalene π-electron density so that electronic repulsion is decreased and thereby the stacked arrangement, Figure 2.20, is favoured. Conversely, electron donating substituents in the naphthalene unit (e.g. **14**) increase its π-electron density so that electronic repulsion would dominate in the 'face to face' stacked geometry and hence destabilise it. In support of this analysis the complex of (**14**) and N-butyl-thymine (**12**) has a crystal structure which shows the naphthalene ring to be almost perpendicular to the 'thymine–pyridine plane', i.e. an 'edge to face' or 'T' shaped geometry, Figure 2.23(b).

(**14**)

References

Aakeroy, C. B. and Seddon, K. R. (1993). The hydrogen bond and crystal engineering. *Chem. Soc. Rev.*, pp. 397–407.

Abraham, M. H., Duce, P. P., Prior, D. V., Barratt, D. G., Morris, J. J. and Taylor, P. J. (1989). Hydrogen bonding. Part 9. Solute proton donor and acceptor scales for use in drug design. *J. Chem. Soc. (Perkin Trans. 2)*, pp. 1355–75.

Bernal, J. D. and Fowler, R. H. (1933). A theory of water and ionic solution, with particular reference to hydrogen and hydroxyl ions. *J. Chem. Phys.*, 1, 515–48.

Blokzijl, W. and Engberts, J. B. F. N. (1993). Hydrophobic effects, opinions and facts.

Angew. Chem. Int. Edn., **32**, 1545–79.

Bolte, J., Demuynck, C., Lhomme, M. F., Lhomme, J., Barbet, J. and Roques, B. P. (1982). Synthetic models related to DNA intercalating molecules: comparisons between quinacridine and chloroquine in their ring–ring intercalation with adenine and thymine. *J. Amer. Chem. Soc.*, **104**, 760–5.

Brockway, L. O. and Robertson, J. M. (1939). The crystal structure of hexamethylbenzene and the length of the methyl group bond to aromatic carbons. *J. Chem. Soc.*, pp. 1324–32.

Bugg, C. E., Thomas, J. M., Sundaralingam, M. and Rao, S. J. (1971). Stereochemistry of nucleic acids and their constituents. X. Solid state base-stacking patterns in nucleic acid constituents and polynucleotides. *Biopolymers*, **10**, 175–219.

Burley, S. K. and Petsko, G. A. (1985). Aromatic–aromatic interaction: a mechanism for protein stabilisation. *Science*, **229**, 23–8.

Burley, S. K. and Petsko, G. A. (1986). Dimerisation energies of benzene and aromatic amino acid side chains. *J. Amer. Chem. Soc.*, **108**, 7995–8001.

Chothia, C. (1984). The principles that determine the structure of proteins. *Ann. Rev. Biochem.*, **53**, 537–72.

Chothia, C. (1990). The classification and origins of protein folding patterns. *Ann. Rev. Biochem.*, **59**, 1007–39.

Constant, J. F., Laguan, P., Roques, B. P. and Lhomme, J. (1988). Heterodimeric molecules including nucleic acid bases and 9-aminoacridine, spectroscopic studies, conformations and interactions with DNA. *Biochemistry*, **27**, 3997–4003.

Diederich, F. and Smithrud, D. B. (1990). Strength of molecular complexation of apolar solutes in water and organic solvents is predictable by linear free-energy relationships: a general model for solvation effects on apolar binding. *J. Amer. Chem. Soc.*, **112**, 339–43.

Eisenberg, D. and Kauzman, W. (1969). *Structure and Properties of Water*. Oxford University Press: Oxford, London.

Etter, M. C. (1990). Encoding and decoding hydrogen-bond patterns of organic compounds. *Acc. Chem. Res.*, **23**, 120–6.

Etter, M. C. (1991). Hydrogen bonds as design elements in organic chemistry. *J. Chem. Phys.*, **95**, 4601–10.

Farber, G. K. and Petsko, G. A. (1990). The evolution of α/β barrel enzymes. *Trends Biochem. Sci.*, **15**, 228–34.

Frank, H. S. (1970). The structure of ordinary water. *Science*, **169**, 635–41.

Frank, H. S. and Evans, M. W. (1945). Free volume and entropy in condensed systems. III. Entropy in binary liquid mixtures; partial molal entropy in dilute solutions; structure and thermodynamics in aqueous electrolytes. *J. Chem. Phys.*, **13**, 507–32.

Frank, H. S. and Wen, W.-Y. (1957). Structural aspects of ion–solvent interactions in aqueous solutions. A suggested picture of water structure. *Disc. Farad. Soc.*, **24**, 133–40.

Franks, F. (1972–80). *The Chemistry and Physics of Water*. Volumes 1–7, Plenum Press: New York.

Gill, S. J., Dec, S. F., Olafsson, G. and Wadso, I. (1985). Anomalous heat capacity of hydrophobic solvation. *J. Phys. Chem.*, **89**, 3758–61.

Gurka, D. and Taft, R. W. (1969). Studies of hydrogen bonded complex formation with *p*-fluorophenol. IV. Fluorine NMR method. *J. Amer. Chem. Soc.*, **91**, 4794–801.

Hamilton, A. D., Muehldorf, A. V., van Engen, D. V. and Warner, J. C. (1988). Aromatic–aromatic interactions in molecular recognition. A family of artificial receptors for thymine that show both face to face and edge to face orientations. *J. Amer. Chem. Soc.*, **110**, 6561–2.

Hardy, A. D. U. and MacNichol, D. D. (1976). Crystal and molecular structure of an –O–H ⋯ π hydrogen bond system: 2,2-bis(2-hydroxy-5-methyl-3-*t*-butylphenyl)-propane. *J. Chem. Soc.* (*Perkin Trans. 2*), pp. 1140–00.

Hendrickson, W. A. and Teeter, M. M. (1981). Structure of the hydrophobic protein crambin determined directly from the anomalous scattering of sulphur. *Nature*, **290**, 107–13.

Henn, R. A. and Kauzman, W. (1989). Equation of state of a random network, continuum mode of liquid water. *J. Phys. Chem.*, **93**, 3770–83.

Hunter, C. A. (1991). Molecular recognition of *p*-benzoquinone. *J. Chem. Soc. Chem. Commun.*, pp. 749–51.

Hunter, C. A. (1993). Sequence dependent DNA structure. The role of base stacking interactions. *J. Mol. Biol.*, **230**, 1025–54.

Hunter, C. A. (1994). The role of aromatic interactions in molecular recognition. *Chem. Soc. Rev.*, **23**, 101–9.

Hunter, C. A. and Sanders, J. K. M. (1990). The nature of π–π interactions. *J. Amer. Chem. Soc.*, **112**, 5525–34.

Hunter, C. A., Hanton, L. R. and Purvis, D. H. (1992). Structural consequences of a molecular assembly that is deficient in hydrogen-bond acceptors. *J. Chem. Soc. Chem. Commun.*, pp. 1134–6.

Jeffery, G. A. (1969). Water structure in organic hydrates. *Acc. Chem. Res.*, **2**, 344–52.

Jeffery, G. A., Gress, M. E. and Takagi, S. (1977). Some experimental observations on H ⋯ O bond lengths in carbohydrate crystal structures. *J. Amer. Chem. Soc.*, **99**, 609–11.

Jeffery, G. A. and Takagi, S. (1978). Hydrogen bond structures in carbohydrate crystals. *Acc. Chem. Res.*, **11**, 264–70.

Jeffery, G. A. and Lewis, L. (1978). Cooperative aspects of hydrogen bonding in carbohydrates. *Carbohydrate Res.*, **60**, 179–82.

Jeffery, G. A. and Mitra, J. (1983). The hydrogen-bonding patterns in the pyranose and pyranoside crystal structures. *Acta Cryst.*, **39B**, 469–80.

Jeffery, G. A., Matias, P. M. and Ruble, J. R. (1988). Structures of the *E, Z* (*cis–trans*) isomer of diacetamide and the 1:1 complex with acetamide at 123 K. *Ab initio* molecular orbital calculations on the *Z, Z* (*trans–trans*), *E, Z* (*cis–trans*) and *E, E* (*cis–cis*) isomers of diacetamide. *Acta Cryst.*, **44B**, 516–22.

Jeffery, G. A. and Saenger, W. (1991). *Hydrogen Bonding in Biological Structures*, Berlin, Heidelberg, New York: Springer-Verlag, p. 569.

Jernigan, R. L., Sarai, A., Mazur, J. and Nussinov, R. (1988). Origin of DNA helical structure and its sequence dependence. *Biochemistry*, **27**, 8498–502.

Jorgensen, W. L. and Severance, D. L. (1990). Aromatic–aromatic interactions: free energy profiles for the benzene dimer in water, chloroform and liquid benzene. *J. Amer. Chem. Soc.*, **112**, 4768–74.

Kamlet, M. J. and Taft, R. W. (1976a). The solvatochromic comparison method. 1. The β-scale of solvent hydrogen bond acceptor (HBA) basicities. *J. Amer. Chem. Soc.*, **98**, 376–83.

Kamlet, M. J. and Taft, R. W. (1976b). The solvatochromic comparison method. 1. The α-scale of solvent hydrogen bond donor (HBD) acidities. *J. Amer. Chem. Soc.*, **98**, 2886–94.

Kamlet, M. J. and Taft, R. W. (1979). Linear solvation free energy relationships. Part 4. Correlations with and limitations of hydrogen bond donor acidities. *J. Chem. Soc. (Perkin Trans. 2)*, pp. 1723–9.

Kauzman, W. (1959). Some factors in the interpretation of protein denaturation. *Adv. Protein Chem.*, **14**, 1–63.

Keegstra, E. M. D., Spek, A. L., Zwikker, J. W. and Jenneskens, L. W. (1994). The crystal structure of 2-methoxy-1,4-benzoquinone: molecular recognition involving intermolecular dipole–dipole and C-H $\cdots$ O-hydrogen bond interactions. *J. Chem. Soc. Chem. Commun.*, pp. 1633–4.

Kennard, O., Taylor, R. and Versichel, W. (1984a). Geometry of the NH $\cdots$ O=C hydrogen bond. 2. Three-centred ('bifurcated') and four centred ('trifurcated') bonds. *J. Amer. Chem. Soc.*, **106**, 244–8.

Kennard, O., Taylor, R. and Versichel, W. (1984b). Geometry of the NH $\cdots$ O=C hydrogen bond. 3. Hydrogen bond distances and angles. *Acta Cryst.*, **40B**, 280–8.

Klotz, I. M. (1958). Protein hydration and behaviour. *Science*, **128**, 815–22.

Kopple, K. D. and Marr, D. H. (1967). Conformations of cyclic peptides. The folding of cyclic dipeptides containing an aromatic side chain. *J. Amer. Chem Soc.*, **89**, 6193–200.

Latimer, W. M. and Rodebush, W. H. (1920). Polarity and ionisation from the standpoint of the Lewis theory of valence. *J. Amer. Chem. Soc.*, **42**, 1419–33.

Legon, A. C. and Millen, D. J. (1987). Angular geometries and other properties of hydrogen-bonded dimers: a simple electrostatic interpretation of the success of the electron pair model. *Chem. Soc. Rev.*, **16**, 467–98.

Legon, A. C. and Millen, D. J. (1993). In *Principles of Molecular Recognition*, eds. A. D. Buckingham, A. C. Legon and S. M. Roberts, Blackie Academic: London, pp. 17–41.

Leonard, N. J. (1979). Trimethylene bridges and synthetic spacers for the detection of intramolecular interactions. *Acc. Chem. Res.*, **12**, 423–9.

Levitt, M. and Perutz, M. F. (1988). Aromatic rings as hydrogen bond acceptors. *J. Mol. Biol.*, **201**, 751–4.

Miller, K. W. and Hildebrand, J. H. (1968). Solutions of inert gases in water. *J. Amer. Chem. Soc.*, **90**, 3001–4.

Mulliken, R. S. (1952). Molecular compounds and their spectra II. *J. Amer. Chem. Soc.*, **74**, 811–24.

Nemethy, G. and Scheraga, H. A. (1962a). Structure of water and hydrophobic bonding in proteins. I. A model for the thermodynamic properties of liquid water. *J. Chem. Phys.*, **36**, 3382–400.

Nemethy, G. and Scheraga, H. A. (1962b). Structure of water and hydrophobic bonding in proteins II. A model for the thermodynamic properties of aqueous solutions of hydrocarbons. *J. Chem. Phys.*, **36**, 3382–400.

Nemethy, G. and Scheraga, H. A. (1964). Structure of water and hydrophobic bonding in proteins. IV. A model for the thermodynamic properties of liquid deuterium oxide. *J. Chem. Phys.*, **41**, 680–9.

Newcomb, L. F. and Gellman, S. H. (1994). Aromatic stacking reactions in aqueous solution: evidence that neither classical hydrophobic effects nor dispersion forces are important. *J. Amer. Chem. Soc.*, **116**, 4993–4.

Pauling, L. (1940). *The Nature of the Chemical Bond and the Structure of Molecules and Crystals – An Introduction to Modern Structural Chemistry*. Second edition, Oxford University Press: London.

Pauling, L. and Corey, R. B. (1951). Configurations of Polypeptide chains with favoured orientations around single bonds: two new pleated sheets. *Proc. Natl. Acad. Sci. (U.S.A.)*, **37**, 729–40.

Pauling, L., Corey, R. B. and Branson, H. R. (1951). The structure of proteins: two hydrogen bonded helical conformations of the polypeptide chain. *Proc. Natl. Acad. Sci. (U.S.A.)*, **37**, 205–11.

Perutz, M. F. (1992). What are enzyme structures telling us? *Faraday Disc.*, **93**, 1–11.

Pimentel, G. C. and McClellan, A. L. (1960). *The Hydrogen Bond*. San Francisco, U.S.A.: Freeman.

Pople, J. A. (1951). Molecular association in liquids II. A theory of the structure of water. *Proc. R. Soc. Lond.*, **205A**, 163–78.

Prout, C. K. and Wright, J. D. (1968). Observations on the crystal structures of electron donor–acceptor complexes. *Angew. Chem. Int. Edn.*, **7**, 659–67.

Quiocho, F. A. (1986). Carbohydrate binding proteins. *Ann. Rev. Biochem.*, **55**, 287–315.

Rice, S. A. and Sceats, M. G. (1981). A random network model for water. *J. Phys. Chem.*, **85**, 1108–19.

Rommerts, F. F. G. (1995/1996). Molecular love – lust. *The biochemist*, pp. 13–15.

Rosen, L. S. and Hybl, A. (1971). The crystal structure of 1,3-bis-(8-theophylline)-propane. *Acta Cryst.*, **27B**, 952–60.

Saenger, W. (1984). *Principles of Nucleic Structure*. Springer-Verlag: New York.

Stillinger, F. H. (1980). Water Re-visited. *Science*, **209**, 451–7.

Stoddart, F. J., Anelli, P. J., Ashton, P. R., Spencer, N., Slavin, A. M. Z. and Williams, D. J. (1991). Self assembling [2]-pseudorotoxanes. *Angew. Chem. Int. Edn.*, **30**, 1036–45.

Symons, M. C. R. (1981). Water structure and reactivity. *Acc. Chem. Res.*, **14**, 179–87.

Taft, R. W., Gurka, D., Joris, L., von R. Schleyer, P. and Rakshys, J. W. (1969). Studies of hydrogen bonded complex formation with *p*-fluorophenol. V. Linear free energy relationships. *J. Amer. Chem. Soc.*, **91**, 4801–8.

Taylor, R. and Kennard, O. (1984). Hydrogen bond geometry in organic crystals. *Acc. Chem. Res.*, **17**, 320–6.

Teeter, M. M. (1984). Water structure of a hydrophobic protein at atomic resolution: pentagon rings of water molecules in crystals of crambin. *Proc. Natl. Acad. Sci. (U.S.A.)*, **81**, 6014–18.

Teeter, M. M. and Whitlow, M. D. (1987). Hydrogen bonding in the high resolution structure of the protein crambin. *Trans. Am. Cryst. Assoc.*, **22**, 75–88.

Tse, Y. C. and Newton, M. D. (1977). Theoretical observations on the structural consequences of cooperativity in H ... O bonding. *J. Amer. Chem. Soc.*, **99**, 611–13.

Tso, P. O. P. (1974). Bases, nucleosides and nucleotides. In *Basic Principles in Nucleic Acid Chemistry*, Volume I, Academic Press: London and New York, pp. 453–584.

Tso, P. O. P., Broom, A. F. and Schweizer, M. P. (1967). Interaction and association of bases and nucleosides in aqueous solution. V. Studies of the association of purine nucleosides by vapour pressure osmometry and proton magnetic resonance. *J. Amer. Chem. Soc.*, **89**, 3612–22.

Tso, P. O. P., Broom, A. D., Kondon, N. S. and Robins, R. K. (1969). Interaction and association of bases and nucleosides in aqueous solution. VI. Properties of 7-methylinosine as related to the nature of the stacking interactions. *J. Amer. Chem. Soc.*, **91**, 5625–31.

Walkinshaw, M. D., Taylor, P., Gray, A. M. and Gould, R. O. (1985). Crystal environments and geometries of leucine, isoleucine, valine and phenylalanine provide estimates of minimum non-bonded contact and preferred van der Waals interaction distances. *J. Amer. Chem. Soc.*, **107**, 5921–7.

Watson, J. D. and Crick, F. H. C. (1953). Genetic implications of structure of deoxyribonucleic acid. *Nature*, **171**, 964–7.

Wilcox, G. S., Paliwal, S. and Geib, S. (1994). Molecular torsion balance for weak molecular recognition forces. Effects of 'tilted-T' edge to face aromatic interactions on conformational selection and solid state structure. *J. Amer. Chem. Soc.*, **116**, 4497–8.

Williams D. H. (1991). The molecular basis of biological order. *Aldrichimica Acta*, **24**, 71–80.

Williams, D. J., Rzepa, H. S., Webb, M. L. and Slawin, A. M. Z. (1991). π Facial hydrogen bonding in the chiral resolving agent (S)-2,2,2-trifluoro-1-(9-anthryl)-ethanol and its racemic modification. *J. Chem. Soc. Chem. Commun.*, 765–7.

Wolfenden, R. (1978). Interaction of the peptide bond with solvent water: a vapour phase analysis. *Biochemistry*, **17**, 201–4.

Wolfenden, R. (1983). Waterlogged molecules. *Science*, **222**, 1087–93.

Wright, J. D. (1987). *Molecular Crystals*. Cambridge University Press: Cambridge.

3

Molecular recognition – phenols and polyphenols

3.1 Introduction

The period 1920–50 witnessed great developments in our knowledge of the mechanism of the processes whereby carbon containing compounds undergo chemical change. The study of reaction mechanism lagged well behind that of structure for as long as the organic chemists' depiction of chemical bonds was as 'wooden' as the sticks used to show them then there could be little basis for writing mechanisms (as students from that era may recall) more general than the chalk box or lasso. The understanding of mechanism was largely derived from the application of electronic theories to the chemistry of carbon compounds, from the initial ideas of Lapworth at the beginning of the century and those of Robinson and Ingold in the 1920s and 1930s. With the advent, in the 1950s, of conformational analysis – the determination (when due allowance has been made for the interplay of non-bonded interactions) of the preferred arrangement of atoms in molecules – then the intimate relationship between reaction mechanism and molecular shape emerged. Taken together the theories were of such evident generality that it became possible not only to rationalise and systematise a vast body of seemingly disparate data but also to use the ideas predictively. In turn the subject thrived upon the application of a series of fairly simple inexact models and this wrought a sea change in outlook over the whole of the science, for although it remained essentially experimental (rather than arm-chair), the romantic era was over. It was as though Berg, Boulez and Birtwistle had replaced Schubert, Brahms and Schumann as icons of the subject. A decade ago Prelog, that most genial and gentle of geniuses, admirably summarised the present dilemma and, more importantly, pointed the way forward.

According to Albert Einstein we live in an era of perfect methods and confused aims. For example in organic chemistry the known synthetic methods allow us to prepare an

138

astronomical number of compounds; the gap between the possible and the relevant becomes larger every day. ... the study of the selectivity of reactions becomes of paramount importance. We can learn quite a lot from nature which uses molecular recognition to achieve selectivity in a degree so far unattainable by mere mortals.

(V. Prelog, 1986
18th Solvay Conference)

The desire to reduce and rationalise apparently unconnected facts, to under-stand underlying principles, to work out simple generalisations which may be applied elsewhere are strong, ever present motivations in any experimental subject – none more so than chemistry. Practitioners of the emerging science of molecular recognition should perhaps be grateful that moments of romance, beauty (and even passion) may still be experienced, if only vicariously, from the study of non-covalent inter- and intra-molecular interactions. A comprehen-sive 'set of rules', such as exists in the area of reaction mechanisms, is still some way distant; however, the guidelines enunciated in chapter 2 nevertheless form the basis for the interpretation of phenomena, underlying which are facets of molecular recognition. It is therefore appropriate, and at this point timely, to examine the properties and behaviour of phenols and natural polyphenols in the light of the previous discussions of hydrogen bonding, hydrophobic effects and solvation.

3.2 General observations

3.2.1 Hydrogen bonding

Phenols are amphipathic molecules – the phenolic hydroxyl group is hy-drophilic whilst the aromatic ring is hydrophobic in character. In a statistical study (based upon high resolution X-ray data of 16 non-homologous proteins) Thanki, Thornton and Goodfellow (1988) examined the distribution of water molecules around the 20 different α amino acid residues. They observed that the proportion of residues whose side chain atoms were in contact with water molecules (within 3.5 Å) showed a clear inverse correlation with the hydropho-bicity of the residue, being as low as 14% for leucine and isoleucine but greater than 80% for arginine. Distinct non-random distributions of water molecules were also observed. These hydration patterns were consistent with the ex-pected stereochemistry of the potential hydrogen bonding sites on the polar side chains. As expected, at the distance criterion of 3.5 Å, the hydrophobic aromatic ring of L-phenylalanine accommodates very few water molecules around itself. Similarly the aromatic ring of L-tyrosine also shows the presence of very few water molecules within 3.5 Å (parenthetically the distribution of water around the prolyl residue indicates that it also is strongly hydrophobic in character). In contrast there is a distinct clustering of water molecules around the phenolic hydroxyl group of L-tyrosine. Resonance appears to constrain the proton of the hydroxyl group to the plane of the aromatic ring and prevent free rotation (the C–O bond distance in phenol is thus somewhat shorter than that found in methanol 1.37–1.38 vs 1.43 Å; the experimental torsional barrier to rotation of the hydroxyl group in phenol is $\sim$ 3.1–3.5 kcal mol^{-1}, Momany *et al.*, 1975; Forest and Dailey, 1966; Lowe, 1968). As a consequence the distribution of water molecules around the tyrosine hydroxyl has two clusters which broadly lie in the plane of the ring, Figure 3.1. Therefore, as perhaps expected, very, very few L-tyrosine hydroxyl groups make

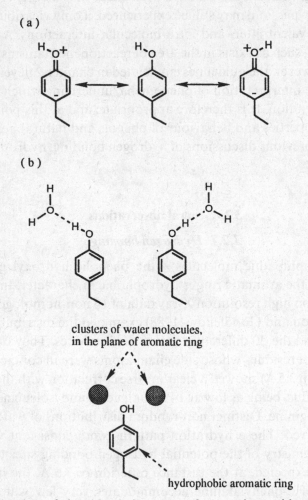

Figure 3.1. Solvation of the side chain atoms of L-tyrosine residues in proteins: (a) hybridisation of the phenolic hydroxyl group; (b) water molecule clusters associated with the phenolic hydroxyl group; few water molecules within 3.5 Å of the aromatic ring atoms.

more than two hydrogen bonds, although roughly one third do make two (one donor, one acceptor). This is again consistent with the hybridisation of the phenolic oxygen atom in L-tyrosine, Figure 3.1. Nevertheless the phenolic hydroxyl group of a tyrosine residue would be predicted to have a strong preference to act as a hydrogen bond donor.

In the earlier discussion of hydrogen bonding attention was drawn to various aspects of '*cooperativity*' in hydrogen bonding and to the favourable energetics of the formation of finite or infinite chains of hydrogen bonds in crystal structures. In this context it is therefore of interest to note some very early observations of Linus Pauling and to connect with these the fact that the two principal groups of polyphenolics (tannins) contain aromatic nuclei with either the 3,4- or the 3,4,5-disposition of phenolic hydroxyl groups. Pauling (1936, 1949) was commenting upon the studies conducted by Wulf and his co-workers on the infra-red absorption spectra of organic compounds containing the hydroxyl group. He put forward some interpretations on the observations. He noted that the C–O bond in phenols has some double bond character (cf. Figure 3.1) and that for molecules such as *ortho*-chlorophenol the two possible configurations (Figure 3.2) were not equivalent. He interpreted the observed two hydroxyl frequencies in the IR spectrum of *ortho*-chlorophenol in terms of the presence of these two conformers; the peak at $7050 \, cm^{-1}$ to the *trans* form, with a frequency very much like that of phenol itself, and the one at $6910 \, cm^{-1}$ to the *cis* form which was the form present in much higher concentration (roughly 10:1). Catechol shows two nearly equal IR peaks with wave numbers of 6970 and $7060 \, cm^{-1}$. This, he argued, showed that catechol, in solution, existed primarily in the form of the *cis–trans* form. Likewise for pyrogallol he suggested the spectroscopic data (two IR peaks at 7050 and $6960 \, cm^{-1}$, approximately in the ratio of 1:2) corresponded to the presence of the internally hydrogen bonded structure for this molecule shown in Figure 3.2. Clearly, if the concepts of '*cooperativity*' in hydrogen bonding and the favourable energetics of the formation of finite or infinite chains of hydrogen bonds have any validity then they would suggest that both pyrogallol and catechol, by virtue of the intramolecular hydrogen bonding which exists between the phenolic hydroxyl groups, would have enhanced monofunctional hydrogen bond donor capacity (–H*, Figure 3.2). Such factors may well influence the capacity of the pyrogallol and catechol nuclei in natural polyphenols to act as hydrogen bond donors, e.g. with the carbonyl groups of prolyl peptides, Figure 3.2.

It is a well established principle that the use of multiple interactions by enzymes often provides specific stabilisation to the transition states of the reactions which they catalyse. Instead of relying upon a single interaction to

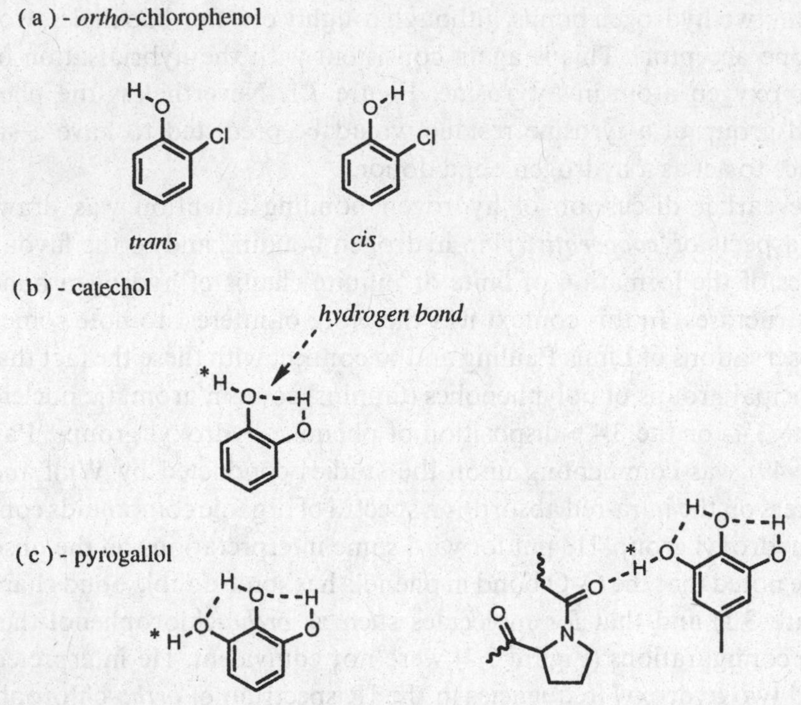

(a) - *ortho*-chlorophenol

trans   cis

(b) - catechol

hydrogen bond

(c) - pyrogallol

Figure 3.2. The structure of some simple phenolic molecules; intramolecular hydrogen bonding (Pauling, 1936, 1949).

generate the bulk of the catalytic power, enzymes use multiple interactions with each one providing a modest contribution to transition state stabilisation. These interactions are often the result of the formation of hydrogen bonds and it has been suggested that the combined energy of multiple hydrogen bonds can be large and nearly additive (Shan and Herschlag, 1996). It should of course always be borne in mind that such non-covalent binding of substrates to enzymes also includes the cost of removing the interacting groups from the aqueous environment in which they are present, cf. Figure 2.12, and the concomitant release of solvent water molecules to the bulk medium.

The association of nucleotide base pairs is a further good example of this particular effect. A triplet of nucleotide bases thus binds strongly to a complementary anticodon of a tRNA, whilst the pairing of isolate complementary bases (e.g. A and U) is not detectable (Fersht, 1977); for example, the association constant between tRNA[Phe] and $tRNA_2{}^{Glu}$ is 2×10^7. Although the hydrophobic stacking of bases is an important feature in the complexation of a complementary pair of base triplets, it is also important to point out that when a single base pair, A and U, associate they gain the energy from the complementary base pairing but at the same time lose the energy of hydrogen bonding to the solvent water. They also forfeit their entropies of independent translation and rotation. In turn there is a gain of entropy as the hydrogen bonded water molecules are released to the bulk medium. However, when a complementary pair of base triplets complex, three times as many molecules of water are freed from solvating these bases but there is still overall the loss of only one set of entropies of rotation and translation. Thus *multiple hydrogen bonding derived from multi-point attachment between substrates may well lead to strong binding.*

Polyphenols (tannins) are typically replete with phenolic groups – a natural polyphenol of $M_R \sim 1000$ generally has some 12–16 phenolic hydroxyl groups in the molecule. In complexation reactions with other substrates polyphenols therefore have the in-built structural facility to form, where this is possible, multiple hydrogen bonds with other substrates. In this way the entropic penalty that must be paid in forming the first hydrogen bond is minimised, if not entirely compensated. This important principle and the importance of multi-point binding is most eloquently and beautifully illustrated in the text *Hydrogen Bonding in Biological Structures* by Jeffery and Saenger (Springer-Verlag, 1991) with the drawing of Gulliver from *Gulliver's Travels* and his 'capture' by the citizens of Lilliput, Figure 3.3.

There is little doubt that this general property of multi-dentate binding to

Figure 3.3. Gulliver: a giant constrained by a multitude of weak bonds (from Jeffery and Saenger, *Hydrogen Bonds in Biological Structures*, Springer-Verlag, 1991).

the co-substrate is responsible, at least in part, for the effectiveness of natural polyphenols in their association with other macromolecules. It also goes a considerable way towards explaining why, in for example the complexations of polyphenols with proteins, strong association results from the property of conformational mobility in both substrates. The binding of natural poly-phenols to compact globular proteins is thus generally much weaker than that with proteins which have a loose, random coil conformation (e.g. salivary proline rich proteins, PRPs). In such cases the complexation is both time dependent and dynamic as the substrates mutually explore routes to engage the 'best fit' for multi-dentate binding, Figure 3.4.

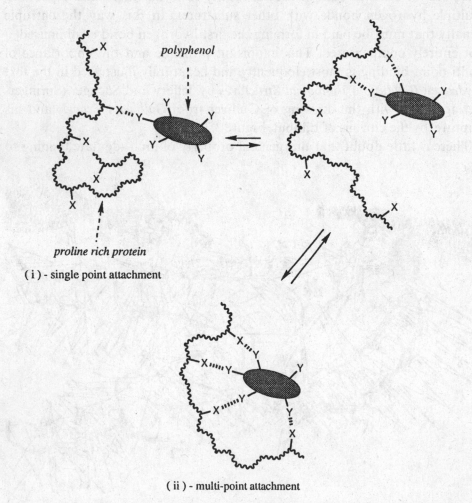

Figure 3.4. Pictorial representation of a polyphenol binding to a proline rich protein (PRP): importance of conformational flexibility in both substrates to attain multi-dentate binding (X ⋯ Y).

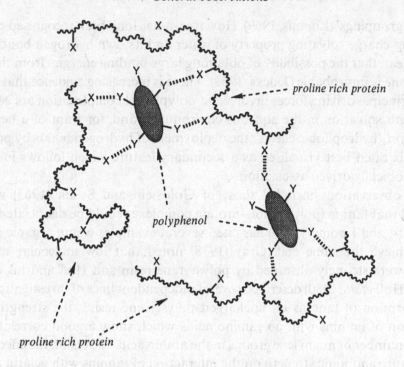

Figure 3.5. Multi-point attachment by polyphenols. Cross-linking of different molecular species, leading to aggregation and precipitation.

This same characteristic for multi-point attachment to the co-substrate is
also responsible for the distinctive property which polyphenols possess to
cross-link different molecular species and molecules of a co-substrate, leading
to large molecular aggregates and ultimately to precipitation of the complex
from solution, Figure 3.5.

3.2.2 *Solvation and 'hydrophobic effects'*

With a molecule such as a polyphenol which possesses a multiplicity of
phenolic groups that can act as both proton donor and acceptor in the
formation of hydrogen bonds and with such a wealth of circumstantial evidence available to underline the general importance of such non-covalent
interactions in molecular recognition processes, historically the first reaction
was therefore to ascribe the driving force for intermolecular complexation
involving polyphenols simply to hydrogen bonding. This emphasis on hydrogen bonding also derived from the fact that tannins (polyphenols) were bound
by modified collagen, gelatin and synthetic polymers such as nylon and
polycaprolactam, the latter containing only –CO–NH– functionalities as

'reactive groupings' (Loomis, 1969). However, it has long been recognised that the strong charge solvating property of water and its own hydrogen bonding ability mean that the possibility of obtaining large binding energies from these forces *alone* is improbable (Jencks, 1969). There is increasing evidence that the initial, principal driving forces involved in polyphenol complexation are associated with solvation in the aqueous environment and, for want of a better description, 'hydrophobic effects'; the deployment of hydrogen bonds by polyphenols is often best visualised as a secondary feature which follows initial hydrophobically driven association.

Early observations included those of Goldstein and Swain (1963) who observed that tannin (polyphenol)–protein complexes could be dissociated by detergents, and Loomis (1969) has cited several examples where organic solvents achieve the same end. Gray (1978) noted that low molecular mass phenols were strongly absorbed by polystyrene resin and Hoff and his colleagues (Hoff *et al.*, 1980) described several independent lines of investigation – the adsorption of tannins on uncharged polystyrene resins; the strength of interaction of tannins with polyamino acids which show a good correlation with the number of methylene groups in the amino acid side chain; the effect of temperature and ionic strength on the interaction of tannins with gelatin and poly-L-proline – all of which supported the view that hydrophobic interactions were of crucial importance in the formation of tannin (polyphenol)–protein complexes.

Polyvinylpyrrolidone (PVP) is interesting from the biological point of view since it has structural features and properties similar to those of proteins; it is soluble in water and it may be precipitated by ammonium sulphate, trichloroacetic acid and polyphenols. It has an amphiphilic character and contains a highly polar tertiary amide group (molecular models show that the carbonyl oxygen is exposed to solvent but the nitrogen atom is buried in the surrounding methylene and methine groups) with a dipole moment close to 4 D from which its hydrophilic properties derive. The apolar methylene and methine groups endow it with hydrophobic properties. In aqueous solution PVP has a loose, random-coil type of conformation analogous to that of the salivary proline rich proteins (PRPs). Polyvinylpyrrolidone (PVP) in aqueous solution displays a strong affinity towards dissolved aromatic compounds, including phenols (Molyneux and Frank, 1961). The binding constants increase with increasing size of the aromatic system and also with the number of phenolic groups (phenol, resorcinol, phloroglucinol). However, in addition to the formation of substrate–PVP hydrogen bonds it was considered that the binding process involved, as its principal factors, an entropy gain due to the disordering of the so-called 'iceberg' water structure in the vicinity of the solute molecules – the so-called hydrophobic effect – and an exothermic interaction between the PVP and the π electron system of the aromatic substrate.

Hansch and his collaborators (Hansch *et al.*, 1964, 1965) have employed a constant π to estimate the hydrophobic bonding of various substituent groups in a particular molecule. They defined π as $\pi = \log P_X/P_H$, where P_H is the partition coefficient of a parent compound between octan-1-ol and water and P_X is that of the derivative X and represents the free energy of transfer of the substituent from an aqueous to a lipophilic phase. They studied the adsorption of a number of phenols (~ 20) by bovine serum albumin and found that binding depends on the lipophilic character of the substituent; it was shown that a linear relationship exists between the logarithm of the binding constant and π. The authors concluded that since the adsorption of the phenols onto the protein very closely parallels the transfer of the phenols from a water phase to one of octan-1-ol, the binding is very probably promoted in large measure by the lipophilic groups and that the phenolic group itself may not play a specific role.

The significance of solvation and 'hydrophobic effects' is most vividly illustrated by the example of three natural polyphenols – β-1,2,3,4,6-penta-*O*-galloyl-D-glucose, and the diastereoisomeric pair vescalagin and castalagin derived from *Quercus* and *Castanea* species. The molecule of β-1,2,3,4,6-penta-*O*-galloyl-D-glucopyranose has, in its most favoured conformation, the shape of a circular disc; molecular models clearly reveal that the periphery of the molecule is hydrophilic, by virtue of the presence of the phenolic groups whilst the upper and lower faces of the molecule (disc-shape) are in contrast largely hydrophobic in character, Figure 3.6.

β-1,2,3,4,6-Penta-*O*-galloyl-D-glucose is amorphous and has a limited solubility in water (~ 1.0 mM at 20 °C). Solutions of higher concentration obtained by heating to 50–60 °C readily gel on cooling to ambient temperature. Such gels probably arise from the ability of the polyphenol to form extensive

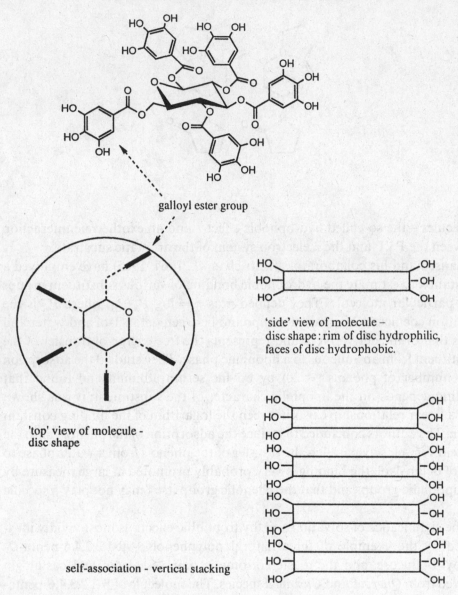

Figure 3.6. β-1,2,3,4,6-Penta-*O*-galloyl-D-glucose: structure and shape. Self-association by vertical stacking of molecules.

three-dimensional lattices, by vertical stacking of molecules (rather as a pile of coins) with their hydrophobic faces brought into juxtaposition to minimise water contacts (Figure 3.6) and by intermolecular cross-linking of stacks by hydrogen bonding. The gel is readily disrupted by addition of solutes such as 1-β-octyl-D-glucopyranoside which presumably functions by intercalation of the hydrophobic hydrocarbon chain in the stacks. The molecule β-1,2,3,4,6-penta-*O*-galloyl-D-glucose has, in principle, a wide range of conformations available to it. The molecule binds strongly to proteins and is a very effective 'tannin'. In partition experiments when bis-*N*-butyl ether is the organic phase β-1,2,3,4,6-penta-*O*-galloyl-D-glucose partitions exclusively to the aqueous phase. However, when the organic phase is octan-1-ol the value of K [octan-1-ol/water] is 32 (Beart *et al.*, 1985; Spencer *et al.*, 1990).

In contrast vescalagin and castalagin, which are unique open-chain derivatives of D-glucose, are both nicely soluble in water but are not extracted therefrom by ethyl acetate and show a K[octan-1-ol/water] of ~ 0.1. Both molecules in the conventional tests for tannins react very poorly indeed and quantitative studies show that they bind very weakly to proteins (Beart *et al.*, 1985). Although these results thus very clearly support the contention of Hansch and his colleagues that solvation and 'hydrophobic effects' are the major driving forces favouring phenol (and hence polyphenol) complexation with proteins they do not answer the question why such dramatic differences should exist. These comparative properties are certainly not ones that would have been readily predicted *a priori*. Thus although both vescalagin and castalagin have formally just six hydrogen atoms less than β-1,2,3,4,6-penta-*O*-galloyl-D-glucose, all three molecules have a similar molecular mass, possess five aromatic nuclei and 15 phenolic hydroxyl groups. The origin of these startling differences presumably lies in the shape, conformation and solvation of these molecules. Thus whilst β-1,2,3,4,6-penta-*O*-galloyl-D-glucose has, in its most favoured conformation, the shape of a circular disc, Figure 3.6, it also possesses considerable conformational mobility. On the other hand vescalagin and castalagin are compact and conformationally quite inflexible. They have the shape of a thin distorted propeller, Figure 3.7, and are well solvated.

At the moment one can only speculate as to the nature of the solvation shell around polyphenol molecules such as β-1,2,3,4,6-penta-*O*-galloyl-D-glucose. Presumably there will be significant clusters (a cloud) of water molecules in the vicinity of the three phenolic groups of each galloyl ester function. The distribution of water molecules around the remainder of each aromatic ring is in present circumstances far more difficult to define. Some pointers may, however, be obtained from X-ray studies of protein structures. Thus analysis of the water structure of γB-crystallin, a structural protein of the eye lens, shows extensive three-dimensional cages of highly ordered solvent molecules around exposed non-polar groups such as the pyrrolidine ring of proline and the phenyl ring of phenylalanine (Kumaraswamy *et al.*, 1996), screening the hydrophobic patches. In the case of the phenylalanine residue three of the closest water molecules lie very close to the plane of the aromatic ring and approxi-

castalagin

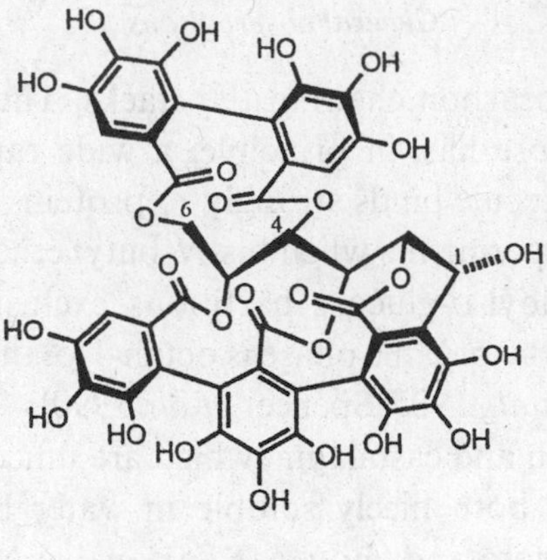

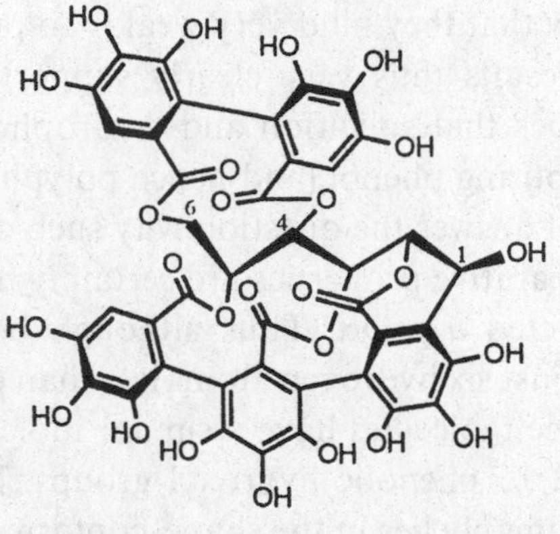

vescalagin

Figure 3.7. Comparison of the molecular shapes of β-1,2,3,4,6-penta-O-galloyl-D-glucose (below; two views) and vescalagin and castalagin.

mately perpendicular to the centre of the C–C bonds, observations which are in agreement with those of Walshaw and Goodfellow (1993). They are also consistent with the earlier studies of Thomas (Thomas *et al.*, 1982) who showed that protein oxygen atoms packing around the phenyl ring of phenylalanine in protein crystals prefer edge interactions with the ring and disfavoured those at ring faces. This is in qualitative agreement with a picture for benzene having electron-rich faces and an array of partially positively charged hydrogen atoms around the periphery (a model which may be used to anticipate other interactions with the aromatic ring, most notably the 'edge to face' arrangement in crystals of benzene – the 'herring bone' pattern). Singh and Thornton (1990) in an analysis of protein structures showed a similar preference for oxygen atoms to pack in the plane of the tyrosine ring, and avoid its faces.

Solvation of the galloyl ester group will doubtless be determined preferentially by the association of water molecules with the three phenolic hydroxyl groups, which in turn will influence the disposition of additional solvent molecules in the vicinity of the more hydrophobic regions of the phenolic ester. However it seems possible that, superimposed upon this broad distribution, there will also be preferred locations in these regions for solvent molecules. The evidence from protein structure analysis discussed above suggests such positions may well occur around the edges of the aromatic ring rather than on its faces. Similar considerations may well apply to the solvation encountered with other phenolic nuclei found in natural polyphenols.

3.2.3 Crystal structures of simple phenols

The crystal structures of several simple phenolic molecules may be found within the Cambridge Crystallographic Database. Analysis of these structures provides some interesting pointers to the interactions which give rise to the preferred spatial packing arrangements in these molecules – principally aromatic–aromatic interactions and hydrogen bonding – and hence to their possible roles in polyphenol intermolecular complexation. Some of those phenolic structures most pertinent to the subject of naturally occurring polyphenols and structural elements apposite to the crystal of each phenol are shown in Figures 3.8–3.12. Intermolecular hydrogen bonding is a ubiquitous feature of all the crystal structures; the arrangement of different aromatic nuclei varies significantly. Hydroquinone (quinol), Figure 3.8, has six molecules linked by a cyclic network of hydrogen bonds. Each oxygen atom in this array is joined by hydrogen bonding to *two* hydrogen atoms of phenolic groups in two different molecules in the cyclic array. Each proton takes part formally in *two* hydrogen bonds (bifurcated) to two different oxygen atoms in

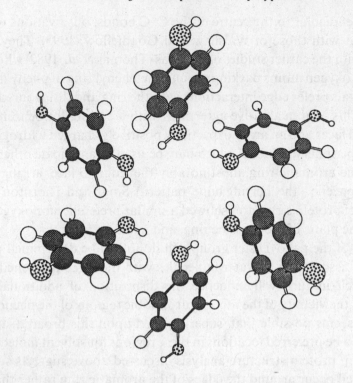

Figure 3.8. Crystal structure of hydroquinone (quinol). Ball and stick models: hydrogen – white; carbon – grey; oxygen – black dots in white circles.

the array. Outside this central core of six molecules, other hydroquinone molecules are related to these by a 'staggered-offset', 'edge to face' relationship of the aromatic nuclei, in a familiar herringbone arrangement similar to that observed with benzene. The crystal of catechol (1,2-dihydroxybenzene) shows a repeating pattern of pairs of catechol molecules ('dimers') linked by hydrogen bonds, Figure 3.9. These are related to other pairs of catechol molecules in a parallel 'off-set' stacking arrangement with the aromatic protons (H*) above or below the face of another aromatic nucleus.

In the crystal of resorcinol (1,3-dihydroxybenzene), Figure 3.10, the structure is again extensively cross-linked by hydrogen bonding. Two 'edge to face' *T* shaped arrangements of aromatic nuclei (see chapter 2) are also clearly discernible in the structure. A cyclic array of four phloroglucinol (1,3,5-trihydroxybenzene) molecules linked by hydrogen bonding is observed in the crystal structure of this phenol, Figure 3.11. Three of the aromatic rings are coplanar whilst the fourth is orthogonal to this plane. In addition stacking arrangements of aromatic nuclei are present; these show the optimal geometry for such structures (see Figure 2.21) with hydrogen atoms poised roughly over ring

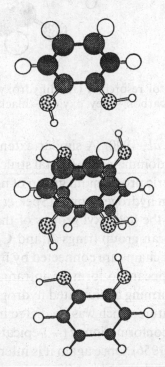

' pairs of catechol molecules '

' stacking ' arrangement

Figure 3.9. Crystal structure of catechol (1,2-dihydroxybenzene). Ball and stick models: hydrogen – white; carbon – grey; oxygen – black dots in white circles.

centres, Figure 3.11. This arrangement may also be facilitated by interaction of the phenolic hydrogen of one ring with the π-system of the other.

The crystal structure of methyl gallate is characterised by extensive hydrogen bonding which involves the phenolic hydroxyl groups and the ester

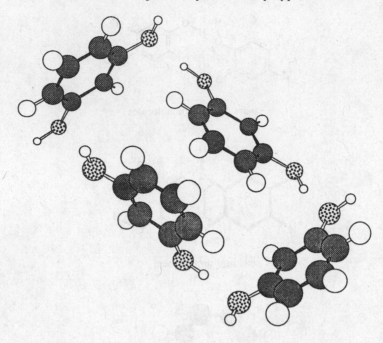

Figure 3.10. Crystal structure of resorcinol (1,3-dihydroxybenzene). Ball and stick models: hydrogen – white; carbon – grey; oxygen – black dots in white circles.

carbonyl group (Spencer *et al.*, 1990). A similar extensive three-dimensional network of hydrogen bonds dominates the crystal structure of (−)-epicatechin – a prototype of the oligomeric procyanidin group of natural polyphenols. All oxygen atoms are involved in hydrogen bonds (Spek *et al.*, 1984; Mattice *et al.*, 1984). Interactions between the hydroxyl groups of the 3,4-dihydroxy-ring B and the 3,5,7-trihydroxy-pyran group (rings A and C) connect molecules to form infinite chains. Parallel chains are connected by hydrogen bonds and the aliphatic hydroxyl group appears to form an intramolecular contact to the heterocyclic oxygen atom, forming a bifurcated hydrogen bond in the process. This is in fact a structural feature which was predicted in earlier work when the absolute configuration and conformation of (−)-epicatechin and (+)-catechin were still debated (Whalley, 1956). Once again it is interesting to note the 'edge to face', '*T* shaped', stereochemical relationship of adjacent 3,4-dihydroxyphenyl rings in the parallel infinite chains of (−) epicatechin molecules, Figure 3.12.

' cyclic array of four phloroglucinol molecules '

' stacking ' arrangement of phloroglucinol nuclei

3.2.4 Model studies with caffeine

The formation of weak intermolecular complexes between caffeine and poly-phenols has been known for some time. Mejbaum-Katzenellenbogen and colleagues (Mejbaum-Katzenellenbogen *et al.*, 1959, 1962) demonstrated that caffeine competes effectively with proteins for polyphenolic substrates and, further, that it is possible to regenerate a wide range of proteins, in a biologically active state, from insoluble protein–tannin (polyphenol) complexes by treat-ment with caffeine. Hydroxycinnamic acids are ubiquitous in the plant king-dom and are usually found as esters or glycosides; the most familiar member of this class is chlorogenic acid, the 5-*O*-caffeoyl ester of quinic acid. Raw coffee beans are a rich source (5–8%) and potassium chlorogenate was first isolated from green coffee beans (*Coffea arabica*) by Gorter (1907, 1908) as a 1:1 crystalline complex with caffeine. Gorter showed that the complex could be recrystallised unchanged from aqueous ethanol. Horman and Viani (1972) concluded from an NMR study that, in aqueous media, the caffeine–potassium chlorogenate complex was best described as a 'hydrophobically bound π molecular complex'. They suggested, percipiently, that the time averaged solution conformation of the complex was one in which the planes of the caffeine

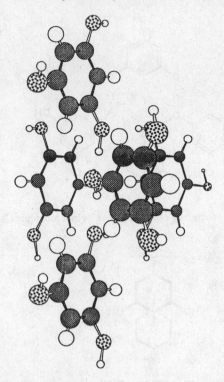

Figure 3.11. Crystal structure of the phloroglucinol (1,3,5-trihydroxybenzene). Ball and stick models: hydrogen – white; carbon – grey; oxygen – black dots in white circles.

molecule and the caffeoyl ester group were parallel and that both the five- and six-membered rings of the nitrogen heterocycle were involved in the complex.

Potassium chlorogenate   caffeine

The crystal structure of the complex was determined in 1987 (Martin *et al.*, 1987). Extensive hydrogen bonding and oxygen coordination around the potassium ion are two factors which serve to stabilise the complex. The third contributing factor is the layer lattice structure adopted in the complex with the phenolic nuclei of the chlorogenate ion alternating with the caffeine molecule in a 'face to face' π–π stacking arrangement (see chapter 2). Within each stack the phenolic rings are parallel and alternating caffeine molecules lie inclined at 3.6° to these rings. The centres of the adjacent planes are separated

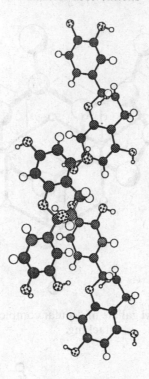

Figure 3.12. Crystal structure of (−)-epicatechin (Spek *et al.*, 1984; Mattice *et al.*, 1984). Ball and stick models: hydrogen – white; carbon – grey; oxygen – black dots in white circles.

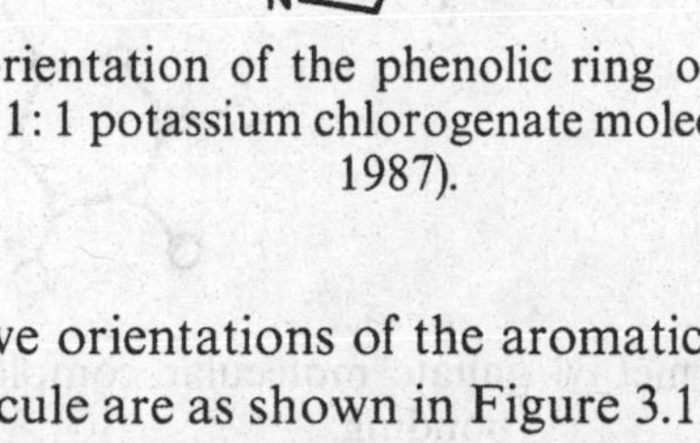

Figure 3.13. Relative orientation of the phenolic ring of the caffeoyl ester and the caffeine molecule in the 1:1 potassium chlorogenate molecular complex (Martin *et al.*, 1987).

by 3.78 Å. The relative orientations of the aromatic ring of the caffeoyl ester and the caffeine molecule are as shown in Figure 3.13.

A very similar layer lattice structure is adopted in the 1:1 molecular complex between methyl gallate and caffeine, Figures 3.14 and 3.15. In this array caffeine and methyl gallate molecules are arranged in alternating layers, approximately parallel, with an interplanar separation of 3.3–3.4 Å (Martin *et al.*, 1986; Cai *et al.*, 1990). This 'stacking' structure is complemented by an exten-

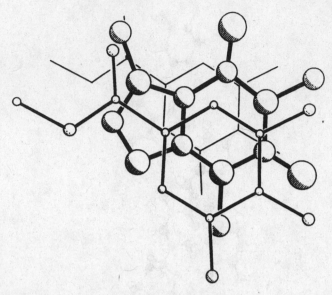

Figure 3.14. The caffeine – methyl gallate molecular complex: vertical 'face to face' π–π stacking.

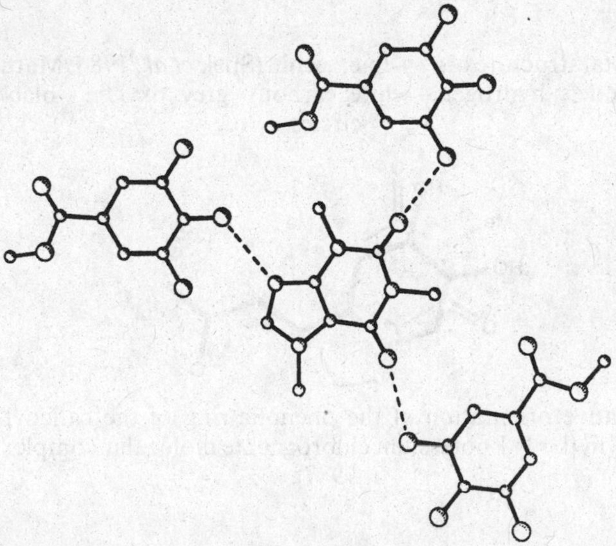

Figure 3.15. The caffeine – methyl gallate molecular complex: in-plane hydrogen bonding.

sive in-plane system of hydrogen bonding in which both ketoamide carbonyl groups and N-9 of caffeine act as proton acceptors and each of the galloyl ester phenolic groups act as proton donors.

Whilst it is not possible to directly extrapolate to behaviour in solution these X-ray crystallographic results confirm the importance of (i) apolar hydrophobic interactions and (ii) hydrogen bonding as primary non-covalent intermolecular forces which operate in caffeine–polyphenol complexation. Circumstantial evidence strongly suggests that this tendency of caffeine to form apolar 'face to face' π–π stacking arrangements with polyphenols also occurs in aqueous media. This is observed in the anisotropic shielding of the protons of the three –N–Me groups and proton H-8 of caffeine and of the aromatic protons of the polyphenol. Typical results are shown in Figure 3.16 for the two flavan-3-ol substrates (−)-epigallocatechin and (−)-epigallocatechin-3-*O*-gallate. This property may be exploited to determine the association constants for the formation of 1:1 complexes between caffeine and various polyphenols in aqueous media, Table 3.1 (Cai *et al.*, 1990). Particularly striking in all these results is the strong dependence of the association constants upon the 'free' galloyl ester content and the tendency (Figure 3.16), for the caffeine molecule to associate preferentially at or near one or more of the galloyl ester groups in the polyphenolic substrate.

The importance of hydrophobic effects in the association of caffeine with polyphenols is demonstrated strikingly by the diminution of the association constant (extrapolated from data in Table 3.1) for the system β-1,2,3,4,6-penta-*O*-galloyl-D-glucose:caffeine by the addition of methanol to the medium. Parenthetically it may also be noted that the various theaflavins which contribute to the colour and taste of black tea liquors associate strongly with caffeine – theaflavin digallate has an affinity for the nitrogen heterocycle which is comparable to that of β-1,2,3,4,6-penta-*O*-galloyl-D-glucose.

As infusions of black tea cool (< 60°C) the originally clear liquor becomes turbid and a colloidal precipitate – commonly referred to in the tea trade as the *tea cream* – ultimately separates; the process is usually reversible and the tea cream may be re-dissolved by heating once again. This 'creaming-down' of the tea liquors is generally taken as an indication of qualities such as 'strength' and 'briskness' in the infusion. The tea cream is principally a series of complexes of the tea polyphenols – theaflavins, thearubigins and phenolic flavan-3-ols – with caffeine. This phenomenon is a general one and polyphenols which form strong molecular complexes with caffeine may usually be precipitated from solution by it – soluble complexes are first generated and these then aggregate and lead to precipitation. Usually the ratio of caffeine to the polyphenolic substrate is higher with lower initial concentrations of the polyphenol. The

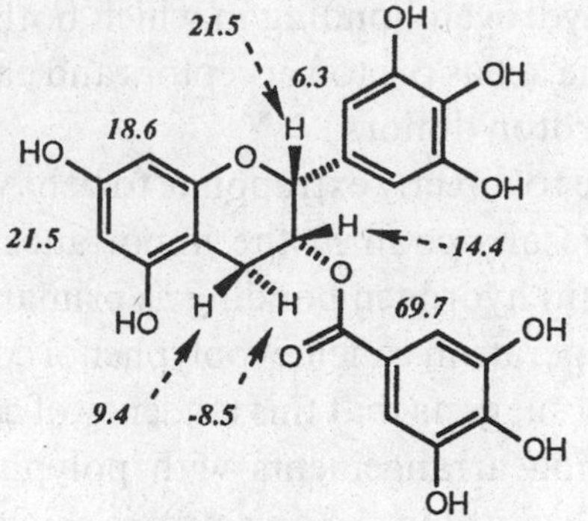

(-)-epigallocatechin-3-O-gallate

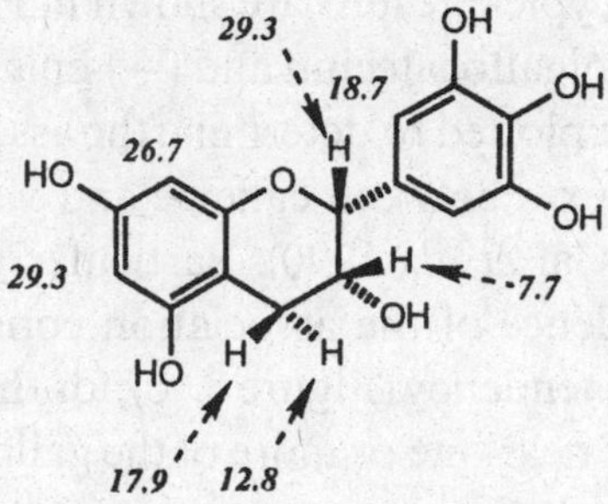

(-)-epigallocatechin

Figure 3.16. Polyphenol – caffeine complexation in solution. ^{1}H NMR chemical shift changes [$\Delta\delta = \delta^0 - \delta^{caff}$] in Hz (250 MHz) induced by caffeine in D$_2$O at 45°C; δ^0 = chemical shift of phenol in absence of caffeine; δ^{caff} = chemical shift of phenol in presence of caffeine; phenol = 8.0 × 10^{-3} M; caffeine = 2.0 × 10^{-3} M.

extent of polyphenol precipitation is not only dependent upon the affinity of a particular polyphenol for caffeine but also on the overall polyphenolic content of the solution. Thus the presence of a good precipitant of caffeine has a synergistic effect upon the extent of precipitation of other phenols present. Extrapolating from the X-ray crystallographic data (Figures 3.13–3.15) the nature of the molecular aggregates which form in solution of theaflavin digallate and caffeine might be of the general form shown in Figure 3.17. In this model it is postulated that the caffeine links separate polyphenol molecules, rather as the latter themselves cross-link proteins in the analogous protein–polyphenol precipitation processes, Figure 3.5.

In the context of polyphenol complexation with caffeine it is also of interest to note that polyphenols complex with other electron deficient aromatic systems such as anthocyanins (e.g. malvin, see chapter 6, copigmentation), the alkaloid berberine and the dyestuff methylene blue, all of which formally bear a

Table 3.1. *Association constants (K, dm³ mol⁻¹) for the formation of 1:1 complexes between caffeine and natural polyphenols in D₂O. Determined by ¹H NMR spectroscopy using the chemical shift changes for H-8 of caffeine (Cai et al., 1990; Liao, Lilley and Haslam, unpublished results)*

Polyphenol	M_R	$K_{(25°C)}$	$K_{(45°C)}$	$K_{(60°C)}$
(−)-Epicatechin	290		34.5	
(−)-Epigallocatechin	306		35.6	
(−)-Epigallocatechin-3-O-gallate	458		52.8	34.6
(+)-Catechin	290		26.1	
(+)-Catechin-3-O-gallate	442		38.2	26.1
β-1,2,3,4,6-Penta-O-galloyl-D-glucose	940	331		81.6
β-1,2,4,6-Tetra-O-galloyl-D-glucose	788			53.4
β-1,3,6-Tri-O-galloyl-D-glucose	636	100		37.1
β-1,2,3,4,6-Penta-O-galloyl-D-glucose	940		58.0*	
Theaflavin	564		35.2*	
Theaflavin monogallates	716		42.3*	
Theaflavin digallate	870		54.7*	

*Association constants determined in D₂O containing 10% d₄ MeOH.

positive charge on the delocalised π-system. Okuda and his colleagues (Okuda, Mori and Hatano, 1985) have employed the ability of polyphenols to precipitate methylene blue from solution as a colorimetric test for the determination of tannins in plant materials. In addition they correlated the relative affinity of a wide range of polyphenols for methylene blue (RMB) and demonstrated that this value had, in general, a good comparative relationship with values for relative astringency (RA) as determined by the precipitation of haemoglobin, (Bate-Smith, 1973a,b).

berberine

malvin

methylene blue

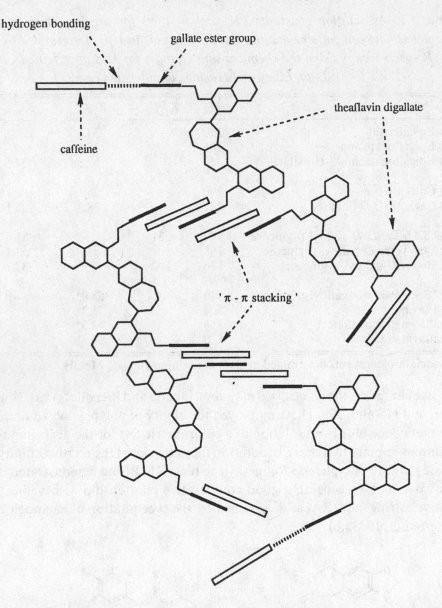

Figure 3.17. Pictorial representation of typical molecular aggregates as they might be formed between caffeine and theaflavin digallate.

This evidence, limited as it may be at present, nevertheless does suggest that the electron rich aromatic nuclei of phenols and polyphenols have the propensity to complex with electron deficient aromatic molecules and that they do so preferentially in apolar 'face to face' π–π stacking arrangements (see chapter 2). A further general feature of interest which emerges from the X-ray crystallographic studies of the caffeine–phenol complexes is the relative orientation of the caffeine and the phenolic nuclei in the layer lattice structures. The term 'polarisation bonding' has been used to describe the generally weak interactions between the polar groups of one component in a complex and the second polarisable component (Wallwork, 1961). For weak non-covalent bonding of this type the principal feature to be expected is the juxtaposition of the polarising groups of one component and the polarisable regions of the other molecule. It is clear from the geometry of the caffeine–phenol complexes that the phenolic groups and associated nuclei are generally stacked above, and below, the six-membered ring of the caffeine molecule. This suggests that in this form of complexation the two components (for example the caffeoyl ester and caffeine, Figure 3.13) develop complementary polar characteristics of the type shown in Figure 3.18.

3.2.5 Cyclodextrins and polysaccharides

Studies of the reversible complexation of polyphenols with proteins have a long history, which, in the past two decades has been considerably facilitated by the availability of both groups of substrate in pure, structurally defined forms. Similar studies of polyphenol–carbohydrate complexation have, in comparison, been hampered by the lack of water-soluble oligo- and polysaccharides with clearly defined structures and molecular masses. Although the significance of polysaccharide/oligosaccharide association with polyphenols has often been inferred from experimental data (see chapter 5) quantitative data are still lacking. The most satisfactory quantitative data have been obtained using polysaccharides in the solid state – the various chromatographic gels Sephadex G-25, G-50 and LH-20 and cellulose tri-acetate in membrane form (Cai *et al.*, 1989). Dextran is a polymeric glucan derived microbiologically from sucrose by the action of *Leuconstoi mesenteroides*. It consists of α-D-(1-6)-linked units to an extent of 90–95%. Cross-linking by epichlorhydrin yields gels such as Sephadex. The affinity of aromatic compounds for dextran gels is well documented (Brook and Munday, 1970; Haglund, 1978; deLigny, 1979; Bywater and Marsden, 1983) and it is noteworthy that the addition of hydroxyl groups to the aromatic nucleus – particularly in the 1,3 and 1,3,5 positions – enhances that affinity (Haglund and Marsden, 1980). The origins of this affinity are uncertain. Several suggestions have been preferred:

(i) interaction (hydrogen bonding) between the phenolic hydroxyl groups and the ether oxygen atoms of the cross-linking chains,

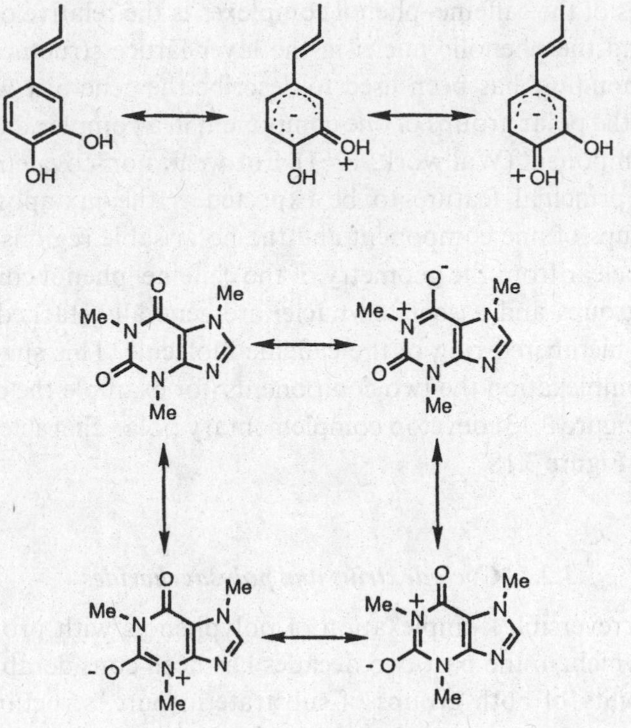

Figure 3.18. Development of complementary polar characteristics in caffeine and caffeoyl ester molecules during complexation (see Figure 3.13).

(ii) the phenyl rings acting as electron donors to the hydroxyl groups of the Sephadex framework (hydrogen bonding), and

(iii) sequestration, inclusion, of the aromatic substrate within the pores of the gel.

Cai and his colleagues determined the rank order of affinity of a range of naturally occurring phenols for Sephadex using the apparent partition coefficient, K_{av}, (Cai *et al.*, 1989), Table 3.2. Subtle changes in polyphenol structure result in marked changes in affinity for the polysaccharide gel. However, molecular size (and with this the possibilities of enhanced cooperative effects in association), conformational flexibility, and water solubility and solvation appear to be the three principal criteria which most strongly influence retention of polyphenols by the polysaccharide matrix.

These studies indicated that the ability of the polysaccharide to generate shapes which are able, either wholly or in part, to encapsulate the polyphenolic substrate may well be an additionally important factor promoting binding to the polysaccharide. Cyclodextrins, first isolated in 1891, are cyclic oligosaccharides composed of (1-α-4) linked D-glucosyl residues; α- and β-cyclodextrins possess respectively six and seven glucose units. They have the shape of a

Table 3.2. *Chromatography of polyphenols on Sephadex gels: K_{av} at 25°C.*

Polyphenol	M_R	(a)	(b)
Phenol	94	1.1	1.13
(−)-Epicatechin	290	1.83	1.96
Procyanidin B-3	578	2.36	3.04
β-1,3,6-Tri-*O*-galloyl-D-glucose	636	4.43	2.49
β-1,2,4,6-Tetra-*O*-galloyl-D-glucose	788	16.25	6.17
β-1,2,3,4,6-Penta-*O*-galloyl-D-glucose	940	*	8.56
Vescalagin	934	2.5	2.28
Rugosin D	1874	*	56.7

(a) Sephadex G-25, Cl⁻ buffer; * not eluted.
(b) Sephadex LH-20, MeOH/H_2O (9:1, v/v).

doughnut with all the D-glucopyranose units in substantially undistorted ($^4C_1$, C-1) conformations. The cavities are slightly 'V' shaped; the secondary hydroxyl groups at C-2 and C-3 lie on the upper side of the torus and the primary hydroxyl groups (C-6) on the lower face. The interior of the cyclodextrin cavity consists of the glucosidic oxygen atoms and two concentric rings of C–H groups (C-3 and C-5), Figure 3.19. The cavity is generally thought to be more accessible from the face bearing the secondary hydroxyl groups and, compared to an aqueous environment, is apolar and relatively hydrophobic.

One of the most important properties of the cyclodextrins is their ability to sequester substrates in the hydrophobic cavity. The nature of the binding and the driving forces that lead to inclusion remain uncertain although several suggestions have been put forward, including: (i) hydrogen bonding, (ii) van der Waals interactions, (iii) the release of strain energy in the cyclodextrin, and (iv) the release of solvent water molecules from the cavity to the bulk medium. The ^1H NMR resonances of the protons H-3 and H-5, located inside the cavity of the cyclodextrins, show substantial changes of chemical shift upon complexation with phenols and polyphenols. The question of the preferred orientation and location of the aromatic nuclei of phenolic substrates complexed within the α- and β-cyclodextrin cavity was determined by Harata (1977) and Inoue and his colleagues (Inoue *et al.*, 1984, 1985). In all the cases examined the phenyl rings were inserted into the cavity from the secondary hydroxyl group side with the *para*-substituent (X) in the lead and with the phenolic hydroxyl group remaining in the aqueous medium, Figure 3.20.

The Sheffield group (Cai *et al.*, 1989, 1990) investigated the importance of 'cavity sequestration' in relation to polyphenol complexation with the cyclodextrins. Several features of their results command immediate attention. Thus compared to the various natural galloyl esters the binding of the phenolic flavan-3-ols is quite strong; the latter give good binding curves typical of the

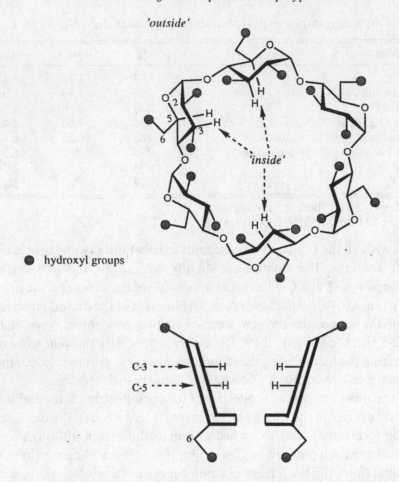

Figure 3.19. Pictorial representation of the molecule of α-cyclodextrin.

formation of 1:1 complexes. Binding is dependent upon the stereochemistry of the C-3 hydroxyl group and is diminished by hydroxyl substitution in the aryl ring B (mono > di > tri substitution). Complexation is also enhanced by galloylation at the C-3 position; modestly (2×) in the case of (+)-catechin, but substantially (9×) in the case of (−)-epigallocatechin. Significantly substitution at C-4 by a further flavan-3-ol molecule (to give a typical proanthocyanidin) suppresses almost completely the ability to complex with the cyclodextrins. A wide range of parameters were measured and probed in an attempt to discover the nature of the complexation process. The anisotropically induced ^{1}H NMR chemical shift changes observed for H-3 and H-5 were such as to suggest complexation by means of cavity sequestration. Although a great deal more work remains to be done it was tentatively suggested that, in the case of the phenolic flavan-3-ols, several modes of association are probably

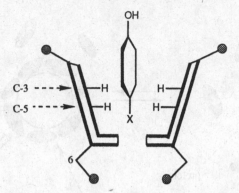

Figure 3.20. Inclusion of *para*-substituted phenols in the cavity of α- and β-cyclodextrin (Harata, 1977, and Inoue *et al.*, 1984, 1985).

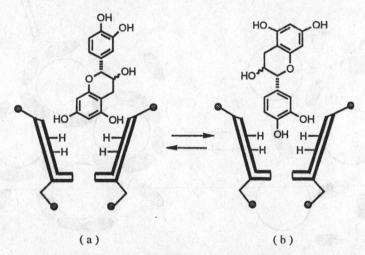

(a)   (b)

Figure 3.21. Possible modes of complexation of phenolic flavan-3-ol substrates with the cyclodextrins (Cai *et al.*, 1989, 1990).

in operation involving the insertion of both rings A and B into the cyclodextrin cavity, Figure 3.21. The enhancement of complexation by galloylation at C-3 was envisaged as due to the pendant gallate ester group acting as an anchor by hydrogen bonding to the peripheral hydroxyl groups at C-2 and C-3 on the top surface of the cyclodextrin cavity.

At the present it seems reasonable to conclude (McManus *et al.*, 1985) that phenols and polyphenols appear to have an affinity for preformed oligosaccharide 'holes', such as are provided by the cyclodextrins, and for 'holes, pores or crevices' in polysaccharide structures, whether these are preformed, such as in the Sephadex gels, or whether they are developed during the processes of complexation in solution (see chapter 5). However, further work, preferably

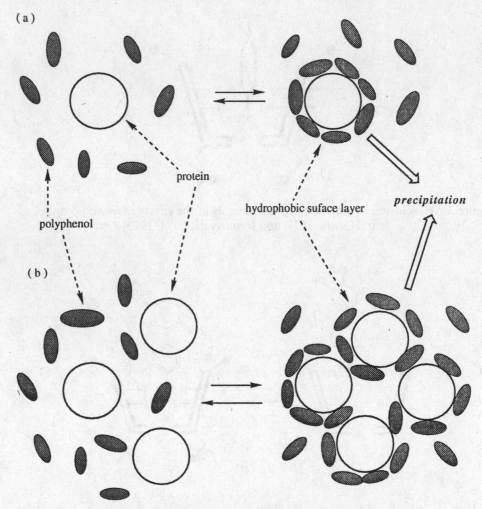

Figure 3.22. Polyphenol–protein precipitation processes: (a) *low* initial protein concentrations; (b) *high* initial protein concentrations.

with structurally defined oligosaccharides, is necessary to determine more precisely the nature, the strength and hence the significance of this associative behaviour.

3.2.6 *Studies with peptides and proteins*

Studies of the reversible association of polyphenols with proteins have a very long history; one of the first scientific papers on this subject is that of Sir Humphry Davy (1803) nearly two hundred years ago. Davy's fame rests principally on his electrochemical studies. He was born in Penzance, Cornwall, in 1778 and was initially apprenticed to a surgeon but took up chemistry in 1795. At the Royal Institution he was persuaded by the Directors to undertake studies of the age-old process of vegetable tannage whereby 'astringent vegetable matter' (Davy's own description) converts animal skins to durable leathers. Much of the early work in this field is covered in the book by Nierenstein – *The Natural Organic Tannins* – published in 1934.

More recent work has utilised a variety of physical techniques – NMR spectroscopy, microcalorimetry, equilibrium dialysis, enzyme inhibition, and precipitation methods – to probe the nature of polyphenol complexation with proteins (Haslam, 1989; Hagerman, 1989). The efficacy of polyphenol (vegetable tannin) binding to proteins derives in large part, almost certainly, from the fact that polyphenols are polydentate ligands able to bind simultaneously/consecutively at more than one point to the protein structure (McManus *et al.*, 1981, 1985; Gustavson and Holm, 1952). In general when polyphenols bring about precipitation of proteins from solution, two *extreme* situations may be envisaged. At *low* protein concentrations the polyphenol associates at one or more sites on the protein surface, ultimately this gives a monolayer which is less hydrophilic than the protein itself, Figure 3.22a. Aggregation and precipitation then ensue. Conversely when the protein concentration is **high** the relatively hydrophobic surface layer is formed by complexation of the polyphenol onto the protein with concommitant cross-linking of different protein molecules by the multi-dentate polyphenols, Figure 3.22b. Precipitation then ensues as above. This proclivity to cross-link protein molecules at higher protein concentrations explains the changing stoicheiometry of the aggregates with changing protein concentrations – an observation first indicated by Sir Humphry Davy! More polyphenol is thus required to precipitate polyphenols from dilute solutions than from concentrated ones (van Buren and Robinson, 1969).

The precipitation of protein by polyphenols may be reversed often by the addition of further protein, and during his observations of the precipitation of astringent vegetable infusions using gelatin Sir Humphry Davy remarked:

In ascertaining the proportions of tannin in astringent infusions, great care must be taken to prevent the presence of any excess of gelatine for when this excess exists, I have found that a small portion of the solid compound formed is re-dissolved.

In a similar way protein–polyphenol complexes may be dissociated, without denaturation of the protein, with solvents such as acetone (Hestrin *et al.*, 1955), with caffeine (Mejbaum-Katzenellenbogen *et al.*, 1959, 1962), urea (Gustavson, 1956), polyvinylpyrrolidone (Hulme *et al.*, 1965), various polyethylene glycols and detergents (Goldstein and Swain, 1965; Jones and Mangan, 1977). Likewise various precipitation methods have been exploited to measure the relative affinities of different proteins and polyphenols for each other. Most familiar of these are polyphenol precipitation by gelatin (Calderon *et al.*, 1968), the enzyme β-glucosidase (Haslam, 1974), by the blood protein haemoglobin (haemanalysis, Bate-Smith, 1973a,b) and the procedure of Hagerman and Butler (1981) using radioiodinated protein (BSA). Results may be expressed in several ways; in haemanalysis, for example, the relative affinity of different polyphenols for haemoglobin is expressed as relative astringency (RA) and compared to an arbitrary standard polyphenolic substrate – tannic acid or β-1,2,3,4,6-penta-*O*-galloyl-D-glucose. Under these conditions procyanidin B-2 had an RA of 0.10–0.11 and the RA values of galloyl glucose esters increased steadily towards a value of 1.0 as the number of galloyl groups increased from 1 to 5. Tannin (polyphenol)–protein precipitation is influenced by a variety of factors (Hagerman, 1989; Calderon *et al.*, 1968; van Buren and Robinson, 1969; Beart *et al.*, 1985; Luck *et al.*, 1994; Haslam *et al.*, 1992) including characteristics of the protein (size, amino acid composition, pI), characteristics of the polyphenol (size, structure, water solubility, conformational flexibility) and the conditions under which the reaction takes place (pH, solvent composition, temperature, the presence of salts and other organic molecules such as polysaccharides).

In terms of its findings in relation to polyphenol–protein interactions there is little doubt that the paper by Hagerman and Butler (1981) has probably been the most influential, particularly in the way in which it has directed subsequent work in this area. The Purdue workers showed that the condensed proanthocyanidin polymer (tannin) from *Sorghum bicolor*, at pH 4.9, had a relative affinity for various proteins and synthetic polyamides which ranged over four orders of magnitude – indicating that this polyphenol interacts quite selectively with proteins and protein-like polymers. Proteins were most efficiently precipitated at or near their isoelectric points: tightly coiled globular proteins were not so readily precipitated as those which had a random coil conformation and in particular those which are rich in the amino acid proline (proline rich proteins – PRPs).

Complementary studies of polyphenol–protein interactions in solution using a range of different polyphenols with the protein BSA were conducted by McManus and his colleagues (McManus *et al.*, 1985) using equilibrium dialysis, Table 3.3. They showed that molecular size, conformational flexibility and water solubility were the dominant features in the determination of the strength of binding of a particular polyphenol to BSA. Good water solubility depressed the effectiveness of the polyphenol in protein complexation, whilst increasing molecular size and conformational flexibility enhanced association

Table 3.3. *Protein–polyphenol complexation. Values of the free energy of transfer* $(-\Delta G^{\ominus,\mathrm{tr}})$ *of a protein, BSA, from an aqueous medium to an aqueous medium containing a polyphenol ligand (concentration* $-m_{\mathrm{f}}$*)*

Polyphenol	M_R	$-\Delta G^{\ominus,\mathrm{tr}}/\mathrm{kJ\ mol}^{-1}$
β-1,3,6-Trigalloyl-D-glucose	636	0.9
β-1,2,3,6-Tetragalloyl-D-glucose	788	9.1
β-1,2,3,4,6-Pentagalloyl-D-glucose	940	26.9
Vescalagin/castalagin	934	1.0
Rugosin D	1874	58.7

Determined in aqueous solution at 25°C, pH 2.2 and at a free ligand (polyphenol) concentration (m_{f}) of 4.5 mole kg^{-1}, McManus *et al.*, 1985.

with the protein. The experimental data, which illustrate these points, are shown in Table 3.3. Results were expressed in terms of the free energy of transfer $(-\Delta G^{\ominus,\mathrm{tr}})$ of the protein BSA from an aqueous medium to an aqueous medium containing the polyphenol ligand, (concentration $-m_{\mathrm{f}}$). High values of the quantity $-\Delta G^{\ominus,\mathrm{tr}}$ are indicative of strong complexation; conversely, low values portray poor complexation, Table 3.3.

The principles which underly molecular recognition phenomena may be analysed not only in terms of the composition, structure and conformation of the substrates taking part in the complexation reactions but also in terms of three idealised concepts (Williams, 1988). 'Dye–mould' (jig-saw) matching is essentially static with an exact fit of donor and acceptor molecules together. 'Key–lock' matching is time dependent, since the key (donor) invariably has to be manoeuvered into the lock (acceptor) to achieve the correct fit. Finally 'hand-in-glove' matching of donor and acceptor molecules is both time dependent and dynamic. Both donor and acceptor molecules are mobile and flexible and may assume a variety of subtly different shapes as complexation proceeds. In such situations it is important that both substrates are able (if necessary) to sample a variety of different relative orientations with respect to each other such that ultimately the maximum number of strong contacts are made between donor and acceptor species. Such associative reactions frequently exhibit strong cooperative effects. Present evidence, and particularly that emanating from the work of Hagerman and Butler (1981) and McManus and his colleagues (McManus *et al.*, 1985), suggests that where polyphenol–protein complexation is at its most effective then it is largely of the 'hand-in-glove' type where polyphenol and protein both possess considerable conformational mobility and structures which permit the possibility of bringing about multi-dentate complexation, Figures 3.4 and 3.5.

One of the most intriguing questions emanating from the work of Hagerman and Butler (1981) is whether the high affinity which polyphenols display towards proline rich proteins (PRPs) derives simply from the open random-coil conformations which these proteins display in solution or if there are additional factors at work which enhance complexation. This particular question was addressed by Williamson and his colleagues in a high resolution NMR study of polyphenol association with proline rich peptides (Murray *et al.*, 1994; Charlton *et al.*, 1996; Luck *et al.*, 1994). Two peptides (one 19 and the other 22 amino acids in length) which both contain the repeat sequence of mouse salivary proline rich protein MP5, but which commence and terminate at different points in the sequence, were synthesised and shown to possess an extended random coil conformation in solution (Murray and Williamson, 1994). Two-dimensional ^{1}H NMR studies were carried out using these peptides with β-1,2,3,4,6-pentagalloyl-D-glucose. No major changes were observed in the averaged conformations of the peptides such as to suggest that distortion of the peptides is not a thermodynamically important feature of the binding process. Intermolecular nuclear Overhauser effects and chemical shift changes demonstrated that the principal binding sites on these peptides are the prolyl groups and the preceding amide bond and its associated amino acid, Figure 3.23. It was concluded from these observations that the pyrrolidine rings of the prolyl residues provide a multiplicity of potential binding sites ('hydrophobic sticky patches') on the PRPs and therefore exert a strong and selective influence on the recognition process. The upfield chemical shift changes seen on titration of the β-1,2,3,4,6-pentagalloyl-D-glucose into the peptide solutions were reasonably ascribed to ring current shifts caused largely by hydrophobically driven 'face to face' stacking of the prolyl rings with the galloyl rings of the phenolic substrate. Chemical shift changes, in particular those of prolyl residues, suggested similarly that association of this type occurred preferentially on the least hindered face (α-H face) of the pyrrolidine ring.

Subsequent hydrogen bonding by one or more phenolic groups to the tertiary amide carbonyl group, *N*-terminal to the prolyl residue (Figure 3.23*) as a second phase, would then consolidate this initial association. (There is now good evidence that the carbonyl function in tertiary amides such as in prolyl peptides is a much better hydrogen bond acceptor than that in primary and secondary amides.) It is presumed that steric accessibility is probably responsible for the preferential binding at or near the N-terminal proline in an oligoproline sequence, Figure 3.23, and this point of association would necessarily bring about chemical shift changes in the proximal amino acid side-chain (Figure 3.23, R). A pictorial representation of the probable mode of selective binding of a galloyl ester group to the pyrrolidine ring of a prolyl group is shown in Figure 3.24.

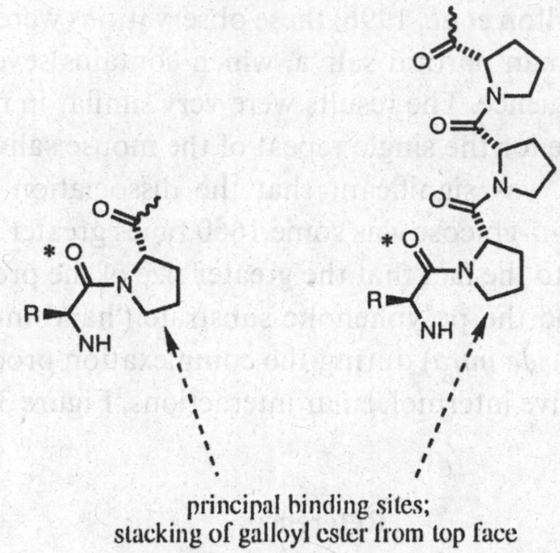

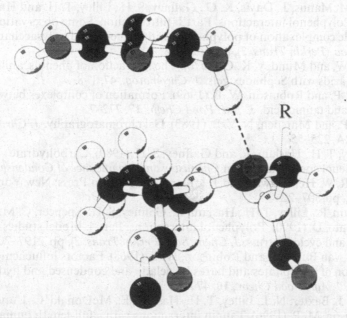

principal binding sites;
stacking of galloyl ester from top face

Figure 3.23. Principal polyphenol binding sites in proline rich peptides and proteins
(Murray *et al.*, 1994; Luck *et al.*, 1994).

Figure 3.24. Suggested mode of preferential association of galloyl ester groups in
β-1,2,3,4,6-pentagalloyl-D-glucose to the prolyl residues in PRPs. The side-chain of the
amino acid N-terminal to the proline is indicated as R; the hydrogen bond between a
phenolic hydroxyl and the carbonyl group is shown as a dashed line. Individual atoms
are drawn as: carbon – black; oxygen – small, dark grey; nitrogen – large, light grey;
and hydrogen – white.

In later work (Charlton *et al.*, 1996) these observations were extended to the protein IB5 from human parotid saliva, which contains several repeats of a short proline rich sequence. The results were very similar in many respects to those described above for the single repeat of the mouse salivary proline rich protein. However, it was significant that the dissociation constant for β-1,2,3,4,6-pentagalloyl-D-glucose was some 1660 times greater. This increase in affinity was ascribed to the fact that the greater size of the protein allows it to fold and wrap around the polyphenolic substrate ('hand-in-glove', dynamic and time dependent, *vide infra*) during the complexation process and thereby increase the cooperative intermolecular interactions, Figure 3.4.

References

Bate-Smith, E. C. (1973a). Haemanalysis of tannins – the concept of relative astringency. *Phytochemistry*, **12**, 907–12.

Bate-Smith, E. C. (1973b). Tannins of the herbaceous Leguminosae. *Phytochemistry*, **12**, 1809–12.

Beart, J. E., McManus, J., Davis, K. G., Gaffney, S. H., Lilley, T. H. and Haslam, E. (1985). Polyphenol interactions. Part I. Introduction. Some observations on the reversible complexation of polyphenols with proteins and polysaccharides. *J. Chem. Soc. (Perkin Trans. 2)*, pp. 1429–38.

Brook, A. J. W. and Munday, K. C. (1970). The interaction of phenols, anilines and benzoic acids with Sephadex gels. *J. Chromatog.*, **47**, 1–8.

van Buren, J. P. and Robinson, W. B. (1969). Formation of complexes between protein and tannic acid. *J. Agr. Food Chem.*, **17**, 772–7.

Bywater, R. P. and Marsden, N. V. B. (1983). Gel chromatography. *J. Chromatog. Libr.*, **22A**, 225–55.

Cai, Y., Lilley, T. H., Haslam, E. and Gaffney, S. H. (1989). Carbohydrate –polyphenol complexation. In *Chemistry and Significance of Condensed Tannins*, editors R. W. Hemingway and J. J. Karchesy, Plenum Press: New York and London, pp. 307–22.

Cai, Y., Martin, R., Lilley, T. H., Haslam, E., Gaffney, S. H., Spencer, C. M. and Magnolato, D. (1990). Polyphenol interactions. Part 4. Model studies with caffeine and cyclodextrins. *J. Chem. Soc. (Perkin Trans. 2)*, pp. 2197–208.

Calderon, P., van Buren, J. and Robinson, W. B. (1968). Factors influencing the formation of precipitates and hazes by gelatin and condensed and hydrolysable tannins. *J. Agr. Food Chem.*, **16**, 479–82.

Charlton, A. J., Baxter, N. J., Lilley, T. H., Haslam, E., McDonald, C. J. and Williamson, M. P. (1996). Tannin interactions with a full-length human salivary proline-rich protein display a stronger affinity than with single proline-rich repeats. *FEBS Letters*, **382**, 289–92.

Davy, H. (1803). An account of some experiments and observations on the constituent parts of some astringent vegetables and on their operation in tanning, (from *Phil. Trans.*, 1803). In *Incunabula of Tannin Chemistry*, ed. M. Nierenstein, Edward Arnold: London, 1932, p. 116.

Fersht, A. (1977). *Enzyme Structure and Mechanism*, Freeman: Reading (U.K.) and San Francisco, pp. 240–2.

Forest, H. and Dailey, B. P. (1966). Microwave spectra of some isotopically substituted phenols. *J. Chem. Phys.*, **45**, 1736–46.

Goldstein, J. L. and Swain, T. (1963). Changes in tannins in ripening fruit. *Phytochemistry*, **2**, 371–83.

Goldstein, J. L. and Swain, T. (1965). The inhibition of enzymes by tannins. *Phytochemistry*, **4**, 185–92.

Gorter, K. (1907). Beitrage zur Kenntniss des Kaffees. *Liebig's Annalen Chemie*, **358**, 327–48.

Gorter, K. (1908). Beitrage zur Kenntniss des Kaffees. *Liebig's Annalen Chemie*, **359**, 217–44.

Gray, J. C. (1978). Adsorption of polyphenols by polyvinylpyrrolidone and polystyrene resins. *Phytochemistry*, **17**, 495–7.

Gustavson, K. H. (1956). *The Chemistry of Tanning Processes*. Academic Press: New York.

Gustavson, K. H. and Holm, B. (1952). The reactions of polyamides with tanning agents. *J. Amer. Leather Trades Chem. Assoc.*, **47**, 700–11.

Hagerman, A. E. (1989). Chemistry of tannin–protein complexation. In *Chemistry and Significance of Condensed Tannins*, editors R. W. Hemingway and J. J. Karchesy, Plenum Press: New York and London, pp. 308–33.

Hagerman, A. E. and Butler, L. G. (1981). The specificity of proanthocyanidin–protein interactions. *J. Biol. Chem.*, **256**, 4494–7.

Haglund, A. C. (1978). Adsorption of monosubstituted phenols on Sephadex G-15. *J. Chromatog.*, **156**, 317–22.

Haglund, A. C. and Marsden, N. V. B. (1980). Hydrophobic and polar contributions to solute affinity for a highly cross-linked water swollen (Sephadex) gel. *J. Polymer Sci. Polymer Lett.*, **18**, 271–9.

Hansch, C., Fujita, T. and Iwasa, J. (1964). A new substituent constant, π, derived from partition coefficients. *J. Amer. Chem. Soc.*, **86**, 5175–80.

Hansch, C., Kiehs, K. and Lawrence, G. L. (1965). The role of substituents in the hydrophobic bonding of phenols by serum and mitochondrial proteins. *J. Amer. Chem. Soc.*, **87**, 5770–3.

Harata, K. (1977). The structure of the cyclodextrin complex. V. Crystal structures of α-cyclodextrin complexes with *p*-nitrophenol and *p*-hydroxybenzoic acid. *Bull. Chem. Soc. Japan*, **50**, 1416–24.

Haslam, E. (1974). Polyphenol–protein interactions. *Biochem. J.*, **139**, 285–8.

Haslam, E. (1989). *Plant Polyphenols – Vegetable Tannins Re-visited*, Cambridge: Cambridge University Press, pp. 167–77.

Haslam, E., Lilley, T. H., Warminski, E., Liao, H., Cai, Y., Martin, R., Gaffney, S. H., Goulding, P. N. and Luck, G. (1992). Polyphenol complexation – a study in molecular recognition. In *Phenolic Compounds in Food and Their Effects on Health. I. Analysis, Occurrence and Chemistry*, A.C.S. Symposium Series **506**, eds C.-T. Ho, C. Y. Lee and M.-T. Huang, American Chem. Soc.: Washington D.C., pp. 8–50.

Hestrin, S., Feingold, D. S. and Schramm, M. (1955). Enzymes of carbohydrate metabolism. Hexoside hydrolases. *Methods in Enzymol.*, **1**, 231–57.

Hoff, J. E., Oh, H. I., Armstrong, G. S. and Haff, L. A. (1980). Hydrophobic interactions in tannin protein complexes. *J. Agric. Food Chem.*, **28**, 394–8.

Horman, I. and Viani, R. (1972). Nature and conformation of the caffeine–chlorogenate complex of coffee. *J. Food Sci.*, **37**, 925–7.

Hulme, A. C., Jones, J. D. and Wooltorton, L. S. C. (1965). Use of poly-

(vinylpyrrolidone) in the isolation of enzymes from apple peel. *Phytochemistry*, **4**, 659–76.

Inoue, Y., Okuda, T., Miyata, Y. and Chujo, R. (1984). NMR studies of cycloamylose–inclusion complexes with substituted phenols. *Carbohydrate Res.*, **125**, 65–76.

Inoue, Y., Hoshi, H., Sakurai, M. and Chujo, R. (1985). Geometry of cycloamylose inclusion complexes with some benzenes in aqueous solution based on carbon-13 NMR chemical shifts. *J. Amer. Chem. Soc.*, **107**, 2319–23.

Jencks, W. P. (1969). *Catalysis in Chemistry and Enzymology*. McGraw-Hill: New York, pp. 393.

Jones, W. T. and Mangan, J. L. (1977). Complexes of the condensed tannins of Sainfoin (*Onobrychis viciifolia*) with fraction I leaf protein and with submaxillary mucoprotein and their reversal by polyethylene glycol and pH. *J. Sci. Food Agric.*, **28**, 126–36.

Kumaraswamy, V. S., Lindley, P. F., Slingsby, C. and Glover, I. D. (1996). An eye lens protein-water structure: 1.2 Å resolution structure of γB-crystallin at 150 K. *Acta Cryst.*, **52D**, 611–22.

deLigny, C. L. (1979). Adsorption of monosubstituted phenols on Sephadex G-15. *J. Chromatog.*, **172**, 397–8.

Loomis, W. D. (1969). Removal of phenolic compounds during the isolation of plant enzymes. *Methods Enzymol.*, **13**, 555–63.

Loomis, W. D. (1974). Overcoming problems of phenolics and quinones in the isolation of plant enzymes and organelles. *Methods Enzymol.*, **31**, 528–544.

Lowe, J. P. (1968). Barriers to internal rotation about single bonds. *Prog. Phys. Org. Chem.*, **6**, 1–80.

Luck, G., Liao, H., Murray, N. J., Grimmer, H. R., Warminski, E. E., Williamson, M. P., Lilley, T. H. and Haslam, E. (1994). Polyphenols, astringency and proline-rich proteins. *Phytochemistry*, **37**, 357–71.

Martin, R., Lilley, T. H., Falshaw, C. P., Bailey, N. A., Haslam, E., Begley, M. J. and Magnolato, D. (1986). Polyphenol–caffeine complexation. *J. Chem. Soc. Chem. Commun.*, pp. 105–6.

Martin, R., Lilley, T. H., Falshaw, C. P., Haslam, E., Begley, M. J. and Magnolato, D. (1987). The caffeine–potassium chlorogenate molecular complex. *Phytochemistry*, **26**, 273–9.

Mattice, W. L., Fronczek, F. R., Gannuch, G., Tobiason, F., Broeker, J. L. and Hemmingway, R. W. (1984). Dipole moment, solution and solid state structure of (−)-epicatechin, a monomer unit of procyanidin polymers. *J. Chem. Soc. (Perkin Trans. 2)*, pp. 1611–16.

McManus, J. P., Davis, K., Lilley, T. H. and Haslam, E. (1981). The association of proteins with phenols. *J. Chem. Soc. Chem. Commun.*, pp. 309–11.

McManus, J. P., Davis, K., Beart, J. E., Gaffney, S. H., Lilley, T. H. and Haslam, E. (1985). Polyphenol interactions. Part I. Introduction. Some observations on the reversible complexation of polyphenols with proteins and polysaccharides. *J. Chem. Soc. (Perkin Trans. 2)*, pp. 1429–38.

Mejbaum-Katzenellenbogen, W., Dobryszychka, W. M., Jaworska, K. and Morawiecka, B. (1959). Regeneration of protein from insoluble protein–tannin compounds. *Nature*, **184**, 1799–800.

Mejbaum-Katzenellenbogen, W. and Dobryszychka, W. M. (1962). Immunochemical properties of the serum proteins after regeneration from the protein–tannin compounds. *Nature*, **193**, 1288–9.

Molyneaux, P. and Frank, H. P. (1961). The interaction of polyvinylpyrrolidone with aromatic compounds in aqueous solution. Part I. Thermodynamics of the binding equilibria and interaction forces. *J. Amer. Chem. Soc.*, **83**, 3169–74.

Momany, F. A., McGuire, R. F., Burgess, A. W. and Scheraga, H. A. (1975). Energy parameters in polypeptides. VII. Geometric parameters, partial atomic charges, non-bonded interactions, hydrogen bond interactions and intrinsic torsional potentials for the naturally occurring amino acids. *J. Phys. Chem.*, **79**, 2361–81.

Murray, N. J. and Williamson, M. P. (1994). Conformational study of a salivary proline-rich protein repeat sequence. *Eur. J. Biochem.*, **219**, 915–21.

Murray, N. J., Williamson, M. P., Lilley, T. H. and Haslam, E. (1994). Study of the interaction between proline-rich proteins and a polyphenol by ^{1}H NMR spectroscopy. *Eur. J. Biochem.*, **219**, 923–35.

Okuda, T., Mori, T. and Hatano, T. (1985). Relationship of structures of tannins to the binding activities with haemoglobin and methylene blue. *Chem. Pharm. Bull.*, **33**, 1424–33.

Pauling, L. (1936). Note on the interpretation of the infra-red absorption of organic compounds containing hydroxyl and imino groups. *J. Amer. Chem. Soc.*, **58**, 94–8.

Pauling, L. (1949). *Nature of the Chemical Bond*, Second Edition, Cornell University Press: Ithaca, New York, pp. 320–7.

Shan, S. and Herschlag, D. (1996). Energetic effects of multiple hydrogen bonds. Implications for enzyme catalysis. *J. Amer. Chem. Soc.*, **118**, 5515–18.

Singh, J. and Thornton, J. M. (1990). SIRIUS. An automated method for the analysis of the preferred packing arrangements between protein groups. *J. Mol. Biol.*, **211**, 595–615.

Spek, A. L., Kojic-Prodic, B. and Labadie, R. P. (1984). Structure of (−)-epicatechin: (2*R*,3*R*)-2-(3,4-dihydroxyphenyl)-3,4-dihydro-2*H*-1-benzopyran-3,5,7-triol, $C_{15}H_{14}O_6$. *Acta Cryst.* **40C**, 2068–71.

Spencer, C. M., Cai, Y., Martin, R., Lilley, T. H. and Haslam, E. (1990). The metabolism of gallic acid and hexahydroxydiphenic acid in higher plants. Part 4; Polyphenol interactions. Part 3. Spectroscopic and physical properties of esters of gallic acid and (*S*)-hexahydroxydiphenic acid with D-glucopyranose ($^4C_1$). *J. Chem. Soc. (Perkin Trans. 2)*, pp. 651–60.

Thanki, N., Thornton, J. M. and Goodfellow, J. M. (1988). Distribution of water around amino acid residues in proteins. *J. Mol. Biol.*, **202**, 637–57.

Thomas, K. A., Smith, G. M., Thomas, T. B. and Feldman, R. J. (1982). Electronic distributions within protein phenylalanine aromatic rings are reflected by the three-dimensional oxygen atom environments. *Proc. Natl. Acad. Sci. (U.S.A.)*, **79**, 4843–7.

Wallwork, S. C. (1961). Molecular complexes exhibiting polarisation bonding. Part III. A structural survey of some aromatic complexes. *J. Chem. Soc.*, pp. 494–9.

Walshaw, J. and Goodfellow, J. M. (1993). Distribution of solvent molecules around apolar side-chains in protein crystals. *J. Mol. Biol.*, **231**, 392–414.

Whalley, W. B. (1956). The stereochemistry of the chromans and related compounds. In *The Chemistry of Vegetable Tannins*, Society of Leather Trades' Chemists: Croydon, pp. 151–60.

Williams, R. J. P. (1988). Biochemical Society, Sheffield, April 15th.

4

Taste, bitterness and astringency

4.1 Food selection – ecological implications

I myself started with the 'information' that tannins were waste products deposited in the wood of trees because the plants had nothing else they could do with them; a horrible thought! Convinced they had a function, I set out to try and find it by way of their systematic distribution linked to the idea of astringency.

E. C. Bate-Smith
(*Personal communication*)

According to the best estimates well over 20 000 secondary metabolites (natural products) have now been isolated and characterised from plant and microbial sources; most, probably at least as many more, remain to be discovered. The view that secondary plant constituents are intimately concerned in the biochemical co-evolution of plants and herbivores was first put forward in the 1950s, notably by Fraenkel (1959) in a paper entitled 'The *raison d'être* of secondary plant substances', in which he drew on the views of the German botanist Stahl expressed some 70 years earlier.

...In the same sense, the great differences in the nature of chemical products and consequently of metabolic processes are brought nearer to our understanding if we regard these compounds as a means of protection, acquired in the struggle with the animal world. Thus the animal world which surrounds plants deeply influenced not only their morphology but also their chemistry.

During the 1960s the topic developed experimentally with explosive effect and secondary metabolites became the cornerstone of a new theory of biochemical co-evolution. This work and the ideas of Feeny (1976) and of Rhoades and Cates (1976) led to a comprehensive view concerning the optimal patterns of chemical defence in plants. The theory had a great impact but its effect on other avenues of enquiry and intellectual endeavour was hypnotic. Some workers, for instance, advocated that the term secondary should now be suppressed on the basis of the clearly perceived biological significance of these compounds and the apparent misconceptions that the term secondary now therefore gave rise to (Beck and Reese, 1976). The following series of quotations illustrates the way in which the theory itself was developed over the intervening years and the general thrust of the ideas which it conveys.

... natural selection serves as a mechanism by which a population of herbivores may, '**call forth de novo**', the evolution of a biosynthetic pathway producing compounds toxic to the herbivore...

and

... The failure to regard such (secondary compounds) as primarily animal (and other plant) toxins renders impossible the explanation of how these products came to be.

(Janzen, 1969)

The evolution of these substances is not, in fact, comprehensible except in an ecological context, including organisms other than the plants producing them.

(Whittaker, 1970)

... much of the purpose of the synthesis of complex molecules of terpenoids, alkaloids and phenolics lies in their use as defense agents in the plants fight for survival.

(Harborne, 1977)

However, it should also be noted that dissenting voices sought to put forward alternative theories to rationalise the occurrence of the extraordinarily diverse range of natural products (secondary metabolites) produced by plants, most notably Muller (1969).

... The secondary or protective role of these toxins may well be their principal role in the ecology of a particular species. They remain, however, primarily metabolic wastes capable of destroying the system that produces them unless they are loosed into the environment or sequestered harmlessly in the plant.

and

... Associated animals are possessed of no mechanism by means of which they can call forth **de novo** the evolution of a specific mechanism in plants.... If, however, a plant species has several alternative and simultaneous metabolic pathways already in operation, producing varying quantities of the numerous by-products characteristic of plants, selective pressure might well increase the proportion of one of these. Thus the toxicity to animals of these metabolic wastes, no matter how important eventually, is subsequent and secondary to their elimination from the protoplasm.

The key element, the *sine qua non*, of the Darwinian view of nature is that the purposeful construction of living matter can be attributed to natural selection. Several assumptions – the *ifs*, Cairns-Smith (1985) – predicate that viewpoint. Most notably; *if* there are random variations in systems which can reproduce their kind; *if* such variations are inherited and confer some advantage to these reproducing systems – then these entities will have an enhanced chance of survival in any competitive situation. Whether or not they are the result initially of random variations the view that secondary plant metabolites (e.g. terpenoids, alkaloids and phenolics) may play a leading role in the ecological interactions between plants and herbivores fits intuitively into this intellectual framework (Swain, 1977). Because of their ubiquitous occurrence in plants more has probably been written (and speculated) about the roles of phenolic natural products in food selection and feeding deterrence than any other group of secondary metabolites. Indeed in the genesis of the general theory of biochemical co-evolution the roles of vegetable tannins (*syn.* plant poly-phenols) were intensively investigated (e.g. Feeny, 1968, 1970) and considerable weight was ultimately placed on the result of these studies. Based on his own and other's work, Bate-Smith (1973a) first succinctly stated the presumed role of vegetable tannins in plant defence.

From the biological point of view the importance of tannins in plants lies in their effectiveness as repellants to predators, whether animal or microbial. In either case the relevant property is astringency, rendering the tissues unpalatable by precipitating proteins or by immobilising enzymes, impeding invasion of the host by the parasite.

Swain was also a natural participant in these developments and in a key paper in 1978 (Swain, 1978) he similarly considered the central proposition and the specific influence of condensed tannins.

Proanthocyanidins or condensed tannins ... are undoubtedly the most useful of all plant chemical defences. Not only are they potent anti-fungal, antibacterial and even anti-viral agents but they bestow on plants which have the ability to synthesise them a powerful feeding deterrent to all herbivores... These phenolic polymers, act through their ability to combine with protein and this inhibits enzymes and reduces nutrition-ally available protein in foods. They are difficult to degrade and because of their molecular size this renders them almost impossible to sequester into vesicles so that their activity may be aborted.

The most prominent characteristic of polyphenols in this context, and as stated by Bate-Smith and Swain, was thus assumed to be their affinity for proteins which deleteriously affected tissue palatability and reduced the avail-ability of food nitrogen to some herbivores (Feeny, 1970). Following the classic experiments of Feeny (1968, 1970) on the detrimental effects of dietary tannins on the feeding of the winter moth (*Operophtera brumata*) on oak leaf, the view was also espoused by Feeny that tannins constitute a unique quantitative defence for plants. Besides their strongly astringent taste (Bate-Smith, 1973a; Haslam and Lilley, 1988) tannins also inhibit virtually every enzyme they are tested with *in vitro*. As a consequence it is also widely assumed that they act as digestion inhibitors by the formation of immutable polyphenol–protein complexes. These, and related observations, gave rise to two well explored theories describing the factors which are thought to influence the distribution and the relative abundance of phenolic metabolites in plant communities. The *resource allocation* theory suggests that the nutrient resources available to plants largely determine the quantity and the types of allelochemical (e.g. polyphenols) that they produce (Bryant, Chapin and Klein, 1983; Coley, 1986; Coley, Bryant and Chapin, 1985). The principal advantage of this theory is its focus upon plant physiology as the determinant of the distribution of allelochemicals rather than their perceived mode of action. This latter factor, however, underpins the *plant apparency* theory which suggests that plants and plant chemistry vary in the degree to which the plants in question are available to, or likely to be discovered by, other organisms such as herbivores. This theory enables predictions to be made concerning the likely defence chemistry to be adopted by abundant long-lived, apparent (often perennial) plants relative to short-lived, rare and herbaceous plants (Feeny, 1976; Rhoades and Cates, 1976).

However, in the intervening period the view has also slowly emerged that the idea of biochemical co-evolution between plants and herbivores as the dominant factor in host range may well have been overemphasised (Bernays and Graham, 1988; Thompson, 1988) at the expense of other potentially crucial interactions. Many workers have therefore moved from the proposition where tannins (polyphenols) were regarded as the ultimate quantitative chemical defence to one where their capacity to deter and their precise mode of action are under further scrutiny (Bernays *et al.*, 1989). Thus the process of tannin–protein complexation which has been thought to be the key element in their mode of action is by no means as immutable as it was once believed to be. Proteases, for example, are poor substrates for complexation with polyphenols (Mole and Waterman, 1987a) and in an *in vitro* simulation of proteolysis it was found possible to achieve rates of proteolysis, either lower or higher than those measured in the absence of tannin, simply by changing the relative concentrations of substrate protein and polyphenols (Mole and Waterman, 1985). Reductions of the rates of proteolysis brought about by the presence of polyphenols may also be alleviated by the addition of surfactants such as cholic acid.

Herbivores, when consuming plant tissues, rarely consume pure natural products but rather an unholy cocktail of metabolites, and Mole and Waterman and their colleagues (1987b, 1987c, 1989) have used this observation as a basis for a critical analysis of the various techniques which may be employed for the measurement of polyphenols (tannins) in ecological studies. However, the results of this extensive analysis simply serve to draw attention to the veritable minefield which this area has become, especially for the naive and unsuspecting recruit; in particular these authors noted the very great difficulty and uncertainty which exist in respect of the problem of establishing a relationship between semi-quantitative measures of polyphenol concentrations and the ability of those polyphenols to precipitate proteins *in vitro*, and the real *in vivo* physiological effects, the ecological impact of these polyphenols as allelochemicals, which ecologists wish ultimately to infer. Butler and his colleagues (1989a,b), whose detailed work in this area has been seminal, have suggested that the question of the mode of action of polyphenols in animal diets is unresolved and that nutritional effects via inhibition of digestion are much less significant than inhibition of the utilisation of digested and absorbed nutrients.

Further doubts and questions concerning the mode of action of polyphenols in mammalian diets and/or the idea of biochemical co-evolution has emerged from a purely physicochemical study of structure–activity relationships in protein–polyphenol complexation (Beart, Haslam and Lilley, 1985; Haslam, 1994). According to the Darwinian paradigm of molecular evolution the structure of a biochemically functional molecule is the result of the processes of selection, and presumes its preformation, either as a product of a biotic process or of a prebiotic one. Biotic preformation might include reactions catalysed by enzymes and also non-enzymic processes. If the question is posed as to which factors have determined the molecular structure of a functional biomolecule then it is usual to argue that this is the result of selection processes directed towards the optimisation of the molecule's biological function. In the case of plant polyphenols the question which must therefore be addressed is whether their structures represent the optimal ones predicted for their role as agents of chemical defence in plants. If their role in the chemical defence of plants is, as has been widely suggested (*vide supra*), their complexation with proteins, then do their structures appear to have been fashioned in the crucible of evolution to maximise their potential to do so? According to Beart *et al.* (1985) the answer is certainly **not** an unequivocal yes! Table 4.1 shows some values of a quantity $-\Delta G^{\ominus,\text{tr}}$, which is a measure of the affinity of a given polyphenolic substrate for a protein, in this particular case BSA (Beart *et al.*, 1985); the larger the negative value the stronger the binding. A number of observations are immediately apparent. Thus in the galloy-D-glucose series the strength of association rapidly declines from that of the key biosynthetic intermediate β-1,2,3,4,6-penta-*O*-galloyl-D-glucose as the number of galloyl groups is decreased. Likewise the substantially increased affinity of the 'dimeric' metabolite rugosin D compared to that of β-1,2,3,4,6-penta-*O*-galloyl-D-glucose is pre-

Table 4.1. *Protein–polyphenol complexation. Values of the free energy of transfer* $(-\Delta G^{\ominus,\mathrm{tr}})$ *of a protein from an aqueous medium to an aqueous medium containing a polyphenol ligand (concentration* m_f*)*

Polyphenol (M_R)	$-\Delta G^{\ominus,\mathrm{tr}}$ kJ mol^{-1}	N	$-\Delta G^{\ominus,\mathrm{tr}}/N$
β-1,2,3,4,6-Penta-*O*-galloyl-D-glucose (940)	26.9	5	5.4
β-1,2,3,6-Tetra-*O*-galloyl-D-glucose (788)	9.1	4	2.3
β-1,3,6-Tri-*O*-galloyl-D-glucose (636)	0.9	3	0.3
Vescalagin/castalagin (934)	1.0	5	0.2
Rugosin D (1874)	58.7	10	5.9

Values of $-\Delta G^{\ominus,\mathrm{tr}}$ determined in aqueous solution at 25°C, pH 2.2 and at a free ligand concentration (m_f) of 4.5 mol kg^{-1}, using equilibrium dialysis. N is the number of 'galloyl' nuclei in the metabolite. Data from Beart *et al.* (1985).

sumably due to the almost doubling of molecular size and with it the enhanced possibility for polydentate binding of the conformationally mobile polyphenol with the protein (see chapter 3). However, it should also be noted that the relative affinity per galloyl ester group ($-\Delta G^{\ominus,\mathrm{tr}}/N$) is very similar for both rugosin D and β-1,2,3,4,6-penta-*O*-galloyl-D-glucose. Conversely the metabolites vescalagin and castalagin which are important polyphenolic compounds in both *Quercus* and *Castanea* spp. and which are in a sense close analogues of β-1,2,3,4,6-penta-*O*-galloyl-D-glucose, are conformationally rigid and are both bound extraordinarily weakly to BSA. Indeed in terms of their affinity for BSA they would hardly warrant the description vegetable tannins.

In the context of the questions posed earlier concerning the theory of biochemical co-evolution and the specific suggestions that the *raison d'être* for the production of plant polyphenols is as a particular form of chemical defence in plants, then these data also throw some doubt on this hypothesis. Thus castalagin and the diastereoisomeric vescalagin are key polyphenolic metabolities of both *Quercus* and *Castanea* spp. and both are formally derived by loss of six hydrogen atoms from the biosynthetic intermediate β-1,2,3,4,6-penta-*O*-galloyl-D-glucose. Both fall within the defined relative molecular mass range for vegetable tannins, but unlike β-1,2,3,4,6-penta-*O*-galloyl-D-glucose neither complexes strongly with gelatin, BSA or other proteins. Based simply on comparisons of the quantity ($-\Delta G^{\ominus,\mathrm{tr}}/N$), Table 4.1, then the structure of β-1,2,3,4,6-penta-*O*-galloyl-D-glucose represents, in the simple galloyl-D-glucose series, the apotheosis of the ability to complex with protein. Although both vescalagin and castalagin possess an identical number of 'galloyl' ester groups as β-1,2,3,4,6-penta-*O*-galloyl-D-glucose ($N = 5$) the biosynthetically (oxidatively) introduced di- and tri-phenyl linkages drastically reduce the affinity of both vescalagin and castalagin for proteins. If, as the theory suggests, the purpose of the synthesis of these metabolites is as agents of chemical defence then it is wholly reasonable to conclude that their structures do not appear, on this evidence, to have been fashioned through the crucible of co-evolution to maximise their potential to bind to proteins. One must there-fore conclude, at this juncture, that either their mode of action is not, as has been widely advertised, to bind and complex with proteins *or* that the 'purpose' of their synthesis is not directly related to plant chemical defence (and hence their ability to associate with proteins) and that the answers surrounding their origins and the circumstances of their metabolic generation must lie elsewhere.

Although it is a precept of modern chemical ecology that energy is not likely to be wasted in the production of secondary metabolites unless there is some compensating advantage to the organism in question, the structural diversity and the proliferation of plant polyphenols is best viewed as a result of their being very little selection pressure on their identity – the products bring no special advantage or disadvantage. An alternative, and in the present situation perhaps more reasonable, proposition, is that the presence of these poly-phenolic metabolites does enhance, however marginally, the prospects for the plant's survival. Nevertheless in making this proposal it is not necessary to presume that this is the *raison d'être* for their formation. Rather the presump-tion is made that their continued metabolism confers a sufficient advantage to balance the metabolic costs of their formation.

One of the alternative extant theories of secondary metabolism suggests that it provides organisms with a means of adjustment to changing circumstances. According to this view the synthesis of enzymes designed to execute the processes of secondary metabolism thus permits the network of enzymes operative in primary/intermediary metabolism to continue to function until such time as conditions are propitious for renewed metabolic activity and growth. Within this framework a general sequence of events leading to the expression of secondary metabolism may be formulated:

 (i) a termination of balanced growth, leading to,
 (ii) an accumulation of intermediates in a primary metabolic pathway, leading to,
(iii) the induced synthesis of secondary metabolites.

In this idealised picture it is suggested that a key secondary metabolite – gallic acid and/or β-1,2,3,4,6-penta-O-galloyl-D-glucose – is first formed and then transformed to a diverse array of secondary metabolites by reactions, which do not have a high substrate specificity, and this results in the production of overlapping patterns of secondary metabolites, Figure 4.1. In this picture it is the *processes* of secondary metabolism rather than the *products* which are

Vescalagin / Castalagin

-6H

β-1,2,3,4,6-Penta-O-galloyl-D-glucose

important – the products themselves bring no great advantage to the plant.

Whatever the final outcome of this debate concerning biochemical co-evolution, the fact remains that plant chemistry and in particular plant phenolics have been implicated as factors in food selection by many animals – herbivores using almost every conceivable nutritional system. One feature to emerge from recent investigations is the wide range in the levels of phenolics that a herbivore may often tolerate (Waterman and Mole, 1994). This can range from essentially no intake of phenolics through to specialisation to feed

β-1,2,3,4,6-Penta-O-galloyl-D-glucose

$2x$, -6H

Rugosin D

on plants with very high levels of phenolic metabolites. However, the more typical result from food selection studies, particularly from those which involve polyphenolics (tannins), is that most herbivores appear to avoid consuming levels of polyphenolics that are in excess of those found in their 'normal diet'. In making this assertion it is nevertheless important to bear in mind that polyphenolics (tannins) constitute just one of the parameters in food selection. Dietary fibre and nitrogen nutrition are likewise other crucial factors and a particular herbivore (e.g. squirrel; Smallwood and Peters, 1986) may feed preferentially on a polyphenolic (tannin) rich food source, such as an acorn, because the benefit to the consumer of the energy rich nut may well offset the nutritional cost of the intake of a high level of polyphenolics.

4.2 Nutritional effects

'An apple a day keeps the doctor away' is an old and familiar aphorism; there are sound medical reasons why man (and woman) should eat a balanced diet including plenty of fresh fruit. Phenolic metabolites are ubiquitous in plants and we often consciously or sub-consciously select fruits and beverages rich in phenolics in our diet. Pierpoint (1986), however, noted that '... food flavonoids

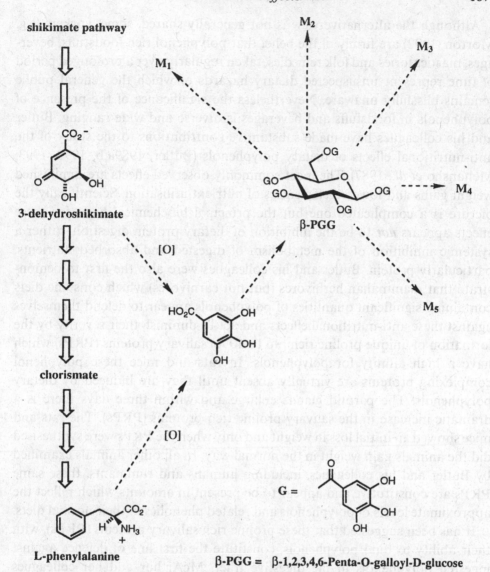

Figure 4.1. Suggested biosynthetic 'scenario' for the derivation of polyphenolic metabolites (M_1, M_2, M_3, etc.) derived from gallic acid *via* the key intermediate β-1,2,3,4,6-penta-*O*-galloyl-D-glucose. It is proposed that gallic acid is generated at the termination of balanced growth from intermediates and/or endproducts of the shikimate pathway; Haslam (1994).

are on the whole beneficial or only harmful in unusual or predictable circumstances.' This view has changed somewhat over the past four or five years with the epidemiological evidence brought forward to sustain the aphorism for the 1990s (and the sales of a particular beverage) that 'A glass of red wine a day keeps the doctor away.'

Although the alternative view is not generally shared, some workers (e.g. Morton, 1992) are firmly of the belief that polyphenol rich foodstuffs, beverages, masticatories and folk remedies, taken regularly over a prolonged period of time represent unsuspected dietary hazards of which the general public remains blissfully unaware. Nevertheless the significance of the presence of polyphenols in foodstuffs and beverages is diverse and wide ranging. Butler and his colleagues have made substantive contributions to the study of the anti-nutritional effects of dietary polyphenols (Butler, 1989a,b, 1992, 1993; Mehansho *et al.*, 1987). The most commonly observed effects are diminished weight gains and lowered efficiencies of nutrient utilisation. Scientifically the picture is a complicated one but the principal biochemical basis for these effects appears *not* to be the inhibition of dietary protein digestion, rather a systemic inhibition of the metabolism of digested and absorbed nutrients, particularly protein. Butler and his colleagues were also the first to demonstrate that mammalian herbivores (but not carnivores) which consume diets containing significant quantities of polyphenols appear to defend themselves against these anti-nutritional effects and act to diminish their severity by the formation of unique proline rich (up to 40%) salivary proteins (PRPs) which have a high affinity for polyphenols. In rats and mice these polyphenol complexing proteins are virtually absent until they are induced by dietary polyphenols. The parotid glands enlarge and within three days there is a dramatic increase in the salivary proline rich proteins (PRPs). The rats and mice showed an initial loss in weight and only when the PRPs were synthesised did the animals gain weight in the normal way. In all other animals examined by Butler and his colleagues, including humans and ruminants, these same PRPs are constitutive and appear to be present in amounts which reflect the approximate levels of polyphenols and related phenolics in their normal diets.

It has been suggested that these proline rich salivary proteins (PRPs), with their ability to bind polyphenols, constitute the first line of defence against ingested polyphenols in the digestive tract. McArthur and her colleagues (McArthur, Sanson and Beal, 1995) have advanced the hypothesis that the ancestral function of PRPs is the maintenance of oral homeostasis (association with calcium salts, lubrication and bacterial aggregation) and that counteracting the reduced digestibility and the toxic costs of consuming dietary polyphenols is a derived one. Thus proline rich salivary proteins (PRPs) are effective in oral homeostasis at low secretion levels, whereas neutralising dietary polyphenolics depends upon high secretion levels.

4.3 Phytotoxicity

In 1981 Singleton comprehensively reviewed the state of knowledge concerning naturally occurring phenolic substances of plant origin as food toxicants. In the opening sentence of his review he freely acknowledges the commonly perceived view that 'The risks to humans of serious poisonous effects caused by natural phenols present in normal foods consumed under usual circumstances seem vanishing small.' Despite the obvious difficulties inherent in drawing together the disparate threads in this multidisciplinary area which bridges nutrition, medicine, pharmacology, toxicology, physiology, chemistry, biochemistry and agriculture, Singleton's review remains an authoritative analysis of this field. However, the review concluded with an important rider.

The majority of the phenols common in foods have very low practical toxicity and are negative in the Ames test, and many have definite or possible dietary or therapeutic effects. We need to know more about their fate in man. Testing both on physiological effects and on metabolism after oral ingestion appears safe and should be expanded... More and better research is needed.

In total contrast to this view that the risk of serious toxic effects to humans by natural phenols is, under normal circumstances, negligible, Morton (1992) has consistently voiced the opinion that 'tannin is often an unsuspected hazard in foods, beverages, folk medicines, or other ingested materials that are considered harmless, enjoyable or even beneficial.' The medical basis of this claim lies, according to Morton, in the carcinogenic action of tannins – taken in quantity, regularly and over a lengthy period of time. In support of this suggestion Morton quotes epidemiological evidence relating to the geographical distribution of oesophageal cancer world-wide and relates these data to the use of beverages of plant decoctions rich in polyphenols. The most significant correlation is in the U.S.A. and Europe with apple-based drinks such as ciders, selected for their high polyphenol content, which lends the beverages both bitterness and astringency.

Whatever the situation in man, evidence of phytotoxicity in other animals is often more clear cut. Thus polyphenolic compounds display clear molluscicidal activity against species of the genus *Schistosoma* – a common cause of illness in tropical countries (Hostettman, Lilley and Haslam, unpublished results). An explanation based upon the alleged phytotoxicity of polyphenolic compounds in acorns has recently been advanced to explain the progressive demise in the United Kingdom of the red squirrel in the face of its more successful North American cousin, the grey squirrel. Since its first well documented introduction from North America in 1890 the grey squirrel, *Sciurus caroliensis*, has displaced the native red squirrel, *Sciurus vulgaris*, from most British deciduous and mixed woodland. South of Cumbria (the Lake District) small red squirrel populations have persisted in discrete isolated pockets and also in some large conifer forests. Recent researches have suggested that the decline of the red squirrel can be almost entirely linked to the available diet. Thus an exclusive diet of acorns (*Quercus* sp.) is poisonous to red squirrels whereas its cousin the grey squirrel thrives on the food. In captivity, compared to the grey, the red squirrel had a very much lower digestive efficiency when fed a diet of acorns. This was attributed to the fact that the red squirrel was far less able to 'neutralise' the effects of the polyphenolics present in acorns. At the same time the growing population of grey squirrels, by consuming hazlenuts before they mature, is also depriving the red squirrel of its natural diet of these nuts which they can readily digest and eat when they are ripe. Although it remains to be seen whether grey squirrels secrete proline rich salivary proteins (PRPs), faecal analyses for the acorn diet were consistent with the view that red squirrels suffered dietary 'poisoning' by polyphenols, whilst grey squirrels were able to overcome their phytotoxic effects (Kenward and Holm, 1993).

4.4 Pasture bloat

Pasture bloat has been described as an insidious disease and remains as a continuing hazard to cattlemen and milk-producers because of its unpredictability and the speed of its onset. It is not a recent disorder but has been recognised for many centuries. Bloat manifests itself in the inability of the animal to expel gases in the rumen arising from normal fermentation and may occur when cattle graze on certain leguminous or graminaceous crops at a particular stage of development. Ruminant animals have the ability to digest large quantities of fibrous feedstuffs and to accomplish this they carry a diverse population of micro-organisms in their fore-stomach. During normal digestion these micro-organisms produce large quantities of gas which collects to form a free gas pocket at the top of the rumen. The continued expansion of this gas pocket is periodically relieved by eructation which expels the gas through the oesophagus. The condition of pasture bloat is characterised by the formation of a frothy complex of partially digested feedstuff that traps the gas in very small bubbles throughout the rumen such that eructation cannot take place. In cases of acute bloat animals may succumb to asphyxiation due to the pressure of the rumen on the diaphragm and the lungs.

Pastures dominated by immature, fast growing legumes such as the highly nutritious alfalfa (*Medicago sativa* L.) and most forage clovers present a high risk of the development of cattle bloat. However, legumes such as sainfoin

(*Onobrychis viciifolia*) and birdsfoot trefoil (*Lotus corniculatus* L.) which contain substantial levels of polyphenolic metabolites in their herbage are known to be non-bloating (Lees, 1992). Similar polyphenolic metabolites do not occur in alfalfa herbage, although they are found in the seed coat of that plant. In both sainfoin (*Onobrychis viciifolia*) and birdsfoot trefoil (*Lotus corniculatus* L.) the principal polyphenolic constituents are of the condensed proanthocyanidin class, and largely those of the type which are difficult to solubilise (Base-Smith, 1973b). The bloat-safe properties of plants containing polyphenolic metabolites were first recognised some 30 years ago by Kendall and confirmed by workers in New Zealand (Mangan *et al.*, 1976). It is generally believed that polyphenols act to prevent bloat by virtue of their ability to precipitate proteins; they inhibit cell-wall degrading enzymes which are secreted by the rumen bacteria and render soluble protein in the plant cytosol immediately unavailable.

Jones and Mangan (1977) conducted studies to determine the nature of the complexes formed between the condensed proanthocyanidins of sainfoin (*Onobrychis viciifolia*) and the major soluble dietary protein of green leaf (~ 80% associated with ribulose diphosphate carboxylase activity). The latter formed insoluble complexes with the condensed proanthocyanidin which were stable in the pH range 3.5–7.0. At pH values below 3.0, polyethylene glycol (M_R 4000) exchanged with the protein in the complexes and released the protein (> 90%) into solution. Parenthetically, in this context, Martin and his colleagues (Martin and Martin, 1984; Martin *et al.*, 1985, 1987), whilst examining the interaction of polyphenols with important insect dietary proteins, concluded that natural insect surfactants are critical factors in the prevention of the polyphenol mediated protein precipitation in insect digestive systems, thereby eliminating protease inhibition. It seems probable in both of these examples that solubilisation of the protein is achieved by disruption of the polyphenol–protein complex rather than by the formation of a soluble ternary complex consisting of polyphenol–protein–surfactant (polyethylene glycol). This, it may be presumed, is achieved by competitive polyphenol–surfactant (polyethylene glycol) complexation.

It is known that bloat may also be controlled by the action of both natural and synthetic anti-foaming agents. In view of the probability (on the basis of the above evidence) that polyphenols complex with surfactants it would be interesting to discover the physico-chemical behaviour of condensed tannins in this same context; do they, for example, lower the surface tension of the foam bubbles and thereby permit the release of gas? Nevertheless in this age where genetic manipulation of plant characters is a realistic possibility (and assuming that condensed tannin biogenesis is actually under genetic control!) the prospect of the availability of a palatable, nutritive alfalfa herbage containing condensed tannin metabolites would clearly be of agronomic benefit world-wide.

4.5 Taste and flavour – fruit and beverages

4.5.1 Introduction

The objective evaluation of the taste and flavour of foodstuffs and beverages is still largely dependent upon sensory perception and the collective observations of the members of experienced taste panels. Although the methodology of testing has been greatly refined and systematised a fundamental understanding of the physiology and chemistry which underlies the sensation of taste is still lacking – eloquently illustrated, for example, by the terms which are used contemporaneously to describe the flavours of wines – nutty, vegetal, mineral, floral, spicy, earthy, woody, burnt and fruity (Robinson, 1994, 1995). In the context of taste and flavour, fruits are of particular interest not only because of their position in the biological cycle of higher plants (and seed dispersal), but also because of their use as foodstuffs, either as the fruits themselves or in processed forms as beverages – wines, beers, lagers, perry, ciders, etc. Wines are thus made by the fermentation of the expressed juices of grapes (*Vitis vinifera*) which have a naturally high sugar content, and although wines and wine-making are particularly important commercially and are rich in folklore and religious symbolism, juices made from many other fruit can be fermented into alcoholic beverages – pear juice into perry, apple juice into cider. Likewise the fermentation of the starchy mass derived from malted barley grain is the basis of the manufacture of beers and lagers. However, the exalted position of wines owes something to the fact that there are thousands of distinctive varieties of vine, and hence grape berries, and this in consequence means that the taste on the palate and the flavours of wines are far more varied than those of beverages derived from other fruit. In turn this gives infinite scope to those influential cognoscenti who write on the subject of wines and wine-making: indeed wine is like sex for no-one will freely admit to not knowing all about it.

A detailed exposition of the complex series of reactions that constitute alcoholic fermentation is beyond the scope of this discussion; suffice it to say that during the numerous changes which lead to the sensory characteristics of a particular beverage, natural phenolic constituents of the original fruit play an important, albeit usually subsidiary, role. A familiar example of a fruit rich in phenolic metabolites is the grape (*Vitis vinifera*) but considerable differences exist between different fruits and their total phenolic contents. Whereas some have a fairly low phenolic content others, such as persimmon and plum, contain quantities often totalling 5–10% of the dry weight. Phenolics are invariably concentrated in particular locations; in the grape, for example, the highest concentrations are found in the skin and its immediate environment and in the pips. With the exception of the anthocyanin pigments, which generally accumulate in those fruit which produce them (e.g. black grapes) as they approach maturity, the highest concentrations of phenolic metabolites are almost invariably found in very young immature fruit during the early growth phase associated with extensive cell division (as opposed to the subsequent period of cell enlargement). During growth, variations in the ratios of different phenolic compounds may be observed. Although external factors may act to modify these changes, genetic control appears to be the ultimate underlying influence (Macheix *et al.*, 1990).

Polyphenols affect a range of characteristics and contribute to the critical assessment of the quality and appearance of many beverages such as teas, coffee, cocoa, beers and lagers and to those beverages derived from fruit such as ciders, perry and wines. Particular emphasis is invariably placed upon their influence on the quality of **astringency**. Whilst each particular fruit or raw material, and hence each beverage, has its own characteristic 'phenolic profile' these often bear very close resemblances to one another. Thus soluble hydroxycinnamoyl esters, anthocyanins and other flavonoids, phenolic flavan-3-ols and related proanthocyanidins are often encountered. Particular examples are given in the ensuing discussion.

4.5.2 *Astringency*

Saliva is produced by the salivary glands which empty their secretions into the oral cavity. The macromolecules in saliva consist almost exclusively of protein, varying from 1.0 to 3.5 mg ml^{-1} of saliva, and this secretion contains a complex mixture of individual proteins. Amino acid analysis of unfractionated human salivary protein reveals an unusually large amount of proline (from 16 to 33%). It is now clear that saliva contains a group of unique proteins which are usually referred to as 'proline rich proteins' (Bennick, 1982). It is characteristic that proline accounts for 25–42% of the amino acids in isolated proline rich proteins (PRPs). In addition there are high contents of glutamine (glutamic acid) and glycine, and these three residues account for 70–88% of all amino acids in these proteins. The phenomenon of astringency is believed to be associated specifically with the interaction of polyphenols with these proline rich proteins.

Astringency is generally recognised as a feeling of extreme puckeriness and dryness in the palate. It is not confined to a particular region of the mouth or tongue, but is experienced as a diffuse stimulus which invariably takes a finite time to develop fully (Joslyn and Goldstein, 1964). A mucous membrane covers all the exposed surfaces of the mouth which are moistened by the secretions of the salivary glands. According to Bate-Smith (1973a) the primary reaction whereby astringency develops is via the precipitation of proteins and muco-polysaccharides in these mucous secretions caused by the astringent principles. Polyphenols (tannins) constitute one of the principal sources of naturally occurring astringent principles. They have a harsh astringent taste and pro-duce in the palate those responses noted above – namely feelings of dryness, constriction and roughness. A pictorial representation of the two-stage recog-nition process whereby polyphenols are thought to act at the molecular level to complex with salivary proline rich proteins (PRPs) and bring about the astringent response is depicted later in Figure 4.11. The general consensus is that an essential feature is the cross-linking of adjacent polypeptide chains of the PRPs by polyphenol molecules (Bate-Smith, 1973a; Joslyn and Goldstein, 1964; Haslam *et al.*, 1994). This gives rise to the formation of aggregates and precipitation of the salivary PRPs and in turn this leads to the loss of lubri-cation in the palate and the sensation of astringency. The whole process is a time dependent and dynamic one and requires a period of time to develop fully. A necessary feature of this model is that the polyphenol should be of sufficient size and composition to bring about, by polydentate binding, the cross-linking process. However, it is now clear that the contribution of much smaller phenolic molecules to the development of astringency in the palate cannot be totally ignored (*vide infra* and Haslam, 1989).

4.5.3 *Grapes and wines*

Changes in the social fabric of countries constitute one of the principal factors which has prompted the rapidly increased popularity of wines as beverages over the past two to three decades. Paralleling this change the markets have witnessed an influx of wines, particularly those from Australia, New Zealand and California, which are produced under 'scientifically controlled' conditions; large retail outlets increasingly demand consistency in the qualities of the wines that they offer for sale. For many of these newer wine makers it is no longer an art in which nature is the driving force but it is a commercial scientifically directed operation. In train with this particular development there has been an increasing commitment on the part of oenologists to the scientific study, not only of the vine and its cultivation, harvesting and fermen-tation, but also to the principal ingredient which shapes the characteristics of the wines, namely the grape itself – to its chemical make-up and to the transformations of these constituents during the various stages of wine-making. Pre-eminent amongst these are a range of closely related phenolic metabolites. The diversity of many wine types and styles is very largely due to the variability in concentration and composition of these phenolic metabolites; in the case of red wines they are intrinsic to the processes of maturation and ageing.

There have been many studies of the total phenolics and polyphenolics (tannins) in grapes. Estimates of total phenolics in red wines from different geographical regions vary from ~ 800 to 4000 mg l⁻¹, but even with prolonged skin contact and maceration, phenolic extraction rarely exceeds 50% of the total amounts present originally in the grape. As a rough guide freshly ripened black grapes, used in the preparation of a typical red wine, would normally give ~ 4.0 g of phenolic compounds per kilo of fresh grapes. Amongst these phenolics one would expect to find about 200 mg of caftaric acid and other non-flavonoids, some 400 mg of anthocyanins, 300 mg of 'monomeric' phenolic flavan-3-ols ((+)-catechin and (−)-epicatechin) and 500 mg of 'oligomeric' proanthocyanidins ('tannins' – principally dimeric, trimeric and tetrameric procyanidins based upon (+)-catechin and (−)-epicatechin), Figure 4.2. Seeds and grape skins account for a very large part of the total phenolic content of the grape; in the case of the phenolic flavan-3-ols and the associated 'oligomeric' proanthocyanidins roughly four-fifths are located in the seeds and the remainder in the skins. As a general rule the quantities of (+)-catechin and (−)-epicatechin are roughly equivalent in the grape seeds but in the skins (+)-catechin usually predominates. Likewise each of the dimeric procyanidins (B-1, B-2, B-3 and B-4) is found in the seeds but the relative concentration of procyanidin B-1 is usually enhanced in the grape skins. Nevertheless these general guidelines should not be allowed to disguise the fact that in practice considerable differences may well be encountered in the analysis of different cultivars.

Modern white wines are usually manufactured by cool fermentation of must (grape juice) drained from de-stemmed crushed grapes, avoiding extraction of the solid pomace (solid mass of crushed grapes including skins and seeds). The problems, as compared to those of red wine manufacture, are less complex as minimal skin contact after crushing ensures the virtual absence of phenolic flavan-3-ols and related proanthocyanidins. Major phenolic components of white wines are caftaric acid (Figure 4.2) and its *p*-coumaroyl analogue, coutaric acid, other non-flavonoids and small quantities of tyrosol (arising from yeast fermentation) (Singleton *et al.*, 1978) Coarseness, bitterness and astringency are the characteristics of white wines which invariably derive from excessive skin contact and the extraction of flavonoids. Temperature control at around 10°C is effective in limiting the solubilisation of phenolic flavan-3-ols and proanthocyanidins during commercial juice preparation (Ramey *et al.*, 1986).

At the other end of the wine spectrum freshly prepared red wines may contain up to 30 × the levels of total phenolics found in white wines, although a figure of 10–12 × is more common for red wines intended for ageing. In red wines ready for consumption this value has dropped to some 7 ×. These differences arise from the fact that red wines are made by several days of fermentation of the whole grape at ∼ 25°C. Only after the alcohol content has reached the desired level are these wines processed to separate the solid pomace and at this stage they may contain up to 40% of the total phenolics in the original grapes.

Polyphenols contribute to a whole range of sensory characteristics of wines, notably flavour, colour and appearance, stability, and bitterness and astringency on the palate. The technical literature on the ageing of red wines is voluminous and remarkably extensive. It remains, however, essentially empirical and speculative in nature because of the great overall physico-chemical complexity of the original red wine composition. Whilst a certain amount of tannin (polyphenols) is both desirable and indeed essential to the making of a good red wine to give 'body, longevity and backbone', it is nevertheless quite possible to have too much tannin. Where vintages have an apparent excess of polyphenols the reason is invariably the weather, particularly when there are extreme variations. Thus, apart from ripening the grapes, a spell of intense summer heat thickens the skins – a major source of both pigments and polyphenols (*vide supra*). If these factors are combined with a relatively small concentrated yield then this produces a claret that embarks on the ageing process with a deep colour and a high polyphenol content. As the wine matures over the next decade the critical question is whether the claret will retain sufficient 'fruit and flavour' to balance the astringent polyphenols or whether it will remain with a hard unyielding backbone of polyphenolics throughout. During this ageing process sediments invariably develop and some of these arise from the interaction of pigments and polyphenols; the mature wine thus has a changed appearance as well as a changed astringency on the palate.

4.5.4 Apples and pears; ciders and perry

There is no kind of fruit better known in England than the apple, or more generally cultivated. It is that use, that I hold it almost impossible for the English to live without it, whether it be employed for that excellent drink we call cider, or for the many dainties

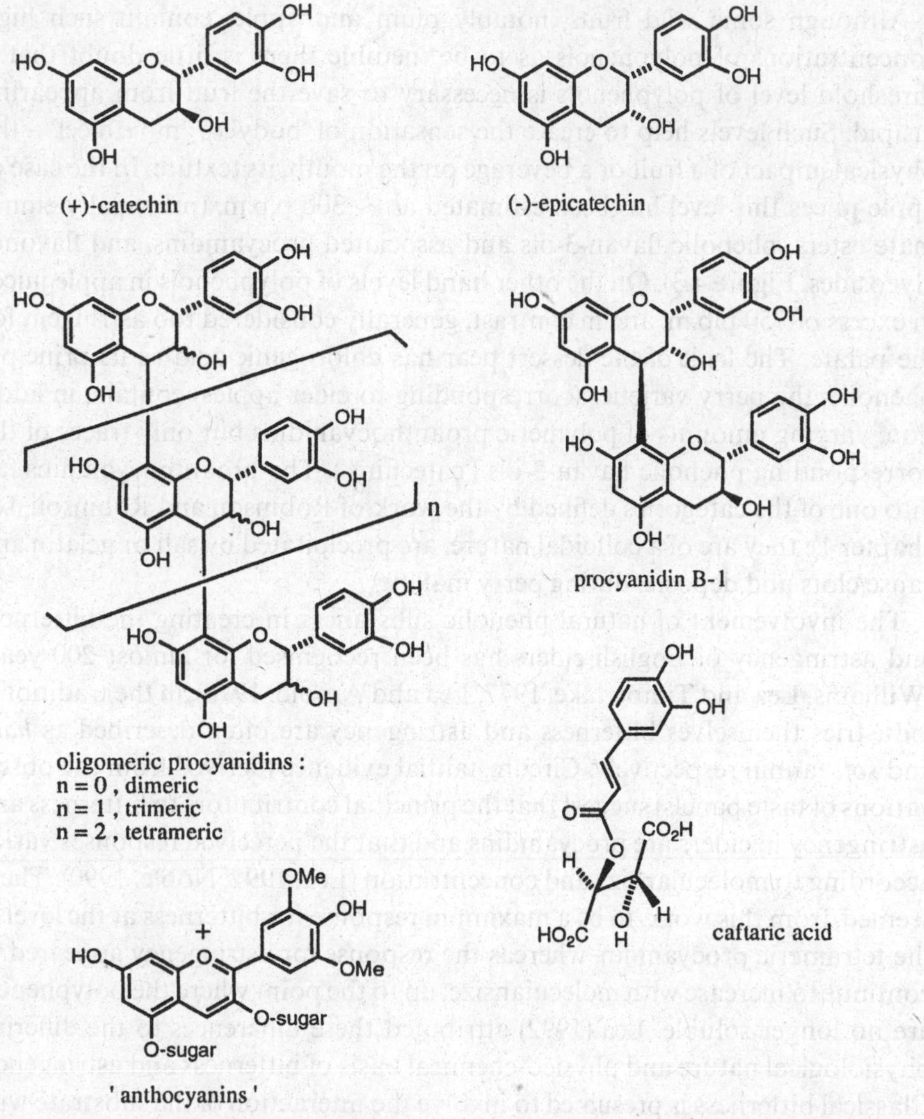

(+)-catechin

(-)-epicatechin

oligomeric procyanidins :
n = 0 , dimeric
n = 1 , trimeric
n = 2 , tetrameric

procyanidin B-1

caftaric acid

' anthocyanins '

Figure 4.2. Some of the principal phenolic constituents of black grapes (Czochanska *et al.*, 1979; Weinges and Piretti, 1971; Singleton, 1992; Singleton and Esau, 1969; Somers and Verette, 1988).

which are made of it in the kitchen. In short were all other fruits wanting us, apples would make amends.

So wrote the horticulturist Richard Bradley in 1718, nearly 1300 years after the first apple is believed to have arrived in Britain in AD 450, brought by the Roman general Ezio. The apple has come to be seen as quintessentially British despite its recorded exotic origins in China's Heavenly Mountains.

Although some wild fruits, notably plum and apple, contain such high concentrations of polyphenols as to be inedible there is little doubt that a threshold level of polyphenols is necessary to save the fruit from appearing insipid. Such levels help to create the sensation of 'body' or 'mouthfeel' – the physical impact of a fruit or a beverage on the mouth, its texture. In the case of apple juices this level has been estimated as ~ 300 p.p.m. (principally cinnamate esters, phenolic flavan-3-ols and associated procyanidins, and flavonol glycosides, Figure 4.3). On the other hand levels of polyphenols in apple juices in excess of 750 p.p.m. are, in contrast, generally considered too astringent for the palate. The fruit of the dessert pear has chlorogenic acid as its principal phenolic: the perry varieties (corresponding to cider apples) contain, in addition, varying amounts of polymeric proanthocyanidins but only traces of the corresponding phenolic flavan-3-ols ('catechins'). The proanthocyanidins fall into one of the categories defined by the work of Robinson and Robinson (see chapter 1); they are of a colloidal nature, are precipitated by salt or gelatin and cause clots and deposits during perry making.

The involvement of natural phenolic substances in creating the bitterness and astringency of English ciders has been recognised for almost 200 years (Williams, Lea and Timberlake, 1977; Lea and Arnold, 1978); in the traditional industries themselves bitterness and astringency are often described as *hard* and *soft* tannin respectively. Circumstantial evidence (derived from the observations of taste panels) suggest that the principal contributors to bitterness and astringency in ciders are procyanidins and that the perceived responses varied according to molecular size and concentration (Lea, 1992; Noble, 1990). There seemed, from this work, to be a maximum response for bitterness at the level of the tetrameric procyanidin, whereas the response for astringency appeared to continue to increase with molecular size, up to the point where the polyphenols are no longer soluble. Lea (1992) attributed these differences to the differing physiological nature and physico-chemical basis of bitterness and astringency. Classical bitterness is presumed to involve the interaction of the substrate with a specific taste receptor and it therefore depends on the ability of molecules to pass through a lipid membrane to bind onto a selected protein receptor. Lea (1992) suggested that the tetrameric procyanidin was of such a molecular size and lipophilicity as to be able to pass through the lipid membrane to act at the selected receptor, whereas soluble oligomeric procyanidins of greater molecular size were unable to do so. Similarly it was presumed that oligomeric procyanidins of smaller molecular size reached the receptor but in turn gave a weaker physiological response.

In the case of ciders the observations of Williams, Lea and Timberlake are consonant with the general idea of the origins of astringency on the palate

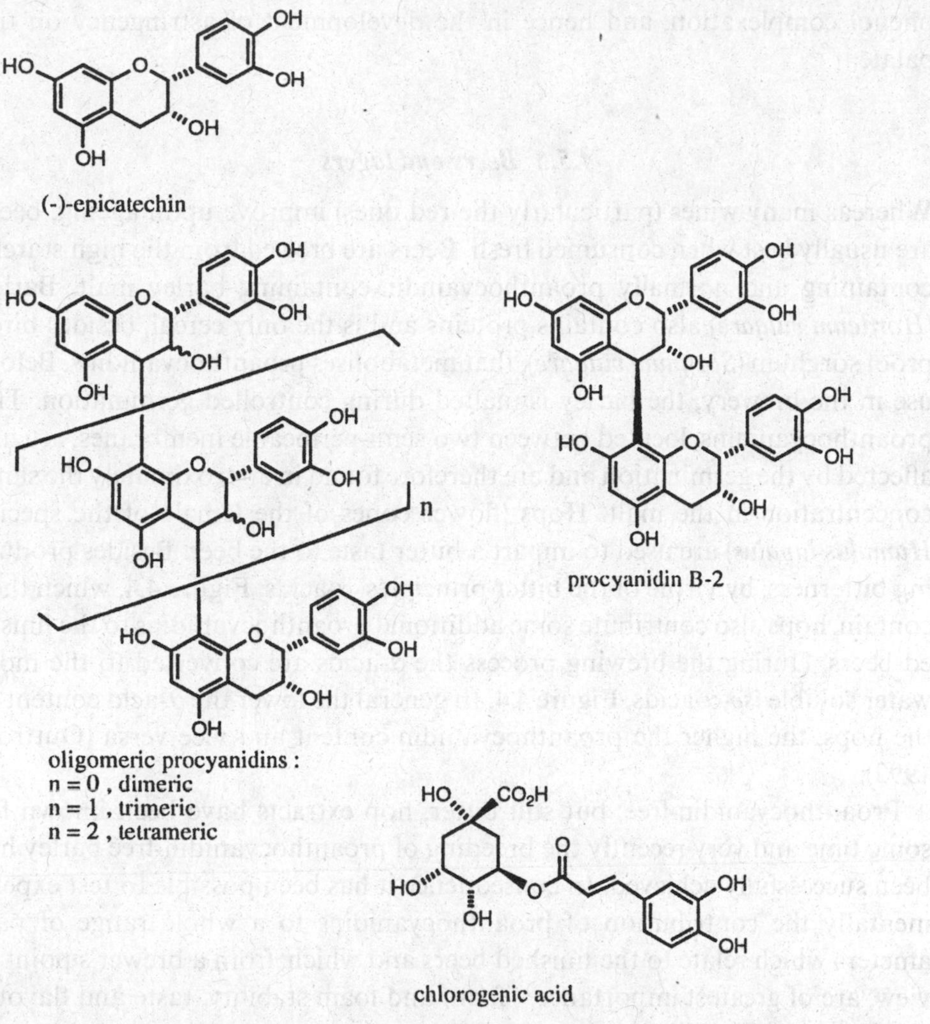

(-)-epicatechin

procyanidin B-2

oligomeric procyanidins :
n = 0 , dimeric
n = 1 , trimeric
n = 2 , tetrameric

chlorogenic acid

Figure 4.3. Principal phenolic constituents of apple (*Malus* sp.) fruit (Williams, 1960; Lea, 1992).

outlined above; the greater the molecular size of the oligomeric procyanidin (assuming that it remains soluble) then the greater its contribution to the astringent response. However, these authors also clearly pointed out that *all* procyanidins contribute to the total bitterness and astringency of the beverage, simultaneously and in concert, and that it was therefore impossible to identify any one specific component as being principally, let alone solely, responsible for either of these two sensory characteristics. Interestingly similar tests carried out in alcoholic media enhanced the sensation of bitterness but depressed the feeling of astringency: an observation which re-enforces the view that '*hydrophobic effects*' are one of the principal driving forces in protein–polyphenol complexation, and hence in the development of astringency on the palate.

4.5.5 Beers and lagers

Whereas many wines (particularly the red ones) improve upon ageing, beers are usually best when consumed fresh. Beers are brewed from the high starch-containing and normally proanthocyanidin-containing barley malt. Barley (*Hordeum vulgare*) also contains proteins and is the only cereal, besides bird-proof sorghum (*Sorghum vulgare*), that metabolises proanthocyanidins. Before use in the brewery, the barley is malted during controlled germination. The proanthocyanidins, located between two semi-permeable membranes, are un-affected by the germination and are therefore found in approximately the same concentration in the malt. Hops (flower cones of the female of the species *Humulus lupulus*) are used to impart a bitter taste to the beer. Besides produc-ing bitterness, by virtue of the bitter principles, α-acids, Figure 4.4, which they contain, hops also contribute some additional proanthocyanidins to the finish-ed beers. During the brewing process the α-acids are converted to the more water soluble *iso*-α-acids, Figure 4.4. In general the lower the α-acid content of the hops, the higher the proanthocyanidin content and vice versa (Outtrop, 1992).

Proanthocyanidin-free, but still bitter, hop extracts have been known for some time and very recently the breeding of proanthocyanidin-free barley has been successfully achieved. In consequence it has been possible to test experi-mentally the contribution of proanthocyanidins to a whole range of par-ameters which relate to the finished beers and which, from a brewer's point of view, are of greatest importance – foam and foam stability, taste and flavour, brightness and stability.

Barley and sorghum are the only two important food grains reported to contain significant quantities of polyphenols – proanthocyanidins – and these metabolites are considered by some to affect flavour and chill haze in beers. Eastmond (1974) was the first to report the isolation and full characterisation of procyanidin B-3 ((+)-catechin – (4α →8) – (+)-catechin), Figure 4.5, from a number of beers and lagers. In later work (Outtrop, 1992; Brandon *et al.*, 1982) a range of proanthocyanidin oligomers were similarly identified in barley containing from two to four phenolic flavan-3-ol units and composed largely of procyanidin and prodelphinidin units, with almost invariably (+)-catechin as the terminal phenolic flavan-3-ol unit, Figure 4.5. The total proan-thocyanidin content of barley is variously reported as from 0.15 to 0.22% of the mass of the dry tissue. The composition of the proanthocyanidins from

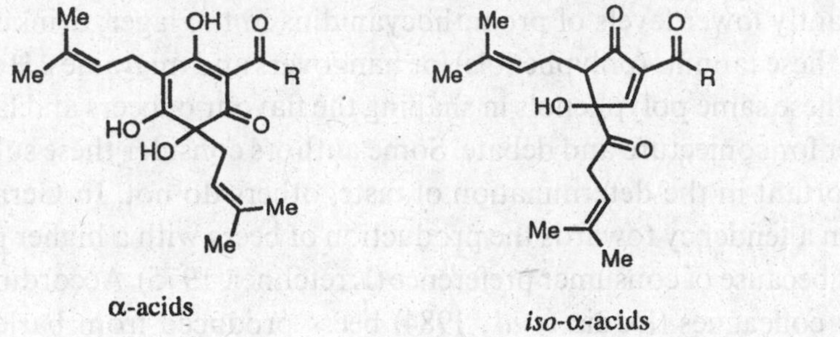

Me
OH
O
R
Me
HO
HO
O
Me
Me
O
O
R
HO
O
Me
Me
α-acids
iso-α-acids

Figure 4.4. Typical α-acids and *iso*-α-acids from hops (*Humulus lupulus*).

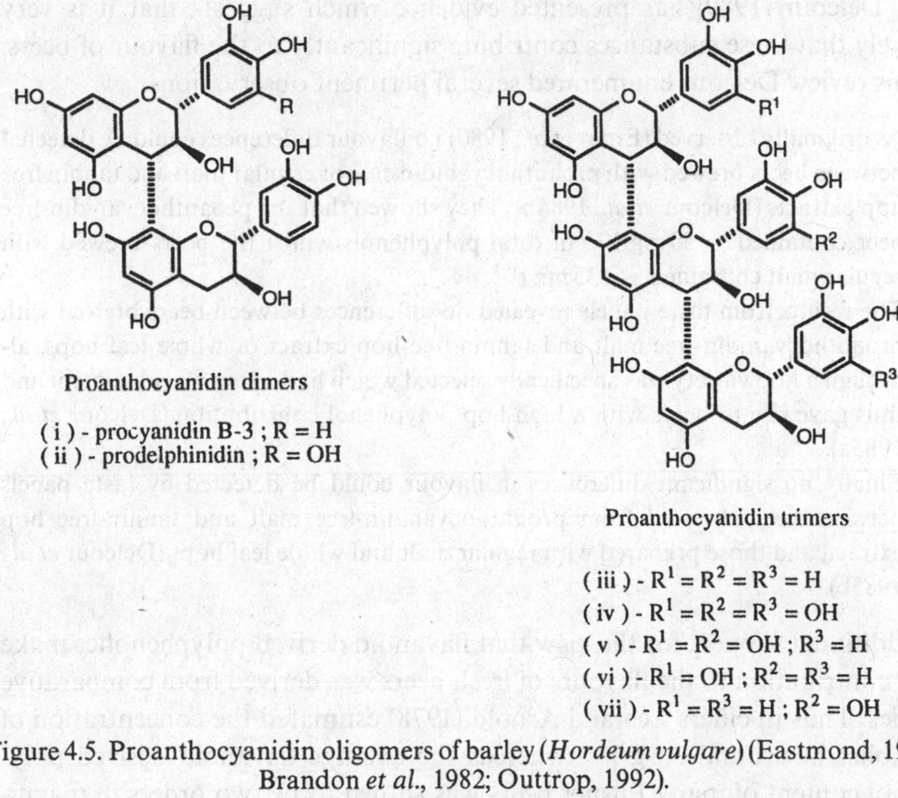

Proanthocyanidin dimers
(i) - procyanidin B-3 ; R = H
(ii) - prodelphinidin ; R = OH
Proanthocyanidin trimers
(iii) - R¹ = R² = R³ = H
(iv) - R¹ = R² = R³ = OH
(v) - R¹ = R² = OH ; R³ = H
(vi) - R¹ = OH ; R² = R³ = H
(vii) - R¹ = R³ = H ; R² = OH

Figure 4.5. Proanthocyanidin oligomers of barley (*Hordeum vulgare*) (Eastmond, 1974; Brandon *et al.*, 1982; Outtrop, 1992).

hops (*Humulus lupulus*) is reported to be more complex and to contain not only procyanidins and prodelphinidins based on flavan-3-ols with the (+)-catechin stereochemistry but also flavan-3-ols with an absolute stereochemistry related to the diastereoisomeric (−)-epicatechin (Outtrop, 1992).

A lager, claimed to be 'hangover free' despite an alcohol content of some 9%, was recently launched on to the U.K. market. The magic ingredient, according to *The Times* (22 September 1995) was said to be a husk-free barley which gave significantly lower levels of proanthocyanidins in the lager; drinking folklore blames these tannins (polyphenols) for hangovers and migraines. However, the role of these same polyphenols in shaping the flavour of beers and lagers is still a matter for conjecture and debate. Some authors consider these substances to be important in the determination of taste, others do not. In Germany there has been a tendency towards the production of beers with a higher polyphenol content because of consumer preference (Kretchner, 1975). According to Dadic and his colleagues (Dadic *et al.*, 1984) beers produced from barley varieties with a higher polyphenol (tannin) content have a better flavour than those produced from barleys which have a lower polyphenol (tannin) content. However, Delcour (1989) has presented evidence which suggests that it is very unlikely that these substances contribute significantly to the flavour of beers. In this review Delcour enumerated several pertinent observations.

(i) As originally observed (Erdal *et al.*, 1980) no flavour differences could be detected between beers brewed with proanthocyanidin-free or regular malt and tannin free hop extracts (Delcour *et al.*, 1984b). They showed that the proanthocyanidin-free beer contained $\sim 50 \, \text{mg} \, l^{-1}$ of total polyphenols whilst the beers brewed with regular malt contained $\sim 135 \, \text{mg} \, l^{-1}$.

(ii) The results from taste panels revealed no differences between beers brewed with proanthocyanidin-free malt and tannin-free hop extract or whole leaf hops, although a hop variety was specifically selected which had a low α-acid content and thus gave rise to beers with a high hop polyphenol contribution (Delcour *et al.*, 1985a).

(iii) Finally no significant differences in flavour could be detected by taste panels between beers brewed from proanthocyanidin-free malt and tannin-free hop extract, and those prepared with regular malt and whole leaf hops (Delcour *et al.*, 1985b).

Additional support for the view that flavanoid derived polyphenolics make little contribution to the flavours of fresh beers was derived from comparative studies. Thus in ciders Lea and Arnold (1978) estimated the concentration of phenolics at around $3.6 \, \text{g} \, l^{-1}$ whereas the average flavanoid derived polyphenol content of many Pilsner beers was shown to be two orders of magnitude less ($\sim 26 \, \text{mg} \, l^{-1}$) (Delcour and Janssens de Varebeke, 1985). Furthermore the 'monomeric' fraction is much more abundant in beers than is the 'dimeric' and 'trimeric' fractions. In de-ionised water the taste thresholds for these 'monomeric', 'dimeric' and 'trimeric' fractions were determined as 46, 17 and $4 \, \text{mg} \, l^{-1}$ (Delcour *et al.*, 1984a), and since these threshold values will be higher in aqueous media containing ethanol (Dadic and Bellau, 1973) then it seems reasonable to conclude that these polyphenolics do not contribute to the flavour of beers.

4.5.6 Teas

A hardened and shameless tea-drinker, who has for twenty years diluted his meals with only the infusion of this fascinating plant; whose kettle has scarcely time to cool; who with tea amuses the evening, with tea solaces the midnight and with tea welcomes the morning.

(Samuel Johnson, Literary Magazine, 1757)

Although verisimilitude does not constitute proof, tea has been regularly described as the world's most popular beverage other than water. Whilst King Charles II's wedding to Catherine of Braganza in 1662 is often credited as the occasion for bringing the tea drinking habit to the English Court, over a 1000 years earlier the Chinese and Japanese peoples had elevated the custom of tea drinking to a social institution approaching a religion in its thoroughness and complexity. Undoubtedly tea owes its popularity, as for Samuel Johnson, to its contribution to the psychological well-being and equanimity of the tea drinker. However, just as in the evolution of a very fine claret it seems the practice of tea production remains as much an art as a science. Thus a great dependence is still placed on the knowledge, skill and judgement of those who select and blend the many types and grades of tea available on the world's markets. The professional tea taster pays attention to a great many factors and has developed a language of his own – principal amongst which are *flavour, colour, strength, quality* and *briskness* – to appraise the characteristics of a black tea liquor (Roberts, 1963; Sanderson, 1972).

The tea plant (*Camellia sinensis*, var. *sinensis* (North China form); *Camellia sinensis*, var. *assamica* (Southern Asia form)) is an evergreen shrub which grows best in tropical and sub-tropical climes. In its natural state the shrub *Camellia sinensis*, var. *sinensis* is a small bush with small leaves; *Camellia sinensis*, var. *assamica* has softer bigger leaves and may grow to a height of 15–30 feet, but the planter usually keeps it pruned to a height of 3–5 feet. New growth (the tea flush) is harvested at intervals of 6–12 days. The apical bud and several new leaves constitute the flush. Generally three principal classes of tea are recognised.

(i) **Fermented or black tea**. The tea leaf is first rolled (partially crushed) and allowed to ferment, before firing which brings the fermentation to a conclusion. This process develops the distinctive colour and aroma of the black tea infusion; many of the compounds responsible for the characteristic aroma of black teas are generated during firing.

(ii) **Unfermented or green tea**. The tea flush is first steamed to inactivate enzymes in the leaf and inhibit fermentation, before drying.

(iii) **Oolong or semi-fermented tea**. This tea is obtained by partial fermentation of the green tea flush. Correspondingly it has some of the characteristics of both green and black tea.

The outstanding characteristic of the chemical composition of the green tea flush is its very high concentration of polyphenolic metabolites – principally phenolic flavan-3-ols, which occur in the cytoplasmic vacuoles of leaf cells, Figure 4.6. Although amounts vary, dependent upon a number of factors such as variety and growing conditions (in Japan for example shading the bushes, over a period of two to three weeks, is employed to cause the level of phenolic flavan-3-ols to drop), phenolic flavan-3-ols *usually* constitute some 20–30% of the dry matter of the leaf. Typically the average content of the major phenolic flavan-3-ols in Sri Lankan tea shoots is shown in Figure 4.6 (Lunder, 1988).

The occurrence of this wealth of phenolic flavan-3-ols in tea leaf would seem to imply the co-occurrence of related proanthocyanidins and their gallate esters. However, it has been noted in earlier work (Haslam, 1989) that where plants metabolise substantial quantities of the gallate esters of flavan-3-ols there is usually found a corresponding diminution in the amounts of the related proanthocyanidins. In an exhaustive study of green tea leaf Nonaka, Kawahara and Nishioka (1983) isolated four dimeric proanthocyanidin gallates (total $\sim 0.07\%$) and two novel dimeric flavan-3-ol gallate esters, theasinensins A and B ($\sim 0.05\%$ and 0.013%, respectively), Figure 4.7. The latter compounds represent a new class of dimeric flavan-3-ols in which the two flavan units are linked by a C–C bond between the two 'B' rings, forming a biphenyl grouping. The chirality of the twisted biphenyl was shown by the Japanese workers to be *S*.

Since almost all the sensory characteristics of manufactured teas derive from the oxidative transformations of the green leaf phenolics a very great deal of attention has been devoted to this topic. E. A. H. Roberts (1962) made pioneering observations in this area and these have been seminal to its subsequent development. The enzymes of principal interest in the tea leaf are those intimately involved in the conversion of the tea flush to the commercially manufactured black and Oolong teas. Pre-eminent amongst these enzymes is tea-leaf polyphenoloxidase. It contains copper (0.32%) and exists as a mixture of several isoenzymes. Its main component has a relative molecular mass of $\sim 140\,000$. Although various mechanisms (free radical, phenoxonium ion) are

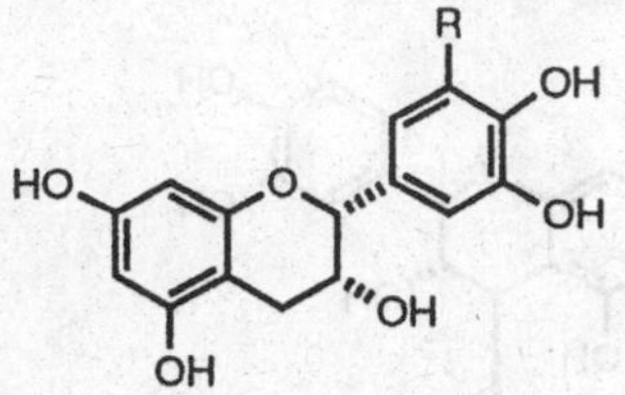

R = H : (-)-epicatechin ; 0.63 %

R = OH : (-)-epigallocatechin ; 2.35 %

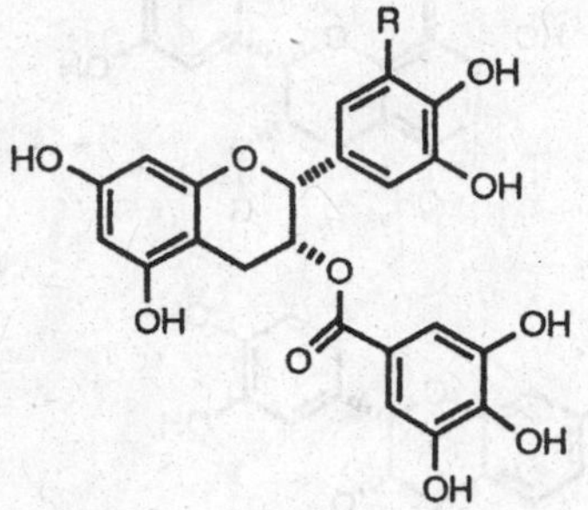

R = H : (-)-epicatechin-3-O-gallate ; 2.75 %

R = OH : (-)-epigallocatechin-3-O-gallate ; 10.55 %

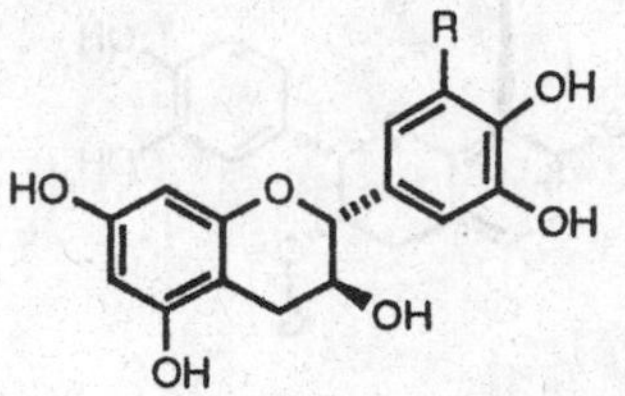

R = H : (+)-catechin ; 0.35 %

R = OH : (+)-gallocatechin ; 0.37 %

Figure 4.6. Principal phenolic flavan-3-ols of a typical Sri Lankan tea flush (Lunder, 1988); % composition of dry weight of leaf.

theoretically possible to initiate the polyphenol oxidase catalysed oxidation, the primary step in tea fermentation is probably best envisaged as the overall conversion of the *ortho*-dihydroxy (*catechin*) and *ortho*-trihydroxy (*gallocatechin*)– phenyl 'B' rings of the substrate tea leaf polyphenolic flavan-3-ols to give the corresponding highly reactive orthoquinone derivatives. Thereafter these orthoquinones may react, under purely chemical control, to give a series of discrete yellow–orange compounds (**theaflavins**) or they may randomly polymerise, and very probably undergo other oxidative reactions such as aryl ring-fission, to give a complex, very poorly defined group of substances known, because of their red-brown colour, as **thearubigins**. Although eventually some 75% of the substrate phenolic flavan-3-ols may be converted to the weakly acidic thearubigins they are the least well characterised of all the components of black tea, notwithstanding that they constitute the largest group of 'compounds' and contribute significantly to the colour, mouthfeel and strength of the subsequent infusion (Roberts, 1963). Approximately 15% of the phenolic

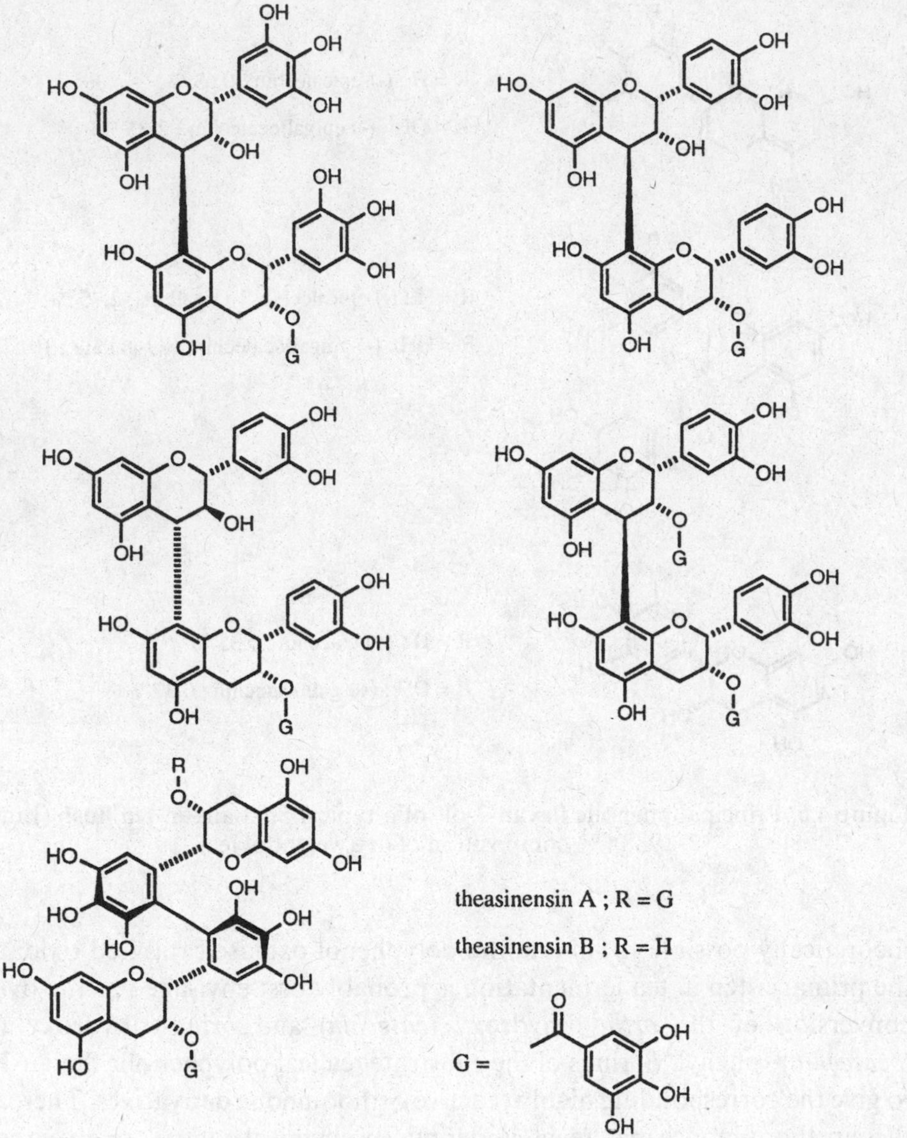

theasinensin A ; R = G

theasinensin B ; R = H

G =

Figure 4.7. Minor polyphenolic components of green tea leaf: proanthocyanidins and theasinensins A and B (Nonaka *et al.*, 1983).

flavan-3-ol substrates may remain unchanged; some 10% are accounted for by the formation of the various theaflavins and associated compounds.

Although his structural proposals were not totally correct, Roberts (1962) correctly identified the fundamental chemical process taking place during the generation of the theaflavins, namely the oxidative condensation of the *ortho*-dihydroxyphenyl 'B' ring of an epicatechin derivative with the corresponding *ortho*-trihydroxyphenyl 'B' ring of an epigallocatechin derivative to give a

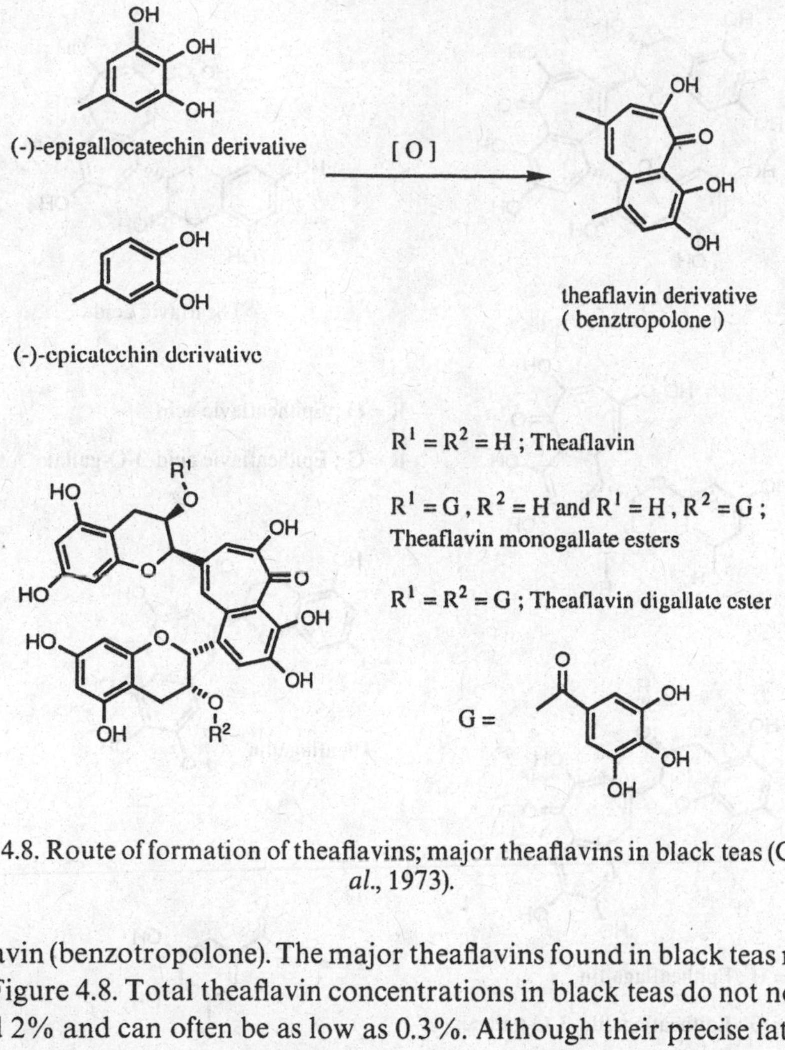

(-)-epigallocatechin derivative

[O]

theaflavin derivative
(benztropolone)

(-)-epicatechin derivative

$R^1 = R^2 = H$; Theaflavin

$R^1 = G$, $R^2 = H$ and $R^1 = H$, $R^2 = G$;
Theaflavin monogallate esters

$R^1 = R^2 = G$; Theaflavin digallate ester

G =

Figure 4.8. Route of formation of theaflavins; major theaflavins in black teas (Collier *et al.*, 1973).

theaflavin (benzotropolone). The major theaflavins found in black teas number four, Figure 4.8. Total theaflavin concentrations in black teas do not normally exceed 2% and can often be as low as 0.3%. Although their precise fate is not known, prolonged oxidative fermentation progressively decreases the theaflavin concentration in the final black tea.

The structure of theaflavin was determined simultaneously by Takino and Ollis and their colleagues (Takino *et al.*, 1965; Ollis *et al.*, 1966). In addition Takino (Takino *et al.*, 1965, 1971) also neatly demonstrated that oxidation (tea polyphenol oxidase or $K_3FeC_6N_6$) of the appropriate pair of phenolic flavan-3-ol substrates gave the predicted theaflavin derivative according to the mechanism of benzotropolone formation put forward earlier by Horner (Horner *et al.*, 1961, 1964); thus oxidation of a mixture of (−)-epicatechin and (−)-epigallocatechin gave theaflavin; similarly that of (−)-epicatechin-3-*O*-gallate and (−)-epigallocatechin-3-*O*-gallate gave theaflavin digallate. Consideration of this mechanism for the oxidative formation of benzotropolones from cate-

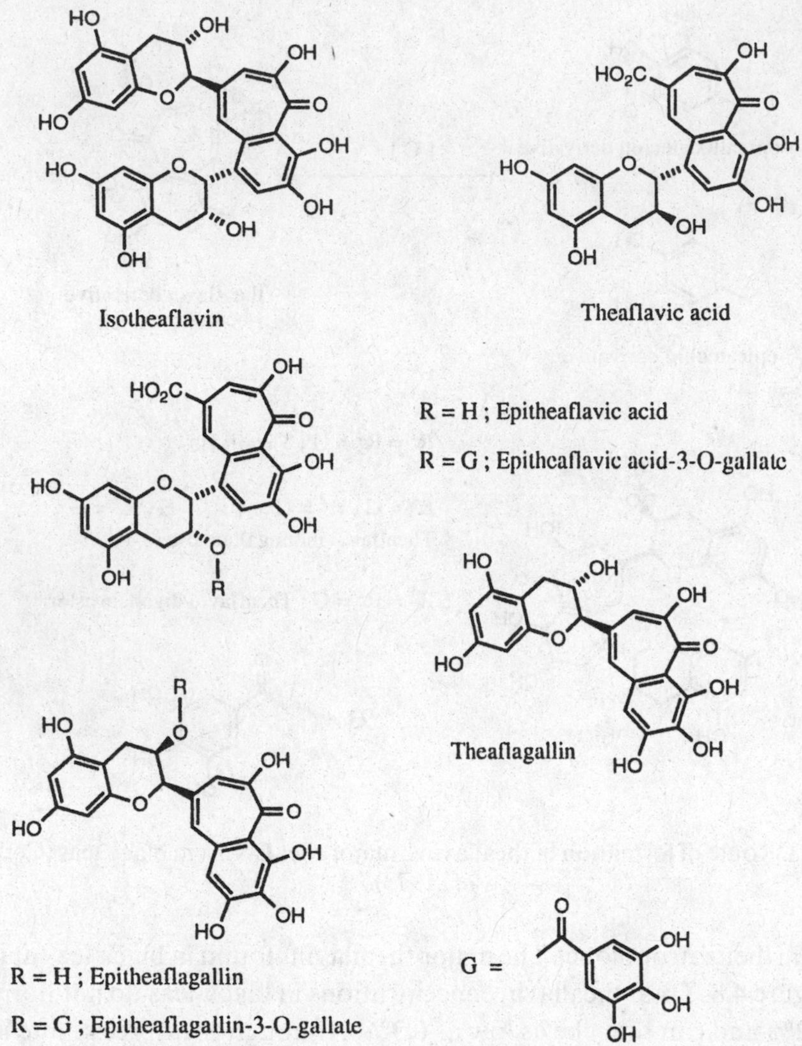

Isotheaflavin

Theaflavic acid

R = H ; Epitheaflavic acid

R = G ; Epitheaflavic acid-3-O-gallate

Theaflagallin

R = H ; Epitheaflagallin

R = G ; Epitheaflagallin-3-O-gallate

G =

Figure 4.9. Some minor phenolic benzotropolone derivatives obtained from black teas (Collier *et al.*, 1973; Nonaka *et al.*, 1986).

chol and pyrogallol derivatives suggests that black teas should also contain other phenolic benzotropolone derivatives in addition to the theaflavins. Several such compounds, derived from the oxidative transformation of phenolic flavan-3-ols and their gallate esters and gallic acid, have now been isolated (Collier *et al.*, 1973; Nonaka *et al.*, 1986) and are shown in Figure 4.9.

There are known to be considerable variations in the concentrations of theaflavins and related benzotropolone derivatives in black teas of different origins, although high theaflavin levels are generally correlated with teas of good quality. It is believed that the quantity and type of benzotropolone derivatives that accumulate during tea manufacture are determined primarily by initial phenolic flavan-3-ol composition in the fresh green leaf, rates of polyphenoloxidase catalysed fermentation and rates of theaflavin oxidative breakdown.

A feature of note, as yet unexplained, is the relative ease of oxidative transformation of (−)-epicatechin and (+)-catechin in model reactions designed to produce benztropolone derivatives; the former is a much more efficient substrate in these reactions yet it only differs in the absolute stereochemistry of the aliphatic hydroxyl group at C-3 in the heterocyclic ring from that of (+)-catechin – a functional group moreover which does not formally participate in the oxidation reactions. The answer probably lies in the earlier elegant and seminal work of Weinges and his collaborators (Weinges *et al.*, 1969, 1971; Van Soest, 1971) who showed that oxidation of (+)-catechin by various oxidative enzymes such as peroxidase, laccase and tyrosinase gave as the major products a polymer ($\sim$ 40%) and ($\sim$ 10% yield) the yellow crystalline dehydro-dicatechin A. Minor products (0.2–1.0% yield) were the dehydro-dicatechins B1, B2, B3 and B4. The dimer dehydro-dicatechin B4 was shown (Weinges and Huthwelker, 1971) to be a C-8 to C-6′ linked dimer, derived presumably by straightforward oxidative coupling between the A-ring of one (+)-catechin molecule and the B-ring of another molecule, Figure 4.10. The formation of the major product dehydro-dicatechin A can be very satisfactorily rationalised by assuming the further oxidation of the B-ring in this dimer and then trapping of the resultant *o*-quinone by the C-3 hydroxyl group, Figure 4.10. This reaction can only readily occur when the C-3 hydroxyl group is juxtaposed correctly to add 1,4 to the *o*-quinone; to do so it must have the absolute stereochemical configuration as in (+)-catechin. Thus when (+)-catechin acts as a substrate in putative reactions to give benzotropolone (theaflavin) derivatives this type of reaction may well partially divert, at the initial coupling of the two phenolic B-rings, intermediates formed by further oxidation by trapping in an analogous manner to that as envisaged in Figure 4.10. Interestingly Guyot and co-workers (Guyot *et al.*, 1996) have isolated various (+)-catechin dimers, including dehydro-dicatechin A, from the oxidation of (+)-catechin using grape polyphenoloxidase and they suggested that such reactions may be lead to changes in astringency, turbidity and colour stability of various fruit juices.

4.5.7 *Astringency of teas*

One of the criteria tea tasters use to describe the quality of teas is their astringent taste. For a great many years attempts have been made to find a

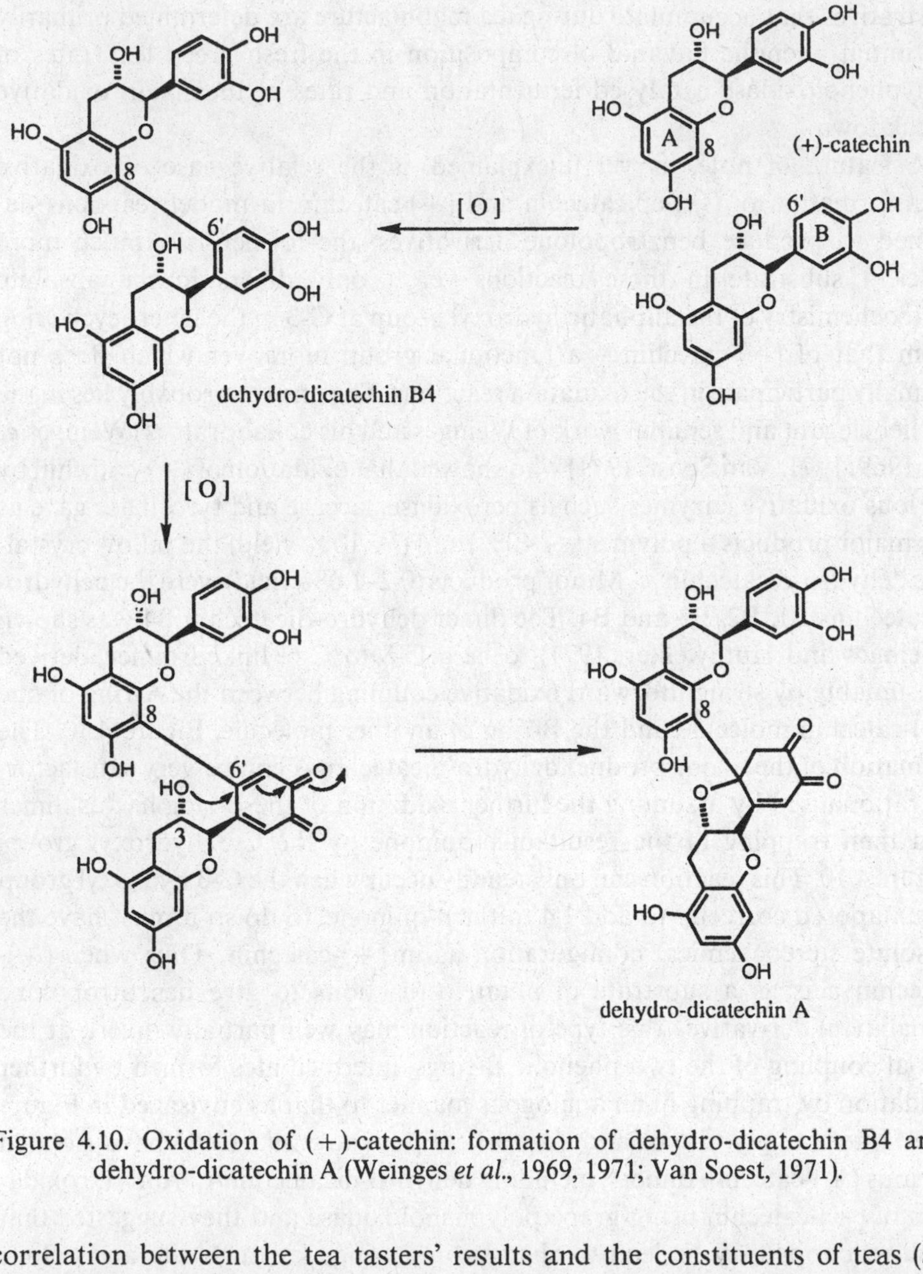

Figure 4.10. Oxidation of (+)-catechin: formation of dehydro-dicatechin B4 and dehydro-dicatechin A (Weinges *et al.*, 1969, 1971; Van Soest, 1971).

correlation between the tea tasters' results and the constituents of teas (as determined by chemical analysis), and their properties. However, with respect to the quality of astringency in teas there remains a difference of emphasis, if not of opinion, a divergence of expectation over observation. Thus for example it has been reported that as one progresses from a green, through oolong, to a black tea the astringency *decreases* as the level of oxidation *increases*. For black teas which are almost completely fermented the quality of astringency is rarely experienced. The explanation lies very probably in the balance between the concentrations of phenolic compounds in the tea brew and the threshold values required for their appreciation by the tea taster. Englehardt, Kuhr and Ding (1992) bring this effect out very clearly in a recent study on the influence of phenolic flavan-3-ols and the theaflavins on the astringent taste of black teas. They thus showed that the correlations between astringency (as recorded by a taste panel) and the phenolics present were generally very good for the phenolic flavan-3-ols, such as (−)-epigallocatechin-3-gallate and (−)-epi-catechin-3-gallate. No correspondingly clear correlation was, however, observed for the theaflavins. According to these workers although the threshold values to experience an astringent taste in the cases of (−)-epigallocatechin-3-gallate and (−)-epicatechin-3-gallate are relatively high their relative concentrations in the tea liquors are also correspondingly high. In the case of the various theaflavins their astringency thresholds are low but so also are their relative concentrations in the tea brew.

Such observations may initially appear to run contrary to the earlier suggestions of, amongst others, Roberts and Sanderson, to the general picture of the origins of astringency noted above, and to the general expectations based upon the comparisons of the chemical and physical properties, and in particular their propensities to bind to proteins, of the theaflavins and the phenolic flavan-3-ols such as (−)-epigallocatechin-3-gallate and (−)-epicatechin-3-gallate. Complexation of polyphenolic substrates with peptides and proteins is driven principally by *'hydrophobic effects'* – the innate compulsion of certain groups or regions in these molecules, because of their anarchic distaste for the surrounding aqueous medium, to congregate together and so decrease the surface area of the hydrophobic groups exposed to water. When the propensity of phenolic flavan-3-ols to bind to salivary proteins (PRPs) is compared to that of the theaflavins on a molar basis then the latter group of substrates are much more effective. Thus, whilst substrates such as (−)-epigallocatechin-3-gallate and (−)-epicatechin-3-gallate bind relatively weakly to such proteins, theaflavin digallate, Figure 4.8, is very effective in this respect and is comparable in its binding to the archetypal tannin/polyphenol β-1,2,3,4,6-penta-*O*-galloyl-D-glucose.

A pictorial representation of the two-stage recognition process ('hydrophobic effects' followed by hydrogen bonding) whereby polyphenols are thought to act at the molecular level to complex with salivary proline rich proteins (PRPs) and bring about the astringent response is depicted in Figure 4.11. An essential feature of this model is the cross-linking of adjacent polypeptide chains of the PRPs by polyphenol molecules (Bate-Smith, 1973a; Joslyn and Goldstein, 1964; Haslam *et al.*, 1994). This gives rise to the formation of aggregates and precipitation of the salivary PRPs and in turn this leads to the loss of lubrication in the palate and the sensation of astringency. The whole process is a time dependent and dynamic one and requires a period of time to develop fully. A necessary feature of this model is that the polyphenol should be of sufficient size and composition to bring about, by polydentate binding, the cross-linking process – such considerations indeed underpin the original definition of vegetable tannins by Bate-Smith and Swain. The theaflavins and their gallate esters fit this picture very well – they are strongly hydrophobic and have several distinct sites for complexation; hence the expectation that they would be astringent to the taste. However, if smaller molecules such as the phenolic flavan-3-ols and their gallate esters, such as (−)-epigallocatechin-3-gallate and (−)-epicatechin-3-gallate are (in high concentration) effective as astringents, as suggested by Englehardt and his colleagues, then this would indicate that the development of astringency may have rather more subtle origins than those suggested in Figure 4.11. For example, saturation binding by the phenolic substrate at specific sites (e.g. prolyl residues) on the PRPs may initiate critical changes in conformation and solubility giving rise to loss of lubrication and sensation of astringency. Clearly there are further important questions to be answered in relation to this phenomena and more detailed work is desirable.

4.5.8 Tea cream and tea creaming

As infusions of black tea cool (< 60°C) the originally clear liquor becomes turbid and a colloidal precipitate – commonly referred to in the tea trade as the *tea cream* – ultimately separates. In a freshly prepared liquor the process is usually reversible and the tea cream may be re-dissolved by heating once again. This 'creaming-down' of the tea liquors has long been exploited in the assessment of tea quality by tea tasters and is generally taken as an indication of qualities such as 'strength' and 'briskness' in the infusion. The tea cream is principally a series of complexes of the tea polyphenols – theaflavins, thearubigins and phenolic flavan-3-ols – with caffeine, although by the very nature of its formation other components of the original tea liquor – carbohydrates (often derived from pectin), proteins, other nitrogenous compounds and metal ions – may well be co-precipitated or occluded in the tea cream. Although the poorly characterised thearubigins are the principal pigmented constituents of tea creams, Roberts (1963) argued (based upon the original concentrations of theaflavins in the tea liquor) that the theaflavins are affected to a much greater extent by the creaming process, i.e. they form stronger complexes with the caffeine than do the thearubigins. Thus in typical examples as much as 62% of

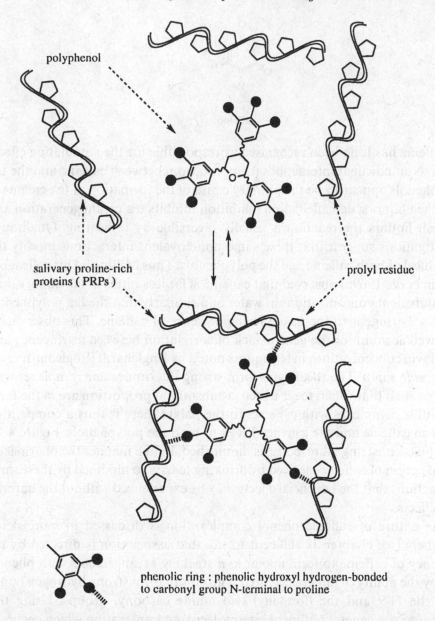

polyphenol

salivary proline-rich
proteins (PRPs)

prolyl residue

phenolic ring : phenolic hydroxyl hydrogen-bonded
to carbonyl group N-terminal to proline

Figure 4.11. Development of astringency. Pictorial representation of polyphenol complexation with salivary proline rich proteins (PRPs): cross-linking and aggregation.

the theaflavins, but only 28% of the thearubigins, in the tea liquor may be precipitated. Later work by Clifford and his group (Clifford *et al.*, 1992) has confirmed these observations and also presented evidence of a synergistic interaction between theaflavins and thearubigins during cream formation, *vide infra*.

caffeine

Caffeine has long been recognised as responsible for the stimulating effects of tea. Non-covalent interactions (complexation) between caffeine and the tea polyphenols appear to be the primary cause of the formation of tea creams in black tea liquors; de-caffeination inhibition inhibits tea cream generation and in fresh liquors the reaction is usually reversible by re-heating. Qualitative investigations suggest that these same non-covalent interactions modify the mouthfeel of both caffeine and the polyphenols. Thus Millin and his colleagues (Millin *et al.*, 1969a) observed that caffeine is far less bitter in tea liquors than the equivalent concentrations in water and similarly that the tea polyphenols are less astringent to the palate in the presence of caffeine. This observation may well account for the general lack of correlation between astringency and theaflavin concentrations in tea liquors noted by Englehardt (Englehardt *et al.*, 1992), *vide supra*. The theaflavins form strong intermolecular complexes with caffeine such that, when tea is tasted, a substantial proportion are in the form of soluble complexes with caffeine. In the palate there is thus a competition between caffeine and the salivary PRPs to bind the polyphenols, Figure 4.12, and thus the astringent response is diminished. In like manner the pharmacological action of caffeine, derived by drinking teas, is so modified by these same interactions that the beneficial effects may be experienced without the harmful side-effects.

The nature of caffeine:phenol complexation is discussed in more detail elsewhere (see chapter 3), suffice it to say that association is directed by the tendency of caffeine to form apolar π–π stacking arrangements with phenols and by the ability of phenolic hydroxyl groups to form strong hydrogen bonds with the N-9 and the (formally) two amide carbonyl groups. Using this behaviour the general nature of intermolecular complexation which occurs in tea creaming can be formulated as shown in Figure 4.13 – namely aggregates formed by extensive apolar π–π stacking arrangements and hydrogen bonds.

Frequently the affinity of a particular substrate, such as caffeine, for different polyphenols has been measured by its ability to precipitate them from solution. Indeed this is often the most convenient experimental technique to employ with substrates like proteins. However, the two phenomena of (i) chemical association and complexation and (ii) aggregation leading to precipi-

caffeine   PRPs

[theaflavins . caffeine] ⇌ [theaflavins] ⇌ [theaflavins . PRPs]

Figure 4.12. Complexation reactions in the palate: modification of the astringent response by partitioning of the tea polyphenols between caffeine and the salivary proline rich proteins (PRPs).

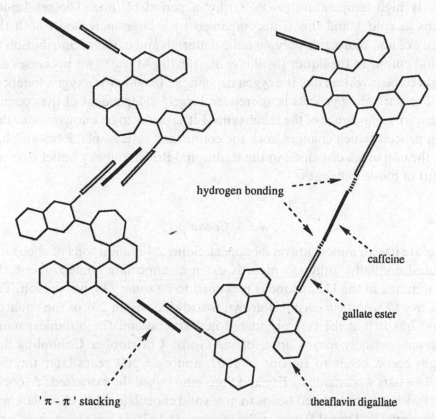

hydrogen bonding

caffeine

gallate ester

' π - π ' stacking   theaflavin digallate

Figure 4.13. Pictorial representation of modes of non-covalent interaction between tea polyphenols (theaflavin digallate) and caffeine; formation of complexes, cross-linking and aggregation to form precipitates (tea cream).

tation, are physically quite distinct and care should be taken in directly extrapolating results from such experiments. In the case of caffeine the extent of precipitation of different polyphenols from a mixture is broadly related to their association constants for the formation of soluble 1:1 complexes with caffeine (Haslam *et al.*, 1992). However, it is also quite clear that the extent of precipitation is also invariably dependent upon the overall composition of the polyphenol solution. Thus the presence of a very good polyphenolic precipitant of caffeine enhances the extent of precipitation of other polyphenols present; there is a synergism between the different polyphenolic components (Haslam, Lilley and Liao, unpublished data).

Finally from government ministers to housewives, aficionados of the relatively strong brew of English breakfast tea will recognise the derogatory description 'stewed' as applied to tea. This is a state of the black tea liquor frequently reached in railway station buffets when the liquor has been held at a relatively high temperature ($\sim 90°C$) for a period of time. The tea liquor darkens in colour and this is accompanied by a large increase in both the quantity of non-dialysable polyphenolic materials and in their contribution to the total colour of the liquor (Millin *et al.*, 1969b). At least two processes are thought to be involved; one is oxygen dependent, the other is oxygen independent. The darkening process is, moreover, largely independent of the concurrent loss from the liquor of the theaflavins. Little seems to be known about this ageing process which changes both the colour and taste of black tea and has made the cup of tea dispensed in the traditional British railway buffet so often the butt of music hall jokes.

4.5.9 Cocoa

Chocolate has an almost universal appeal. Some 2.4 million tons of cocoa are produced annually, although most is eaten as chocolate confectionery; the annual intake in the U.K. alone is reckoned to be some 7 kg per person. The cocoa tree (*Theobroma cacao*) is grown worldwide within 20° of the equator. Legend has it that the tree originated in Paradise and the botanical name *Theobroma* actually means 'food of the Gods'. Christopher Columbus first brought cocoa beans to Europe in 1502, and over 300 years later the first chocolate bars were made by Fry and Sons who mixed the extracted fat (cocoa butter) with ground roasted beans to give solid chocolate. Milk chocolate was later invented by Daniel Peters in Switzerland in 1876. It is interesting to note that in the early years of this century chocolate production in the U.K. was largely in the hands of family firms such as Fry, Cadbury and Rowntree, renowned for their philanthropic work and their devotion to the seriously disadvantaged in society.

The principal constituent of cocoa is cocoa butter, which accounts for over half of the dry weight of the bean. Acids such as citric, lactic, oxalic and succinic acids contribute to the sour (acidic) taste of cocoa and bitterness is generally attributed to the alkaloids theobromine and caffeine (the former is the predominant alkaloid) and to various diketopiperazines derived from the thermal degradation of proteins. Polyphenols in cocoa (mainly flavan-3-ols and procyanidins) are responsible for its astringency. Three processes are necessary to develop the unique flavour and aroma of cocoa. Once harvested the ripe pods are normally stored for several days before **fermentation** in which sucrose is degraded to its constituents, storage proteins are hydrolysed to oligopeptides and free amino acids, and the enzyme polyphenoloxidase converts some or all of the polyphenols present to quinones, which then react in a familiar, if little understood, manner with peptide and protein substrates (quinone tanning, see chapter 8). Following fermentation the beans are then *dried*. The action of the enzyme polyphenoloxidase continues well into this stage and the beans assume a brown colour as quinone tanning proceeds. As it does so the astringent character of the beans declines. Finally *roasting* converts the flavour precursors to compounds which contribute to the intense cocoa flavour normally associated with chocolate – well over 400 distinctive volatile constituents have thus far been detected. Almost all the reducing sugars and approximately half of the amino acid content of the bean is lost upon roasting (MacDonald, 1993).

theobromine

Very early important work on the polyphenolic constituents of the cocoa bean was carried out by Forsyth and his colleagues (Forsyth, 1953, 1955; Forsyth and Roberts, 1960). If the eventual structural conclusions of Forsyth were incorrect his contributions to this area were nevertheless seminal for his work was the very first of a number of investigations which showed that the characteristic so-called '*leucoanthocyanidin*' reactions of many plant tissues were due to the presence of flavan-3-ol dimers, trimers and higher oligomers (procyanidins) and that these same metabolites constituted up to 60% of the polyphenol content of the fresh cocoa bean. Haslam and his colleagues (Haslam *et al.*, 1972) established that the major lower molecular mass procyanidins of *Theobroma cacao* were (−)-epicatechin, the dimeric procyanidins B-2 and B-5 and the trimeric procyanidin C-1, Figure 4.14. Very minor components were shown to be (+)-catechin and procyanidin B-1. In rather later work Porter, Ma and Chan (1991) showed that fresh unfermented cocoa beans contain ~ 2% w/w (−)-epicatechin and almost exclusively (−)-epicatechin based procyanidin oligomers and polymers, largely concentrated in the flesh of the beans. They also identified, as additional minor components, two proanthocyanidin glycosides related to proanthocyanidin A-2.

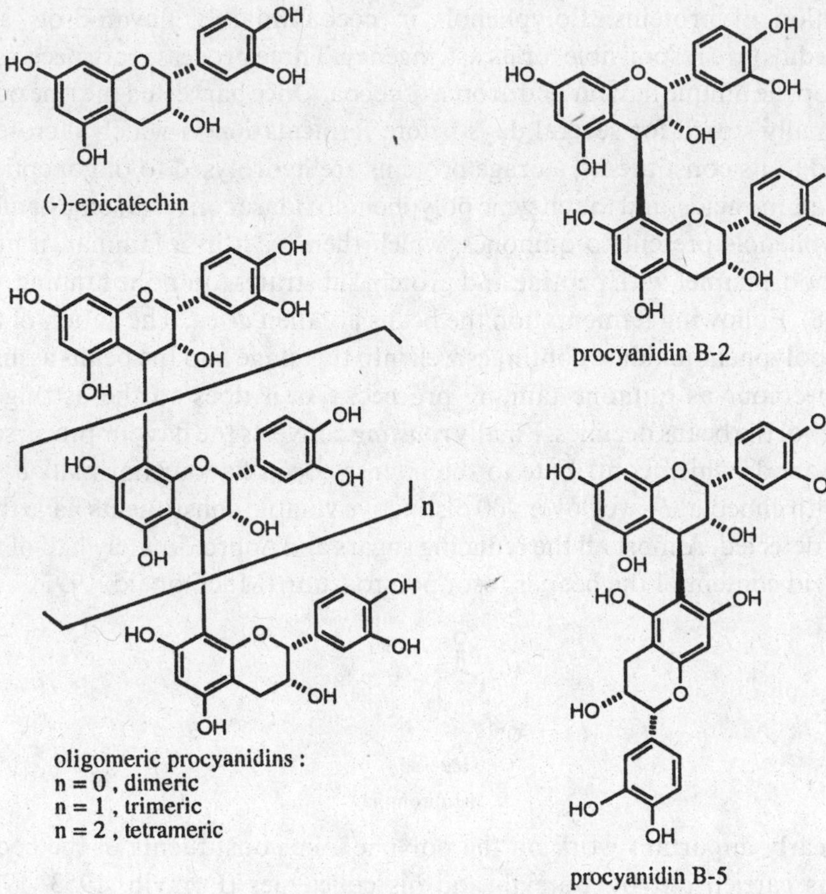

(-)-epicatechin

procyanidin B-2

oligomeric procyanidins :
n = 0 , dimeric
n = 1 , trimeric
n = 2 , tetrameric

procyanidin B-5

Figure 4.14. Major procyanidin oligomers of cocoa bean (*Theobroma cacao*) (Porter *et al.*, 1991; Haslam *et al.*, 1972).

Cacao bean fermentation is not only essential for the formation of chocolate flavour precursors but it also gives rise to the conditions in which polyphenols are converted to the largely insoluble red-brown materials, probably via processes of quinone tanning and which result in the characteristic colour of chocolate (Pettipher, 1986). Kim and Keeney (1984) demonstrated that there is a dramatic fall in soluble polyphenol (flavan-3-ols and procyanidins) content (~ 90%) during fermentation and these results were broadly corroborated by Porter and his colleagues (Porter *et al.*, 1991). Extensive studies of the precursors of chocolate flavour have led to the conclusion that amino acids, reducing sugars and flavonoids are the principal precursors of the primary aroma (Darsley and Quesnel, 1972).

Whilst chocolate is found highly desirable by many people, for others it can become an object of intense craving; most notably some women, around the period of menstruation, find they **must** have chocolate. For others there may be an allergic reaction and it can precipitate migraine attacks. Chocolate is a complex mixture of pharmacologically active nitrogenous materials (Cockroft, 1993) principally:

(i) methyl xanthines, of which theobromine ($\sim 1\%$ w/w) is the major constituent,
(ii) exorphin peptides (exogenous opioid–active peptides present in added cows milk solids in milk chocolate), and
(iii) several types of amines including tetrahydroisoquinolines and monoamine neuro-transmitters.

The major components are serotonin, phenylethylamine, theobromine, sal-solinol and several exorphins. Parethentically it is worth noting, in passing, that theobromine (like caffeine) forms complexes with polyphenolic substrates, although less strongly than caffeine itself. The exorphins analogously contain amino acid residues (proline, arginine, phenylalanine and tyrosine) which are known to promote the affinity of peptide substrates for polyphenols; likewise the aromatic biological amines serotonin, salsinol and phenylethylamine. It therefore seems very probable that the physiological effects which result from the presence of these compounds in chocolate may well be modified by the level of free polyphenolic compounds present in the finished product.

theobromine · phenylethylamine

serotonin salsinol

References

Bate-Smith, E. C. (1973a). Haemanalysis of tannins – the concept of relative astringency. *Phytochemistry*, **12**, 907–12.

Bate-Smith, E. C. (1973b). Tannins of the herbaceous Leguminoseae. *Phytochemistry*, **12**, 1809–12.

Beart, J. E., Haslam, E. and Lilley, T. H. (1985). Plant polyphenols – secondary metabolism and chemical defence: some observations. *Phytochemistry*, **24**, 33–8.

Beck, S. D. and Reese, J. D. (1976). Insect plant interactions: nutrition and metabolism. *Rec. Adv. Phytochem.*, **10**, 41–92.

Bennick, A. (1982). Salivary proline-rich proteins. *Molecular and Cellular Biochem.*, **45**, 83–99.

Bernays, E. A., Cooper-Driver, G. and Bilgener, M. (1989). Herbivores and plant tannins. *Adv. Ecol. Res.*, **19**, 263–302.

Bernays, E. A. and Graham, M. (1988). On the evolution of host specificity in phytophagous arthropods. *Ecology*, **69**, 886–92.

Brandon, M. J., Foo, L. Y., Porter, L. J. and Meredith, P. (1982). Proanthocyanidins of barley and sorghum: composition as a function of maturity of barley ears. *Phytochemistry*, **21**, 2953–7.

Bryant, J. P., Chapin, F. S. and Klein, D. R. (1983). Carbon/nutrient balance of boreal plants in relation to vertibrate herbivory. *Oikos*, **40**, 357–68.

Butler, L. G. (1989a). Effects of condensed tannin on animal nutrition. In *The Chemistry and Significance of Condensed Tannins*, editors R. W. Hemingway and J. J. Karchesy, Plenum Press: New York, pp. 391–402.

Butler, L. G. (1989b). New perspectives on the antinutritional effects of tannins. In *Food Proteins*, editors J. E. Kinsella and W. G. Soucie, American Oil Chemists Society: Champaign, Illinois, pp. 402–9.

Butler, L. G. (1992). Antinutritional effects of condensed and hydrolysable tannins. In *Plant Polyphenols: Synthesis, Properties and Significance*, editors R. W. Hemingway and P. E. Laks, Plenum Press: New York, pp. 693–8.

Butler, L. G. (1993). Polyphenols and herbivore diet selection. In *Polyphenolic Phenomena*, editor A. Scalbert, INRA Publications: Paris, pp. 149–54.

Cairns-Smith, A. G. (1985). *Seven Clues to the Origin of Life*. Cambridge University Press: Cambridge.

Clifford, M. N., Powell, C., Opie, S., Ford, M., Robertson, A. and Gibson, C. (1992). Tea cream formation: the contribution of black tea phenolic pigments determined by HPLC. *J. Sci. Food Agric.*, **63**, 77–86.

Cockroft, V. (1993). Chocolate on the brain. *The Biochemist*, **15**, 14–16.

Coley, P. (1986). Costs and benefits of defense by tannins in a neotropical tree. *Oecologia*, **70**, 238–41.

Coley, P., Bryant, J. P. and Chapin, F. S. (1985). Resource availability and plant antiherbivore defense. *Science*, **230**, 895–9.

Collier, P. D., Bryce, T., Mallows, R., Thomas, P. E., Frost, D. J., Korver, O. and Wilkins, C. K. (1973). The theaflavins of black tea. *Tetrahedron*, **29**, 125–42.

Czochanska, Z., Foo, Y. L. and Porter, L. J. (1979). Compositional changes in low molecular weight flavans during grape maturation. *Phytochemistry*, **18**, 1819–22.

Dadic, M. and Bellau, G. (1973). Polyphenols and beer flavour. *Proc. Amer. Soc. Brewing Chem.*, **33**, 158–65.

Dadic, M., Van Ghelawe, G. E. A. and Valyi, Z. (1984). Barley and malt tanninogens and beer quality. *J. Inst. Brewing*, **90**, 558–64.

Darsley, R. R. and Quesnel, V. C. (1972). The production of aldehydes from amino acids in roasted cocoa. *J. Sci. Food Agric.*, **23**, 215–25.

Delcour, J. A. (1989). Structure elucidation of proanthocyanidins: direct synthesis and isolation from Pilsner beers. *Pharm. Tidjs. voor Belgie*, **66**, 7–20.

Delcour, J. A., Vandenberghe, M. M., Corten, P. F. and Dondeyne, P. (1984a). Flavour thresholds of polyphenolics in water. *Amer. J. Enol. Viticulture*, **35**, 134–6.

Delcour, J. A., Vandenberghe, M. M., Dondyne, P., Schrevens, E. L., Wijnhoven, J.

and Moerman, E. (1984b). Flavour and haze stability differences in unhopped and hopped all malt Pilsner beers brewed with proanthocyanidin-free and with regular malt. *J. Inst. Brewing*, **90**, 67–72.

Delcour, J. A., Schoeters, M. M., Meysman, E. W., Dondeyne, P., Schrevens, E. L., Wijnhoven, J. and Moerman, E. (1985a). Flavour and haze stability differences due to hop tannins in all-malt Pilsner beers brewed with proanthocyanidin-free malt. *J. Inst. Brewing*, **91**, 88–92.

Delcour, J. A., Schoeters, M. M., Dondeyne, P., Schrevens, E. L., Wijnhoven, J. and Moerman, E. (1985b). Flavour and haze stability differences in unhopped and hopped all malt Pilsner beers brewed with proanthocyanidin-free and with regular malt. *J. Inst. Brewing*, **91**, 302–5.

Delcour, J. A. and Janssens de Varebeke, D. (1985). A new colourimetric assay for flavonoids in Pilsner beers. *J. Inst. Brewing*, **91**, 37–40.

Eastmond, R. (1974). The separation and identification of a dimer of catechin occurring in beer. *J. Inst. Brewing*, **80**, 188–92.

Englehardt, U. H., Kuhr, S. and Ding, Z. (1992). Influence of catechins and theaflavins on the astringent taste of black tea brews. *Z. Lebensm. Unters. Forsch.*, **195**, 108–11.

Erdal, K., Ahrenst-Larsen, B. and Jende-Strid, B. (1980). Use of proanthocyanidin-free barley in brewing. In *Cereals for Foods and Beverages. Recent Progress in Cereal Chemistry and Technology*. Academic Press: New York, pp. 365–79.

Feeny, P. P. (1968). Effects of oak leaf tannins on larval growth of the winter moth *Operophthera brumata*. *J. Insect Physiol.*, **14**, 805–17.

Feeny, P. P. (1970). Seasonal changes in oak leaf tannins and nutrients as a cause of spring feeding by winter moth caterpillars. *Ecology*, **51**, 565–81.

Feeny, P. (1976). Plant apparency in chemical defence. *Rec. Adv. Phytochem.*, **10**, 1–40.

Forsyth, W. G. C. (1953). Leuco-cyanidin and epicatechin. *Nature*, **172**, 726.

Forsyth, W. G. C. (1955). Cacao phenolic substances. III. Separation and estimation on chromatograms. *Biochem. J.*, **60**, 108–11.

Forsyth, W. G. C. and Roberts, J. B. (1960). Cacao phenolic substances. The structure of cacao leucocyanidin. *Biochem. J.*, **74**, 374–8.

Fraenkel, (1959). The *raison d'etre* of secondary plant substances. *Science*, **129**, 1466–70.

Guyot, S., Vercauteren, J. and Cheynier, V. (1996). Structural determination of colourless and yellow dimers resulting from (+)-catechin coupling catalysed by grape polyphenoloxidase. *Phytochemistry*, **42**, 1279–88.

Harborne, J. B. (1977). *Introduction to Ecological Biochemistry*. Academic Press: London and New York.

Haslam, E. (1989). *Plant Polyphenols: Vegetable Tannins Revisited*, Cambridge University Press: Cambridge, p. 230.

Haslam, E. (1994). Plant polyphenols – a case of biochemical co-evolution? In *Natural Phenols in Plant Resistance. Acta Horticulturae*, **381**, 722–37.

Haslam, E., Thompson, R. S., Jacques, D. and Tanner, R. J. N. (1972). Plant proanthocyanidins. Part 1. Introduction; the isolation, structure and distribution in nature of plant procyanidins. *J. Chem. Soc. (Perkin Trans. I)*, p. 1387–00.

Haslam, E. and Lilley, T. H. (1988). Natural astringency of foodstuffs – a molecular interpretation. *CRC Rev. Food Sci. and Nutrition*, **27**, 1–40.

Haslam, E., Lilley, T. H., Warminski, E., Liao, H., Cai, Y., Martin, R., Gaffney, S. H., Goulding, P. N. and Luck, G. (1992). Polyphenol complexation – a study in molecular recognition. In *Phenolic Compounds in Food and Their Effects on Health. I. Analysis, Occurrence and Chemistry*, A.C.S. Symposium Series, **506**, eds. C.-T. Ho, C. Y. Lee and M.-T. Huang, American Chem. Soc.: Washington D.C., pp. 8–50.

Haslam, E., Luck, G., Liao, H., Murray, N. J., Grimmer, H. R., Warminski, E. E., Williamson, M. P. and Lilley, T. H. (1994). Polyphenols, astringency and proline-rich proteins. *Phytochemistry*, **37**, 357–71.

Horner, L., Durckheimer, W. and Weber, K.-H. (1961). Zur Kenntnis der o-Chinone. XIX. Hydrolysestudien an 2-substitutierten 1.3 Dicarbonylverbindungen als Beitrag zum Mechanismus der Purpurogallinbildung. *Chem. Ber.*, **94**, 2881–7.

Horner, L., Durckheimer, W., Weber, K.-H. and Dolling, K. (1964). Zur Kenntnis der o-Chinone. XXIV. Synthese, Struktur und Eigenschaften von 1′,2′-Dihydroxy-6,7-Benztropolonen. *Chem. Ber.*, **97**, 312–24.

Janzen, D. H. (1969). Co-evolution. *Science*, **165**, 415.

Jones, W. T. and Mangan, J. L. (1977). Complexes of the condensed tannins of sainfoin (*Onobrychis viciifolia*) with fraction I leaf protein and with submaxillary mucoprotein and their reversal by polyethylene glycol and pH. *J. Sci. Food Agric.*, **28**, 126–36.

Joslyn, M. A. and Goldstein, J. L. (1964). Astringency of fruits and fruit products in relation to phenolic content. *Adv. Food Res.*, **13**, 179–217.

Kenward, R. E. and Holm, J. L. (1993). On the replacement of the red squirrel in Britain: a phytotoxic explanation. *Proc. R. Soc. Lond.*, **251B**, 187–94.

Kim, H. and Keeney, P. G. (1984). (−)-Epicatechin content in fermented and unfermented cocoa beans. *J. Food Sci.*, **49**, 1090–4.

Kretchner, K.-F. (1975). Pilsener Biertypus in aller Welt. *Brauwelt*, **115**, 1049–57.

Lea, A. G. H. (1992). Flavour, colour and stability in fruit products: the effects of polyphenols. In *Plant Polyphenols: Synthesis, Properties and Significance*, eds. R. W. Hemingway and P. E. Laks, Plenum Press, New York, pp. 827–47.

Lea, A. G. H. and Arnold, G. M. (1978). The phenolics of ciders: bitterness and astringency. *J. Sci. Food Agric.*, **29**, 478–83.

Lees, G. L. (1992). Condensed tannins in some forage legumes: their role in the prevention of ruminant pasture bloat. In *Plant Polyphenols: Synthesis, Properties and Significance*, editors R. W. Hemingway and P. E. Laks, Plenum Press: New York, pp. 915–34.

Lunder, T. V. (1988). *Tea*. Nestlé, Nestec L[td.] Technical Assistance, p. 11.

Macdonald, H. (1993). Flavour development from cocoa bean to chocolate bar. *The Biochemist*, **15**, 3–5.

Macheix, J. J., Fleuriet, A. and Billot, J. (1990). *Fruit Phenolics*. CRC Press: Boca Raton, U.S.A.

Mangan, J. L., Vetter, R. L., Jordan, D. J. and Wright, P. C. (1976). The effect of condensed tannins of sainfoin (*Onobrychis viciifolia*) on the release of soluble leaf protein into the food bolus of cattle. *Proc. Nutr. Sci.*, **35**, 95A.

Martin, M. M. and Martin, J. S. (1984). Surfactants: their role in preventing the precipitation of proteins by tannins in insect guts. *Oecologia*, **54**, 205–11.

Martin, M. M., Martin, J. S. and Rockholm, D. C. (1985). Effects of surfactants, pH and certain cations on the precipitation of protein by tannins. *J. Chem. Ecology*, **11**, 484–94.

Martin, M. M., Martin, J. S. and Bernays, E. A. (1987). Failure of tannic acid to inhibit digestion or reduce digestibility of plant protein in gut fluids of insect herbivores: implications for theories of plant defense. *J. Chem. Ecology*, **13**, 605–21.

McArthur, C., Sanson, G. D. and Beal, A. M. (1995). Salivary proline-rich proteins in mammals: roles in homeostasis and counteracting dietary tannin. *J. Chem. Ecology*, **21**, 663–91.

Mehansho, H., Butler, L. G. and Carlson, D. M. (1987). Dietary tannins and salivary proline rich proteins: interactions, induction and defense. *Ann. Rev. Nutrition*, **7**, 423–40.

Millin, D. J., Swaine, D. and Dix, P. L. (1969a). Separation and classification of the brown pigments of aqueous infusions of black tea. *J. Sci. Food Agric.*, **20**, 296–302.

Millin, D. J., Swaine, D. and Sinclair, D. S. (1969b). Some effects of ageing on pigments of tea extracts. *J. Sci. Food Agric.*, **20**, 296–302.

Mole, S. and Waterman, P. G. (1985). Stimulatory effects of tannins and cholic acid on tryptic hydrolysis of proteins. *J. Chem. Ecol.*, **11**, 1323–32.

Mole, S. and Waterman, P. G. (1987a). Tannic acid and proteolytic enzymes: enzyme inhibition or substrate deprivation. *Phytochemistry*, **26**, 99–102.

Mole, S. and Waterman, P. G. (1987b). A critical analysis of techniques for measuring tannins in ecological studies. I. Techniques for chemical defining tannins. *Oecoligica*, **72**, 137–47.

Mole, S. and Waterman, P. G. (1987c). A critical analysis of techniques for measuring tannins in ecological studies. II. Techniques for biochemically defining tannins. *Oecologica*, **72**, 148–56.

Mole, S., Waterman, P. G., Hagerman, A. E. and Butler, L. G. (1989). Ecological tannin assays: a critique. *Oecologica*, **78**, 93–6.

Morton, J. F. (1992). Astringent masticatories. In *Plant Polyphenols: Synthesis, Properties and Significance*, editors R. W. Hemingway and P. E. Laks, Plenum Press: New York, pp. 739–65.

Muller, C. H. (1969). Co-evolution. *Science*, **165**, 415–16.

Noble, A. C. (1990). Bitterness and astringency in wines. In *Bitterness in Foods and Beverages*, editor R. Rouseff, Amsterdam: Elsevier, pp. 145–58.

Nonaka, G.-I., Kawahara, O. and Nishioka, I. (1983). Tannins and related compounds. XV. A new class of dimeric flavan-3-ol gallates, theasinensins A and B, and proanthocyanidin gallates from green tea leaf. *Chem. Pharm. Bull.*, **31**, 3906–14.

Nonaka, G.-I., Hashimoto, F. and Nishioka, I. (1986). Tannins and related compounds. XXXVL. Isolation and structure of theaflagallins, new red pigments from black tea. *Chem. Pharm. Bull.*, **34**, 61–5.

Ollis, W. D., Brown, A. G., Haslam, E., Falshaw, C. P. and Holmes, A. (1966). The constitution of theaflavin. *Tetrahedron Letters*, pp. 1193–204.

Outtrop, H. (1992). Proanthocyanidins, the brewing process and the quality of beer. In *Plant Polyphenols: Synthesis, Properties and Significance*, eds. R. W. Hemingway and P. E. Laks, Plenum Press: New York, pp. 849–58.

Pettipher, G. L. (1986). An improved method for the extraction and quantitation of anthocyanins in cacao beans and its use as an index of fermentation. *J. Sci. Food Agric.*, **37**, 289–96.

Pierpoint, W. S. (1986). Flavonoids in the human diet. In *Plant Flavonoids in Biology and Medicine: Biochemical, Pharmacological, and Structure – Activity*

Relationships, eds V. Cody, E. Middleton and J. B. Harborne. A. R. Liss: New York, pp. 125–40.

Porter, L. J., Ma, Z. and Chan, B. G. (1991). Cacao procyanidins: major flavonoids and identification of some minor metabolites. *Phytochemistry*, **30**, 1657–63.

Ramey, D., Bertrand, A., Ough, C. S., Singleton, V. L. and Sanders, E. (1986). Effects of skin contact temperature on Chardonnay must and wine composition. *Amer. J. Enol. Vitic.*, **37**, 99–106.

Rhoades, D. F. and Cates, R. G. (1976). Toward a general theory of plant herbivore chemistry. *Rec. Adv. Phytochem.*, **10**, 168–213.

Roberts, E. A. H. (1962). Economic importance of flavonoid substances: tea fermentation. In *The Chemistry of Flavonoid Compounds*, editor T. A. Geissman, Pergamon Press: Oxford, London and New York, pp. 468–512.

Roberts, E. A. H. (1963). The phenolic substances of manufactured tea. X. The creaming of tea liquors. *J. Sci. Food Agric.*, **14**, 700–5.

Robinson, J. (editor) (1994). *The Oxford Companion to Wine*, Oxford University Press: Oxford and New York.

Robinson, J. (1995). *Jancis Robinson's Wine Course*, BBC Publications: London, pp. 16–17.

Sanderson, G. W. (1972). The chemistry of tea and tea manufacturing. In *Recent Advances in Phytochemistry*, volume **5**, editors V. C. Runeckles and T. S. Tso, Academic Press: London and New York, pp. 247–316.

Singleton, V. L. (1981). Naturally occurring food toxicants: phenolic substances of plant origin common in foods. *Adv. Food Res.*, **27**, 149–242.

Singleton, V. L. (1992). Tannins and the qualities of wines. In *Plant Polyphenols*: *Synthesis, Properties and Significance*, editors R. W. Hemingway and P. E. Laks, Plenum Press: New York, pp. 859–80.

Singleton, V. L. and Esau, P. (1969). *Phenolic Substances in Grapes and Wine, and Their Significance*. Academic Press: New York.

Singleton, V. L., Timberlake, C. F., Bridle, P. and Lea, A. G. H. (1978). The phenolic constituents of white grapes and wines. *Amer. J. Enol. Vitic.*, **30**, 289–300.

Smallwood, P. D. and Peters, W. D. (1986). Grey squirrel food preferences: the effects of tannin and fat concentration. *Ecology*, **67**, 168–74.

Somers, T. C. and Verette, E. (1988). Phenolic composition of natural wine types. In *Modern Methods of Plant Analysis*, New Series Volume **6**, *Wine Analysis*, eds H. F. Linskens and J. F. Jackson, Springer Verlag: Berlin, Heidelberg, New York, London, pp. 219–57.

Swain, T. (1977). Secondary compounds as protective agents. *Ann. Rev. Plant Physiol.*, **28**, 479–501.

Swain, T. (1978), Plant–animal co-evolution; a synoptic view of the Paleozoic and Mesozoic. In *Biochemical Aspects of Plant and Animal Co-evolution*, ed. J. B. Harborne, Academic Press: London and New York, pp. 3–19.

Takino, Y., Ferretti, A., Flanagan, V., Gianturco, M. and Vogel, M. (1965). Structure of theaflavin, a polyphenol of black tea. *Tetrahedron Letters*, pp. 4019–25.

Takino, Y., Hiroo, S., Shigehiko, M. and Fukunaga, C. (1971). *Nogei Kagaku Kaishi*, **45**, 176–83.

Thompson, J. N. (1988). Coevolution and alternative hypotheses on insect/plant interactions. *Ecology*, **69**, 893–5.

Van Soest, T. C. (1971). Aufklarung der Molecularstruktur des Dehydrodicatechins A durch Rontgenstrukturanalyse seines Bromoheptamethyl Ethers. *Liebig's Annalen Chemie*, **754**, 137–8.

Waterman, P. G. and Mole, S. (1994). *Analysis of Phenolic Plant Metabolites*. Blackwell Scientific Publications: London, Edinburgh and Bonn, p. 238.

Weinges, K., Ebert, W., Huthwelker, D., Mattauch, H. and Perner, J. (1969). Oxidative Kuplung von Phenolen II. Konstitution und Bildungsmechanismus des Dehydro-dicatechins A. *Liebig's Annalen Chemie*, **726**, 114–24.

Weinges, K. and Huthwelker, D. (1969). Oxidative Kuplung von Phenolen III. Isolierung und Konstitutionsbeweiss eines 8,6'-Verknupften Dehydro-dicatechins B4. *Liebig's Annalen Chemie*, **731**, 161–70.

Weinges, K. and Piretti, M. V. (1971). Isolierung des $C_{30}H_{26}O_{12}$-Procyanidins B-1 aus Weintrauben. *Liebig's Annalen Chem.*, **748**, 218–20.

Weinges, K., Mattauch, H., Wilkins, C. and Frost, D. (1971). Oxidative Kuplung von Phenolen V. Spectroscopische und Chemische Konstitutionsaufklarung des Dehydro-dicatechins A. *Liebig's Annalen Chemie*, **754**, 124–36.

Whittaker, R. H. (1970). In *Chemical Ecology*, eds. E. Sondheimer and J. B. Simeone, Academic Press: London and New York.

Williams, A. H. (1960). The distribution of phenolic compounds in apple and pear trees. In *Phenolics in Plants in Health and Disease*, editor J. B. Pridham, Pergamon Press: Oxford, pp. 3–7.

Williams, A. H., Lea, A. G. H. and Timberlake, C. F. (1977). Measurement of flavour quality in apples, apple juices and fermented ciders. In *Flavour Quality: Objective Measurement*, editor R. A. Scanlan, A.C.S. Symposium Series No. **51**, American Chemical Society: Washington, U.S.A., pp. 71–88.

5

Maturation – changes in astringency

5.1 Introduction

The word astringent is derived from the Latin *ad* (to) *stringere* (bind); thus astringency is properly defined as a binding reaction. Indeed, astringents in medicine and pharmacology are recognised as substances that bind to and precipitate proteins. They are used, for example, to control haemorrhage and diarrhoea and to inhibit mucous secretions. In this context it is therefore of particular interest to note that many Japanese and Chinese folk medicines frequently employ, as antidiarrhoeic and haemostatic agents, plants rich in polyphenolics (see chapter 7; Haslam and Lilley, 1988; Haslam, 1996).

The sensation of astringency and its origins have been delineated earlier (see chapter 4). Polyphenols (tannins) constitute one of the principal groups of naturally occurring astringent principles; they have a harsh astringent taste and produce in the palate those responses noted earlier – namely feelings of dryness, constriction and roughness. The means whereby these effects are produced at the molecular level are by the highly selective cross-linking by polyphenolic substrates of the proline rich proteins characteristic of saliva. Details of the general processes thought to be involved have been outlined (Haslam *et al.*, 1994) and illustrated earlier in this text (Figure 4.11). Equally important from a fundamental point of view, and because of its practical implications, is an understanding of the various ways in which the astringent response may be modified and ultimately lost, e.g. the loss of astringency in fruits upon ripening, the loss of astringency in wines and other beverages on storage and maturation, and the development of fugal astringency in beverages.

5.2 Ripening of fruit

Loss of astringency is one of the major changes which occurs during the ripening of fruit. Although it is generally agreed that the property of astringency devolves upon the presence of polyphenols (tannins) in the fruit (*vide supra*), according to Goldstein and Swain (1963) some astringent fruits show a decrease in tannins on ripening, others do not. Goldstein and Swain argued that the capacity of polyphenols to form strong cross-links with the salivary proteins and glycoproteins in the palate depends to a large extent on their molecular size and dimensions; low molecular weight phenolic compounds they suggested were too small to effectively cross-link different proteins (parenthetically it is now clear that high concentrations of relatively low molecular weight phenols, such as (−)-epigallocatechin-3-*O*-gallate, *may* give rise to an astringent response (cf. teas, *vide supra*)); highly polymerised polyphenolic molecules are either not sufficiently soluble or are too large, in terms of their molecular size, to fit readily in the salivary protein matrix. Maximum astringency it was argued is therefore most probably shown by polyphenols of an intermediate molecular size (hence the original definition of vegetable tannins by Bate-Smith and Swain). According to Goldstein and Swain in ripening fruits it may be expected therefore that changes in astringency are a reflection of changes in the molecular size of the polyphenols (tannins) present, with a decline in those polyphenols most effective in producing an astringent response in the palate. Examination of several fruits gave support to this view.

It is well known, following the work of Robinson and Robinson (1935), and Swain and Hillis (1959) that the condensed proanthocyanidins (leuco-anthocyanins) of plant tissues can be broadly divided into three groups depending upon their respective solubilities in organic and aqueous/organic solvents. Goldstein and Swain examined the changes in the condensed proanthocyanidins (leuco-anthocyanins) during the ripening of several fruit – banana, persimmon, peach and plum. In each case a diminution, on ripening, of those condensed proanthocyanidins (leuco-anthocyanins) which are extracted by methanol was observed. This was interpreted by the authors as showing that, during maturation and ripening, there is a further polymerisation of the condensed proanthocyanidins (leuco-anthocyanins) present in the fruit rendering them less easily extracted by methanol. Goldstein and Swain concluded that in these cases the loss of astringency which occurs upon ripening is most probably associated with the increased oligomerisation of the condensed proanthocyanidins which not only affects their solubility and extractability but also their astringency. Until comparatively recently there has been little hard scientific evidence to support this eminently plausible theory put forward by Goldstein and Swain. However, Japanese workers examining the loss of astringency in persimmon (*Diospyros kaki*) have now obtained firm scientific evidence to show how this oligomerisation proceeds concomitant upon the processes designed to externally induce de-astringency of this fruit; equally importantly their observations also provide a rational basis, a paradigm, to explain the natural loss of astringency from persimmon and probably other fruits and also the changes which occur upon the maturation of various beverages, particularly red wines.

5.2.1 Persimmon

The immature fruit of persimmon (*Diospyros kaki*) has remarkable astringent properties; some fruit such as 'Hiratanenashi' and 'Yokono' retain this astringency through to the mature stage such that the fruit are only edible after drying or various treatments to attenuate the astringency (Matsuo and Ito, 1982). The strong binding capacity of kaki-tannin with proteins and carbohydrates has led to several practical applications in Japan. Umbrellas used to be made from paper soaked in crude kaki-tannin and similar treatments were employed in the manufacture of fishing nets. Kaki-tannin has also been widely used to remove protein in the brewing of sake, Japanese rice wine. Various studies of the chemical structure of kaki-tannin have been made (Matsuo and Ito, 1978), most recently by Tanaka and his colleagues (1994). Matsuo and Ito (1978) showed that the tannins of persimmon fruit are polymeric proanthocyanidins with a large relative molecular mass ($\sim 1.12 \times 10^4$ Da on average) which give both cyanidin and delphinidin upon acid hydrolysis. Tanaka and his colleagues (1994) carried out thiol (2-sulphanyl-ethanol – HS–CH_2–CH_2–OH) promoted degradation of the aqueous acetone extract of the soluble proanthocyanidin polymers from unripe persimmon fruit. These showed that the principal '*extension*' units of the polymers were (−)-epicatechin, (−)-epicatechin-3-*O*-gallate, (−)-epigallocatechin and (−)-epigallocatechin-3-*O*-gallate, in the approximate ratio of 4:1:22:6, Figure 5.1. In these experiments **no** polymer flavan-3-ol '*terminal*' units could be detected. The authors suggested that this indicated that either the relative molecular mass of persimmon kaki-tannin is too large to permit the detection of the lower '*terminal*' units, or, that these units are quite different from that of the typical proanthocyanidin polymer. This latter suggestion is of some interest in the light of the very early comments of Robinson and Robinson (1935) and the proposal of Haslam and Lilley (1988) that such difficultly soluble proanthocyanidins were in fact anchored to an insoluble carbohydrate matrix within the plant tissues, (see Figure 1.14).

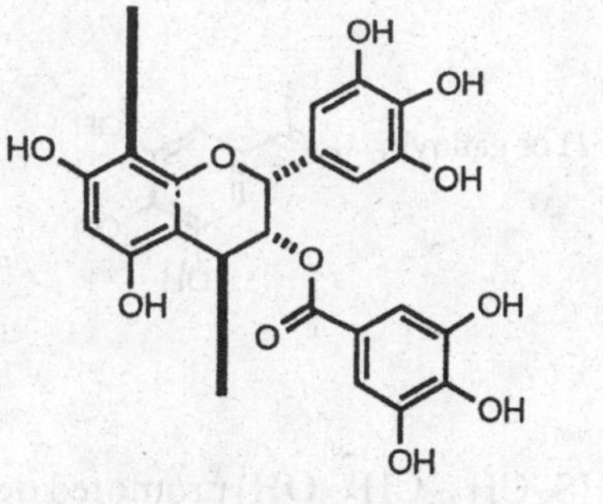

(-)-epigallocatechin-3-O-gallate [6]

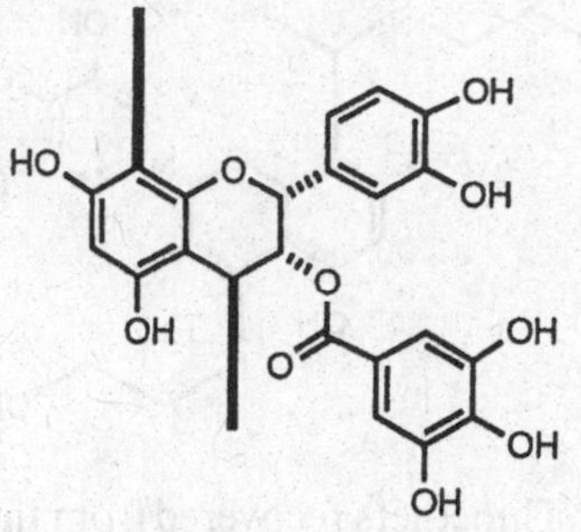

(-)-epicatechin-3-O-gallate [1]

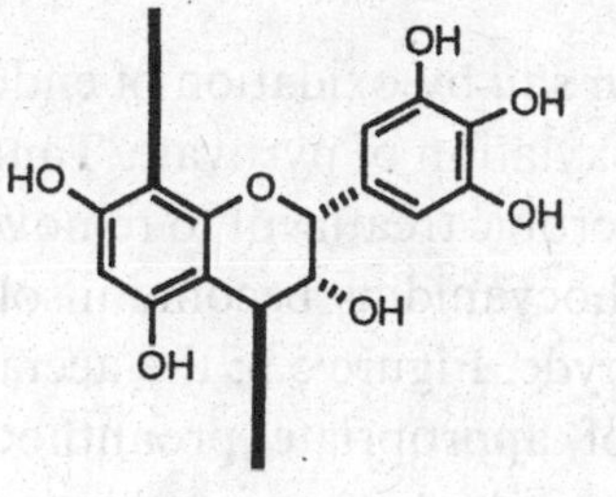

(-)-epigallocatechin [22]

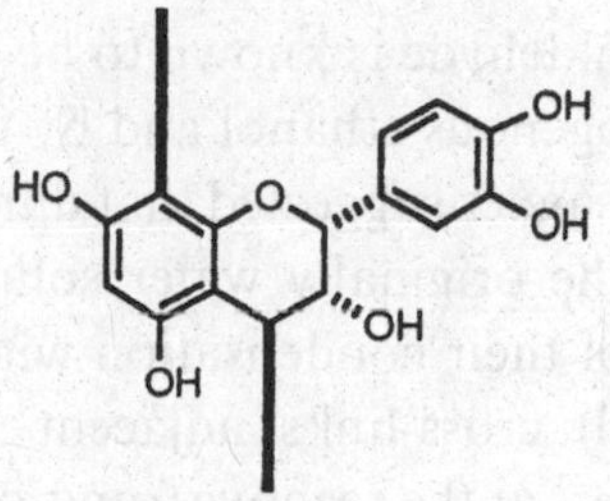

(-)-epicatechin [4]

Figure 5.1. '*Extension*' units of the proanthocyanidin polymers derived from Persimmon fruit (*Diospyros kaki*); numbers in parentheses indicate approximate ratios (Tanaka *et al.*, 1994).

Japanese persimmon cultivars are usually classified into two groups – the astringent and the non-astringent types, depending on the perception of astringency at the mature stage of the fruit. The fruit of the astringent cultivars are usually edible but only after the application of procedures (treatment with alcohol vapour, carbon dioxide or warm water) designed to reduce the astringency. During these anaerobic treatments, acetaldehyde is known to accumulate in the flesh of the fruit and concomitantly the partially water soluble kaki-tannins, responsible for the astringent taste, are steadily transformed into insoluble forms which give a decreased astringent response. Tanaka *et al.* (1994) (after complete removal (ethanol) of the soluble proanthocyanidins), applied the thiol-promoted degradation of proanthocyanidins directly to the insolubilised tannins in the fleshy debris from persimmon fruit which had been subject to external procedures known to decrease their astringency. They recovered thiols typical of those obtained from the original polymer and in addition small amounts of thioethers (Figure 5.2) whose structures suggested the involvement of acetaldehyde in the proanthocyanidin insolubilisation process.

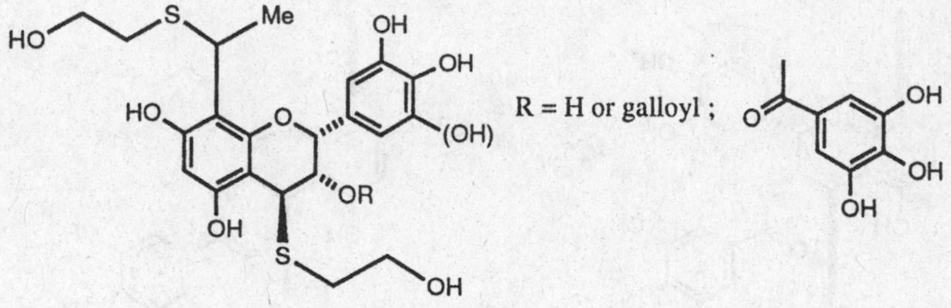

Figure 5.2. Thioethers recovered from the thiol (HS–CH$_2$–CH$_2$–OH) promoted degradation of insolubilised proanthocyanidin polymers from non-astringent persimmon fruit (Tanaka *et al.*, 1994).

Acetaldehyde is known to be generated *in situ* by oxidation of endogenous and exogenous ethanol and by the decarboxylation of pyruvate. Tanaka and his colleagues suggested that during the anaerobic treatment to remove astringency the originally water soluble proanthocyanidins become insoluble by virtue of their condensation with acetaldehyde, Figure 5.3; the acetaldehyde (ethanal) cross-links adjacent 'A' rings of appropriate proanthocyanidin oligomers at the reactive 6 and/or 8 positions. The reaction is analogous to the generation of the familiar phenol–formaldehyde resins and to the initial reaction in the colorimetric determination of catechin derivatives with vanillin. Loss of astringency by treatment of fruit with warm water or by drying, and the natural loss of astringency in such fruit in early winter (contingent with softening) were all shown to be based upon similar origins – the result of the cross-linking of proanthocyanidin oligomers by condensation with acetaldehyde. These experimental observations thus confirm the original ideas of Goldstein and Swain (1963) and the proposal of Matsuo and Ito (1982) who showed that *in vitro* the water soluble proanthocyanidins of persimmon form a gel upon treatment with acetaldehyde.

The non-astringent types of persimmon (*vide supra*), which are astringent when immature, lose their astringency naturally during the course of maturation and ripening. Thiol promoted degradation of the proanthocyanidin polymers from these fruit showed a similar profile of 'extension' units to those present in astringent cultivars, Figure 5.1, with a preponderance of the (−)-epigallocatechin unit. In immature fruit the amount of soluble tannins (~ 4%) was comparable to those of the astringent cultivars (~ 7%). However, as the fruit mature (~ 4 months) the residue of these polymers (~ 0.1–0.4%) is much less than that found in the astringent fruit (~ 1.6%). Over this same period the gross weight of the fruit increases about 10 times. Tanaka and his colleagues (Tanaka *et al.*, 1994) interpreted these observations as showing that the amount of polyphenols in the fruit does not change over

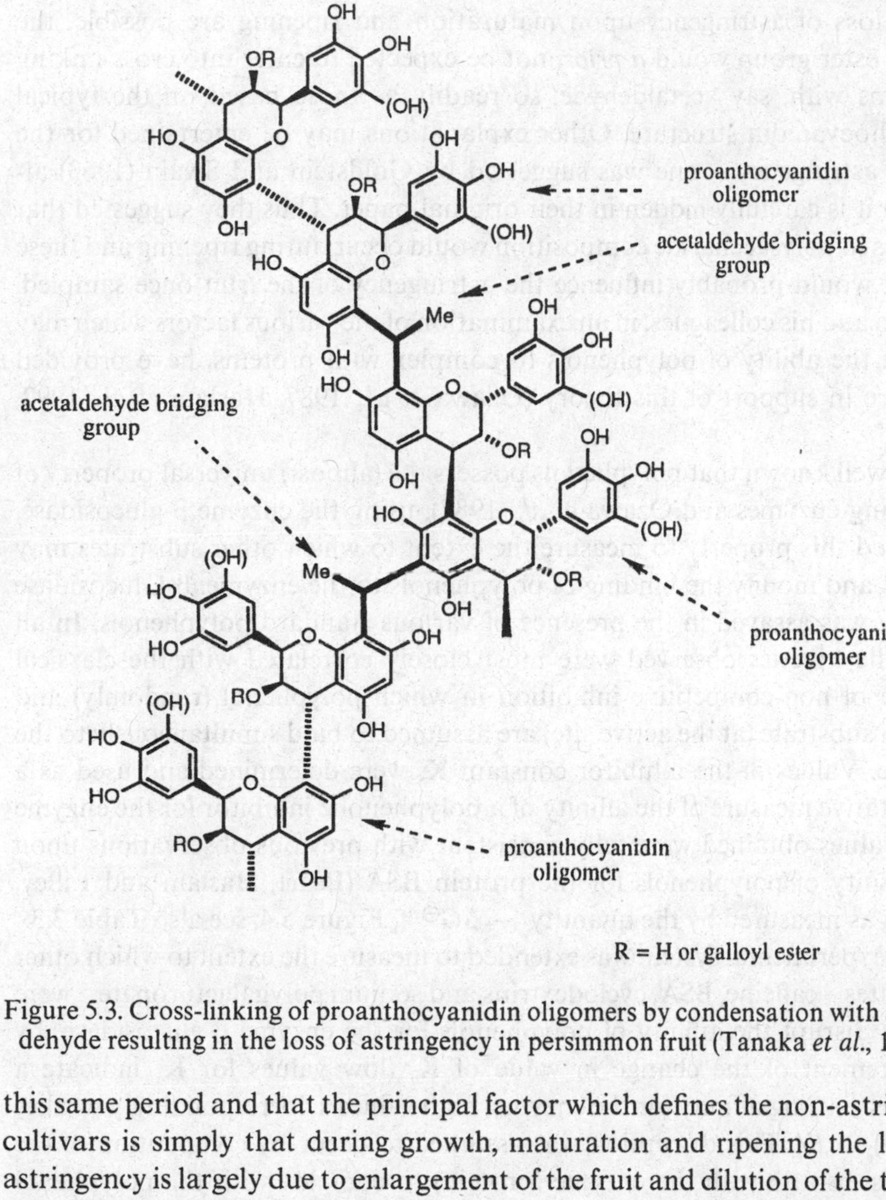

Figure 5.3. Cross-linking of proanthocyanidin oligomers by condensation with acetaldehyde resulting in the loss of astringency in persimmon fruit (Tanaka *et al.*, 1994).

this same period and that the principal factor which defines the non-astringent cultivars is simply that during growth, maturation and ripening the loss of astringency is largely due to enlargement of the fruit and dilution of the tannin concentration. They suggested moreover that tannin coagulation by means of acetaldehyde cross-linking, as described above, is a minor factor in the non-astringent cultivars.

5.3 Moderation of astringency – carbohydrate complexation

What then of other fruit whose polyphenolic constituents are predominantly those based upon gallic acid and its derivatives? Although similar explanations of the loss of astringency upon maturation and ripening are possible, the galloyl ester group would *a priori* not be expected to enter into cross-linking reactions with, say acetaldehyde, so readily as those based on the typical proanthocyanidin structure. Other explanations may be entertained for the loss of astringency – one was suggested by Goldstein and Swain (1963) although it is carefully hidden in their original paper. Thus they suggested that changes in polysaccharide composition would occur during ripening and these in turn would probably influence the astringency of the fruit once sampled. Haslam and his colleagues, in an examination of the various factors which may disrupt the ability of polyphenols to complex with proteins, have provided evidence in support of this theory (Ozawa *et al.*, 1987; Haslam *et al.*, 1992, 1994).

It is well known that polyphenols possess the (almost) universal property of inhibiting enzymes and Ozawa *et al.* (1987), using the enzyme β-glucosidase, exploited this property to measure the extent to which other substrates may disrupt and modify the binding of polyphenols to the enzyme. β-Glucosidase activity was assayed in the presence of various standard polyphenols. In all cases the kinetics observed were most closely correlated with the classical pattern of non-competitive inhibition in which polyphenol (randomly) and normal substrate (at the active site) are assumed to bind simultaneously to the enzyme. Values of the inhibitor constant K_i were determined and used as a quantitative measure of the affinity of a polyphenolic inhibitor for the enzyme – the values obtained were fully consistent with previous observations upon the affinity of polyphenols for the protein BSA (Beart, Haslam and Lilley, 1985a), as measured by the quantity $-\Delta G^{\ominus,tr}$, Figure 5.4; see also Table 3.3.

The experimental system was extended to measure the extent to which other substrates – caffeine, BSA, cyclodextrins and sodium polygalacturonate – were able to disrupt the affinity of polyphenols for the enzyme β-glucosidase, by measurement of the change in value of K_i (low values for K_i indicate a relatively strong affinity for the enzyme, high values a correspondingly weaker affinity) to K_i. This disruption was assumed to occur by competition of the added substrate with the enzyme (protein) to complex with the polyphenol inhibitor and was observed experimentally as a relief of the previous enzyme inhibition (K_i to K_i). This work very clearly demonstrated the ability of each of the above substrates including sodium polygalacturonate to disrupt the binding of typical polyphenols to the enzyme and thus to lessen the inhibition of enzyme activity. The authors suggested that in the case of sodium polygalacturonate this probably resulted from the ability of the polysaccharide to develop a secondary structure in solution with hydrophobic pockets able to encapsulate and complex the polyphenol. It is known that changes of texture

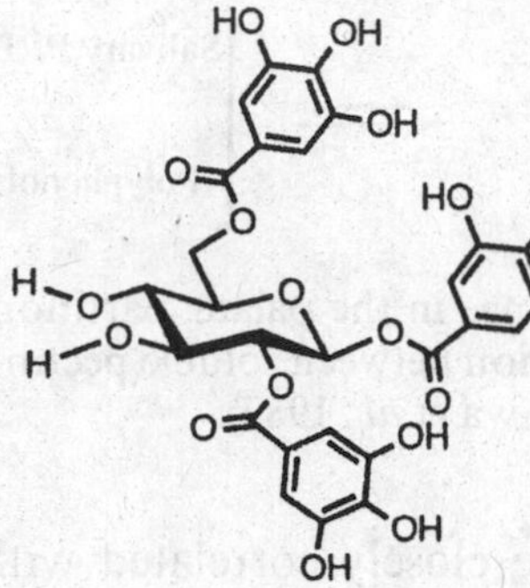

β-1,2,6-Tri-*O*-galloyl-D-glucose

$-\Delta G^{\ominus,\mathrm{tr}} = 0.9\,\mathrm{kJ\,mol^{-1}}$; $K_i = 10.8 \times 10^{-4}\,\mathrm{M}$

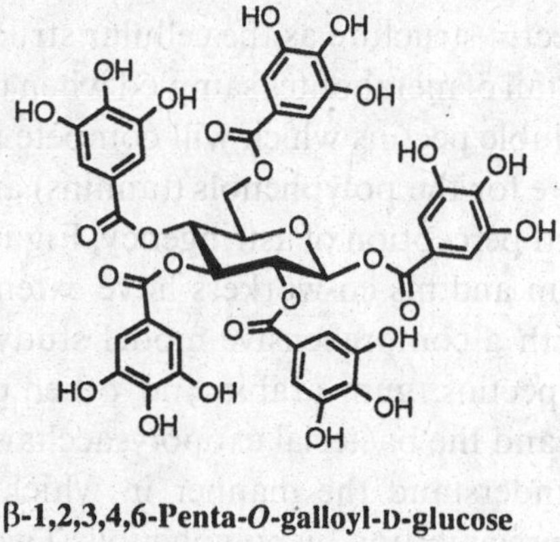

β-1,2,3,4,6-Penta-*O*-galloyl-D-glucose

$-\Delta G^{\ominus,\mathrm{tr}} = 26.9\,\mathrm{kJ\,mol^{-1}}$; $K_i = 0.85 \times 10^{-4}\,\mathrm{M}$

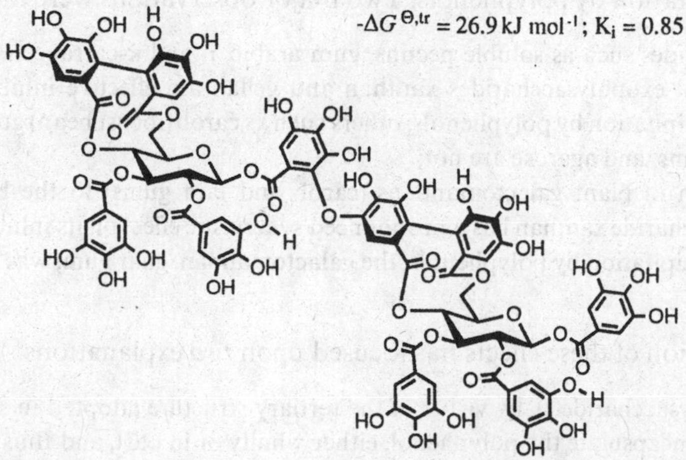

Rugosin D

$-\Delta G^{\ominus,\mathrm{tr}} = 58.7\,\mathrm{kJ\,mol^{-1}}$; $K_i = 0.08 \times 10^{-4}\,\mathrm{M}$

Figure 5.4. Comparative measurements of the affinity of polyphenols for proteins: (i) $-\Delta G^{\ominus,\mathrm{tr}}$, free energy of transfer of the protein BSA from an aqueous solution to an aqueous solution containing the polyphenol; (ii) K_i, inhibition of the enzyme β-glucosidase.

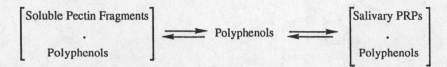

Figure 5.5. Modification of the astringent response in the palate: partitioning of the polyphenol substrates by competitive complexation between soluble pectin fragments and the salivary PRPs (Ozawa *et al.*, 1987).

that result in the softening of the fruit are closely correlated with the depolymerisation of pectin, and it was concluded that these observations supported the view that in some fruit the loss of astringency upon ripening may well be due to the formation of significant amounts of water soluble fragments of the pectin structure as the cellular structure of the fruit breaks down. Thus the cocktail of metabolites sampled when the ripened fruit is tasted will contain these soluble pectins which will compete with the salivary proteins (PRPs) in the palate for the polyphenols (tannins) and would lead to a correspondingly decreased perception of astringency, Figure 5.5.

Haslam and his co-workers have extended this work (Haslam *et al.*, 1992, 1994) with a comprehensive model study of a series of polysaccharides (e.g. soluble pectins, gum arabic and other plant gums, *ι*- and *κ*-carrageenans, agarose and the bacterial exopolysaccharides xanthan and gellan) in order to try to understand the manner in which such macromolecules may inhibit protein precipitation by polyphenols. Two major observations were made:

(i) polysaccharides such as soluble pectins, gum arabic, *ι*- and *κ*-carrageenans, and the bacterial exopolysaccharides xanthan and gellan are effective inhibitors of protein precipitation by polyphenols; others such as carob (locust bean) gum, guar and tara gums, and agarose are not;

(ii) the addition of plant galactomannans (carob and tara gums) to the bacterial exopolysaccharide xanthan has a pronounced synergistic effect on its inhibition of protein precipitation by polyphenols; the galactomannan, guar gum, was without effect.

Rationalisation of these effects has focused upon *two* explanations:

(a) that the polysaccharide(s), by virtue of the tertiary structure adopted in solution, are able to encapsulate the polyphenol, either wholly or in part, and thus prevent its association with the protein, or

(b) that the polysaccharide(s), which are generally polyelectrolytes, may form ternary complexes with the protein · polyphenol complex rendering this soluble in the aqueous medium.

The structure and properties of some of the relevant polysaccharides which are most effective as inhibitors of protein precipitation by polyphenols are discussed below and their possible roles as agents to sequester/encapsulate polyphenols and thus prevent complexation with the protein ((a) above) is considered.

5.3.1 *Algal polysaccharides*

It has been known for some time that polysaccharides may adopt a variety of ordered shapes in the condensed phase – ribbons, double and triple helices – and that these may carry over, fully or in part, to the hydrated environment of the solution (Rees, 1977, 1981; Rees and Welsh, 1977). Double helix conformations have been characterised in the solid state for a number of algal polysaccharides – typically alternating co-polymers of 1,3-linked β-D-galactose and 1,4-linked 3,6-anhydro-D- or L-galactose. Individual members of the family carry variable amounts of sulphate ester substitution, e.g. Figure 5.6. However, the native polymers do not consist entirely of a regular alternating sequence of sugar residues; interruptions occur at various points of the chain which terminate the double helical regions with a 'kink' (in, for example, ι-carrageenan, galactose-2,6-sulphate residues replace a proportion of the 3,6-anhydro-D-galactose-2-suphate groups). The carrageenan chain is then pictured as having regions of regular sequence with the potential, under appropriate conditions, to enter into double helix formation, separated by other regions which act as 'kinks'. A typical chain may thus contain 8–10 helix-forming regions. In the conversion to helical forms in solution each of these regions must nucleate independently and such processes may well not involve the same chain. In this way a cross-linked three dimensional network is created and the solution moves towards a gel, Figure 5.7. The position and frequency of occurrence of the 'kinks' in the polysaccharide chain influence the size and shape of the pores in the gel or gel-like structure. In turn these influence transport through the medium and the encapsulation of other substrates within the gel-like structure.

5.3.2 *Pectin*

Pectin is a major constituent of the cell wall of higher plants and is involved in the maintenance of structural integrity. It occurs as the partial methyl ester of 1,4-linked poly-(α-D-galacturonic acid), Figure 5.8. Just as in the carrageenans the polygalacturonate sequences are interrupted by residues, in this case 1,2-linked L-rhamnose, which introduce severe 'kinks' into the chain. Pectin also usually contains a proportion of branch chains which may limit chain associations. Present evidence suggests that, particularly in pectins of low

repeating helix-forming sub-unit structure of ɩ-carrageenan

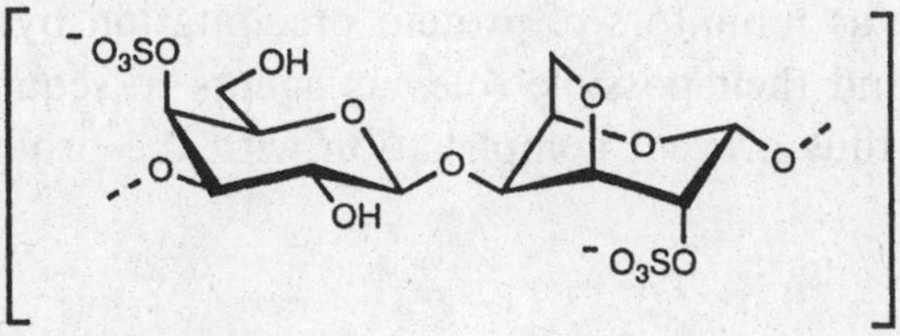

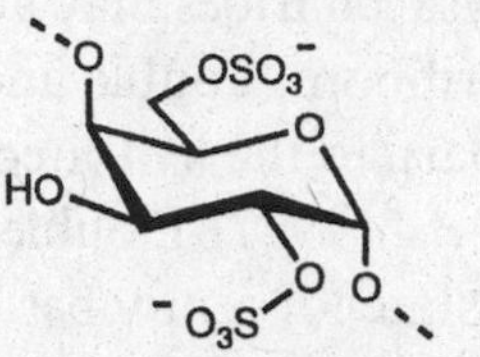

residue introducing 'kinks' into ɩ-carrageenan ;
periodically replaces 4-linked 3,6-anhydro- α-D-galactose

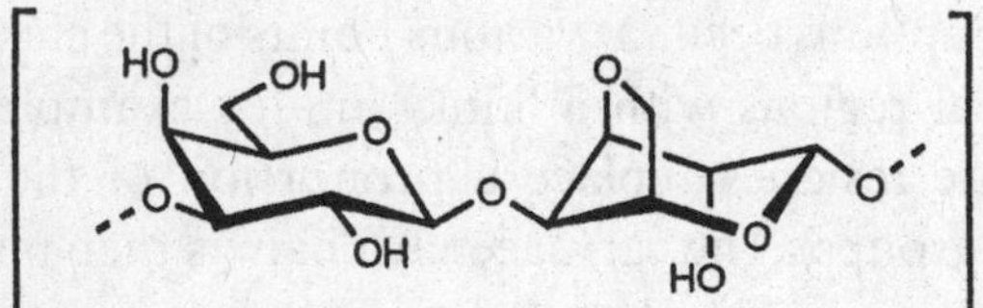

repeating helix-forming sub-unit structure of agarose

Figure 5.6. Structural repeat units of algal polysaccharides: agarose and ɩ-carrageenan: formation of pores and channels.

methyl ester content, chain–chain association probably proceeds by an 'egg-box' type of mechanism (Rees, 1977, 1981; Rees and Welsh, 1977) in which the alignment of two chains creates cavities which accommodate metal ions (often calcium). These, by coordination with the carboxylate and other electronegative oxygen atoms, hold the chains together and induce the formation of gel-like structures in solution, Figure 5.8.

5.3.3 Xanthan

The structure of the bacterial exopolysaccharide xanthan (*Xanthomonas campestris*) is based upon a linear 1,4-linked β-D-glucose backbone, as in cellulose, but solubilised by a charged trisaccharide side-chain attached to every second glucose backbone residue, Figure 5.9. The technological utility of xanthan centres on its unusual solution properties: the polymer has been

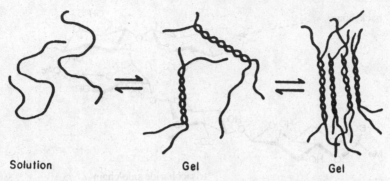

Figure 5.7. Schematic representation of the random coil to double helix conversion in solutions of ι-carrageenan: formation of pores and channels.

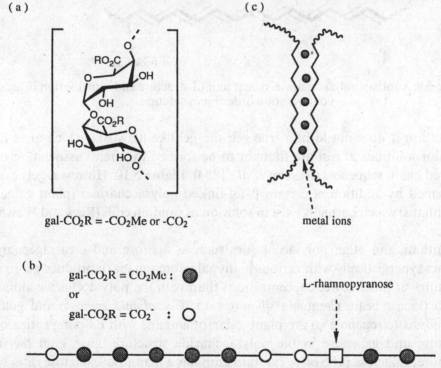

Figure 5.8. (a) Repeat unit of pectin structure; (b) typical sequence of sugar residues in pectin molecule; (c) schematic representation of association of polygalacturonate sequences by chelation of carboxylate groups, ether oxygens and hydroxyl groups and with metal ions: 'egg-box' model.

visualised to possess an ordered conformation in solution in which the main backbone adopts a single helical conformation with the trisaccharide side-chains folded down to align with the main chain and stabilise the overall conformation, Figure 5.9. Although xanthan adopts an ordered conformation

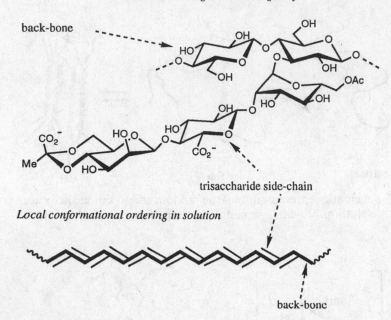

back-bone

trisaccharide side-chain

Local conformational ordering in solution

back-bone

Figure 5.9. Xanthan polysaccharide: repeat unit of structure and visualisation of local conformation ordering in solution.

in solution it does not form a true gel; the gel-like networks which arise in xanthan solutions at rest are thought to be formed by lateral association of ordered chain sequences (Norton *et al.*, 1984), Figure 5.10. However, gels can be formed by addition of certain β-1,4-linked polysaccharides (plant galactomannans) which normally exist in solution as random coils (Rees and Welsh, 1977).

Xanthan, and other polysaccharides such as agarose and κ-carrageenan, interact synergistically with various plant galactomannans to produce gel-like structures at much lower concentrations than with the polysaccharide alone. Carob (locust bean, *Ceratonia siliqua*), tara (*Caesalpinia spinosa*) and guar (*Cyamopsis tetragonoloba*) are plant galactomannans with differing ratios of mannose and galactose in the polysaccharide structure; guar gum has a mannose: galactose ratio of ~ 2:1, tara gum has a mannose: galactose ratio of ~ 3:1 and carob gum has a mannose: galactose ratio of ~ 4:1. Chemically all the structures consist of β-1,4-linked D-mannose units with α-1,6-linked D-galactose units in the ratios indicated. The fine structure indicates that the positions of D-galactose substitution may be regular, random or in regular blocks, Figure 5.11. The degree of complexation of these plant galactomannans decreases with increase in the content of D-galactose, i.e. carob > tara > guar, suggesting that the effective binding occurs principally via sites on the D-mannose backbone. In view of the evidence that these can occur in

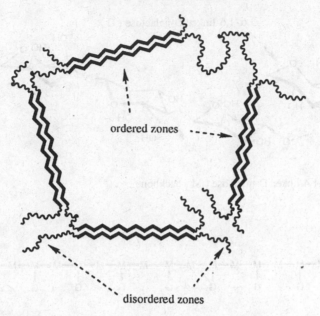

ordered zones

disordered zones

Figure 5.10. Xanthan: lateral association of ordered chain sequences to give, in solution, a weak gel-like network.

regular blocks it has been suggested by Morris (see Rees and Welsh, 1977) that the most probable mechanism for cross-linking, and hence the development of gel-like structures in solution involves helix-ribbon associations as shown in Figure 5.11.

Two possible explanations for the inhibition of protein precipitation by various polysaccharides have been considered, *vide supra*. The polysaccharides which are most effective in this respect – carrageenans, polygalacturonate, xanthan – are all polyelectrolytes and the inhibition of protein–polyphenol precipitation may be due to the formation of ternary complexes – [protein · polyphenol · polysaccharide] – which are solvated by the ionic character of the polysaccharide components. On the other hand there is circumstantial evidence in the data to strongly suggest that for many of the polysaccharides their ability to develop (with or without other polysaccharides) gel-like structures in solution could lead to situations in which the polyphenolic substrates are encapsulated – [polyphenol · polysaccharide] – in pores in the gel-like structure of the polysaccharide solution, which therefore physically prevents their interaction with the protein and thus inhibits [protein · polyphenol] complexation and precipitation, Figure 5.12.

Whatever the final scientific explanation for these phenomena there seems little doubt that modification of the astringent response to polyphenols in the palate may well be achieved by choice of the appropriate soluble polysacchar-

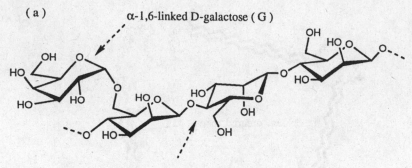

(a)

α-1,6-linked D-galactose (G)

β-1,4-linked D-mannose (M) backbone

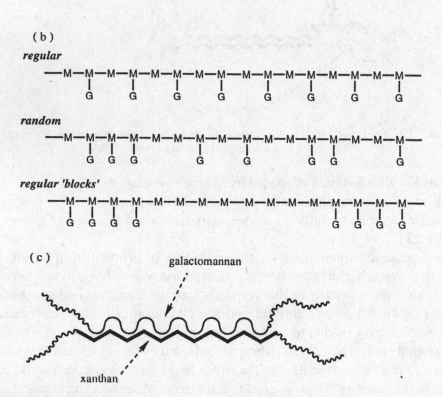

(b)

regular

—M—M—M—M—M—M—M—M—M—M—M—M—M—M—M—M—
 | | | | | | | |
 G   G   G   G   G   G   G   G

random

—M—M—M—M—M—M—M—M—M—M—M—M—M—M—M—M—
 | | | | | | | |
 G G G   G   G   G G   G

regular 'blocks'

—M—M—M—M—M—M—M—M—M—M—M—M—M—M—M—M—
 | | | | | | | |
 G G G G   G G G G

(c)

galactomannan

xanthan

Figure 5.11. Plant galactomannans: (a) general structural motif – β-1,4-linked D-mannose backbone with α-1,6-linked D-galactose substitution; (b) fine structure of galactose (G) substitution of mannan (M)$_n$ backbone; (c) model proposed for the interaction between xanthan helix and the unsubstituted backbone regions of the galactomannan leading to gel-like structures in solution – Morris (see Rees and Welsh, 1977).

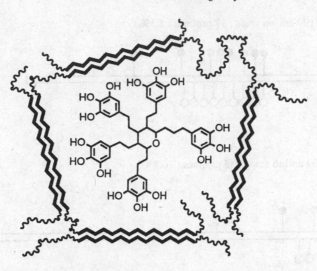

Figure 5.12. Pictorial representation of the encapsulation of a polyphenol within the pores of the gel-like structure developed in an aqueous solution of a polysaccharide.

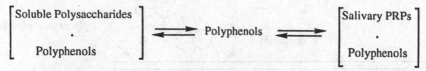

Figure 5.13. Modification of the astringent response in the palate: partitioning of the polyphenolic substrates between soluble polysaccharides and salivary PRPs.

ide to compete with the salivary PRPs for the polyphenolic substrates, Figure 5.13 (cf. Figure 5.5), or to form soluble ternary complexes.

5.4 Teas, casein and astringency

The average citizen of the U.K. drinks 1650 cups of tea each year: almost invariably it is served with lemon or milk. When milk is added to infusions of a 'strong' tea the astringency is moderated substantially (Brown and Wright, 1963).

The caseins are a major group of phosphoproteins synthesised and stored during mammalian lactation and subsequently secreted as stable calcium phosphate complexes (micelles). They constitute some 80% of the proteins of bovine milk and are precipitated at their isoelectric point (~ pH 4.6). 'Isoelectric casein' contains four principal primary proteins, α_{s1}-, α_{s2}-, β- and κ-, plus several minor peptides which originate by post-secretion proteolysis. All the caseins are relatively small (169–209 amino acids) amphipathic molecules

α_{S1} -casein   ; 199 amino acids , 17 prolines (8.5%)

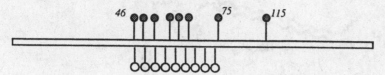

β-casein ; 209 amino acids , 35 prolines (16.8%)

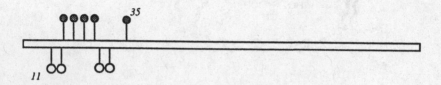

κ-casein   ; 169 amino acids , 20 prolines (11.8%)

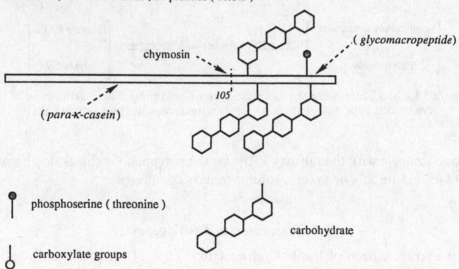

Figure 5.14. Structure of the major bovine casein proteins. Relative hydrophobicity:
β- > κ- > α$_{S1}$- > α$_{S2}$-.

containing high numbers of proline residues (17–35 prolyl groups; 8–17%)
which are fairly evenly distributed throughout the peptide chains, Figure 5.14
(Fox, 1989). Like the salivary PRPs this gives the caseins relatively open,
random coil conformations. Both the α$_{S1}$- , α$_{S2}$- and β-caseins contain phosphoseryl clusters and all the caseins bear a negative charge (from − 21
(α$_{SB}$-casein) to − 4 (κ-casein)) around neutrality. At very low calcium concen-

trations (as in milk) calcium ions are probably bound exclusively to these phosphoseryl groups causing conformational changes in their vicinity (Sleigh *et al.*, 1983). Charge neutralisation facilitates aggregation and precipitation. κ-Casein is also proline rich, contains just one phosphoseryl group but has a water solvating glycosylated C-terminal fragment. It is the most highly structured of the caseins and is an obligate micellar component. Hydrolysis (chymosin) of the Phe_{105}–Met_{106} bond of κ-casein (Fox, 1989; Taborsky, 1974; Mercier *et al.*, 1973) gives rise to two fragments – the glycomacropeptide (hydrophilic C-terminal region) and *para*-κ-casein (the strongly hydrophobic N-terminal region). Using the Bigelow scale (Hil and Wake, 1969) all the caseins (α_{S1}- , α_{S2}- , β- and κ-) are strongly hydrophobic, although the hydrophobic groups are not distributed uniformly throughout the structures of the individual caseins. This property nevertheless promotes strong mutual association of the caseins and therefore facilitates similarly strong complexation with polyphenols in aqueous media.

Casein aggregates (micelles) lend milk its whiteness and its opaque appearance. These particles (averaging ~ 90 nm in diameter) correspond to the aggregation of $\sim 2 \times 10^4$ monomers (Holt, 1985). It is clear (Donnelly *et al.*, 1984; Griffin and Anderson, 1983; Dalgleish *et al.*, 1989) that the micellar surfaces are composed predominantly of κ- and the α_S-caseins with β-casein generally being present to less than 10%. Firm evidence also points to the definitive role of κ-casein as a colloidal protecting agent at the surface of the micelles. It has been suggested that *in vivo* formation of micelles is predominantly the result of deposition of calcium phosphate on α_S- and κ- / α_S-casein complexes and that β-casein then associates with these primary aggregates partly through calcium phosphate bridges but principally through hydrophobically driven complexation to the α_S-caseins. The question of whether caseins form sub-micellar particles is still debated, but Haslam and his group assumed the formation of this sub-structure in the formulation of a model for the moderation of tea polyphenol astringency in the presence of caseins (milk) (Haslam *et al.*, 1994). They showed that each of the caseins – probably via the proline rich and/or hydrophobic regions in the polypeptide chains – forms strong complexes with polyphenols. In salt-free media the complexes are generally soluble. The solubility of the polyphenol complexes involving β- and α_S-caseins is, however, sensitive to the presence of small amounts of calcium ions but these complexes are re-solubilised by the addition of κ-casein.

It is widely, although not universally, held that the micelles are composed of spherical sub-micelles (~ 10–15 nm in diameter) and have a porous structure. A combination of hydrogen and electrostatic bonds, calcium ions and hydrophobic effects holds the sub-micelles together (Figure 5.15). According to Hil

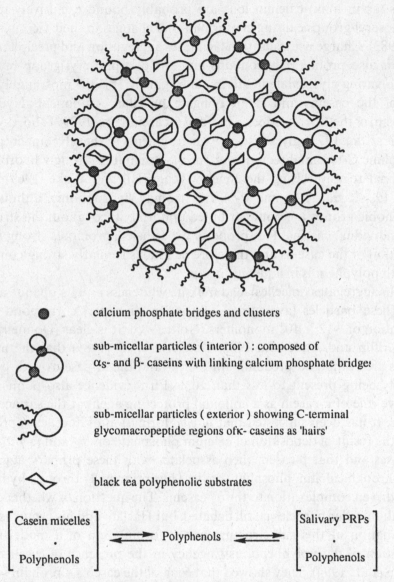

calcium phosphate bridges and clusters

sub-micellar particles (interior) : composed of
α_S- and β- caseins with linking calcium phosphate bridges

sub-micellar particles (exterior) showing C-terminal
glycomacropeptide regions of κ- caseins as 'hairs'

black tea polyphenolic substrates

$$\begin{bmatrix} \text{Casein micelles} \\ \cdot \\ \text{Polyphenols} \end{bmatrix} \rightleftharpoons \text{Polyphenols} \rightleftharpoons \begin{bmatrix} \text{Salivary PRPs} \\ \cdot \\ \text{Polyphenols} \end{bmatrix}$$

Figure 5.15. Pictorial representation of the casein micelle and its sequestration of black tea polyphenols in aqueous media: solubilisation and modification of the astringent response.

and Wake (1969) molecules of κ-casein adopt a radial polarity in the sub-micelles; the hydrophobic N-terminal *para*-region (Figure 5.14) anchors the outwardly directed hydrophilic glycomacropeptide (C-terminal) region (Figure 5.15). With this putative picture of the casein micelle, Haslam and his colleagues rationalised the moderation which the addition of milk has upon the astringency of black tea liquors. In the presence of milk it was suggested that the tea polyphenols were removed from free solution by sequestration within the casein micelles – in the hydrophobic pores, in the hydrophobic regions of the sub-micellar particles and possibly by displacement of the non-essential β-casein structural entities in the micellar structure. The tea polyphenols thus remain solubilised but are 'protected' by encapsulation in the casein micelles from the salivary PRPs. The perceived astringency of the original black tea liquor is thus reduced, Figure 5.15.

5.5 Ageing of red wines

Wines change with time; moreover, they have the capacity to change for the better. However, the perception that all wine improves with age applies perhaps to the top 10% of red wines and to rather less of white wines (Johnson, 1974; Robinson, 1994, 1995). In the case of red wines these changes intimately involve the anthocyanin pigments and the polyphenols (tannins; proan-thocyanidins) which they contain. The processes involved in the ageing of good red wines have been of endless fascination to man for centuries; the mystique underlying these processes was admirably and eloquently summarised by Hugh Johnson (1974).

The making of red wine, which involves the skin and the pips as well as the juice of grapes, leaves extra substances dissolved, above all tannin. This gives the wine the special quality of hardness, of drying up your mouth. These extras need time to resolve themselves to carry out slow and obscure chemical changes which make all the difference in the world to the eventual glass of red wine.

All young red wines, destined for ageing, have a relatively high concentration of polyphenols (tannins; proanthocyanidins) from which they derive (as is the case with ciders) their astringency on the palate. As the wine matures the astringency is modified and generally the wine becomes more acceptable to the consumer. That many of the chemical transformations which occur in such wines over a period of time are relatively slow is not disputed, nor is the fact that progress towards defining the role of each individual substance in the wine, how it participates and changes, has often seemed equally slow (and also painful!) but it is safe to say that over the past 25 years some progress has now been made in lifting the veil of obscurity on the chemical nature of these changes. A freshly made red wine is an extraordinarily complex physico-chemical system with a seemingly infinite number of variables. Pivotal to the 'slow and obscure chemical changes' are the anthocyanin pigments and the polyphenols (tannins; proanthocyanidins); estimates of the total phenolics (largely procyanidins) in a full-bodied red wine range from 800 to 4000 mg l^{-1} and of anthocyanins typically 500 mg l^{-1} (Somers and Verette, 1988). As the wine ages and its sensory characteristics are modified then so also are there concomitant changes in colour and the pigment composition becomes progressively more complex.

5.5.1 Colour changes

The initial colour of red wine is fundamentally due to the extraction of anthocyanin pigments from the black grape skins during vinification. Hplc studies have shown that during conservation residual anthocyanins make decreasing contributions to wine colour (Nagel and Wulf, 1979; Bakker *et al.*, 1986). Concurrent with the progressive decline in anthocyanin pigments there is the formation of new, stable oligomeric/polymeric pigments which maintain the wine colour as it matures (Somers, 1971, 1978; Somers and Evans, 1979, 1986; Somers and Verette, 1988). The rate of decline in the anthocyanin concentration is influenced by such factors as temperature, oxygen access, pH, and free SO_2 content (Somers and Verette, 1988).

Anthocyanin (oenin) self-association and copigmentation (see chapter 6) play major roles in determining 'la robe' of a young red wine and the subtle changes in colour density and hue which take place during early maturation, Figure 5.16. The intensely bluish-red colour of berry juices is largely due to the interplay of these two effects. Much of the initial steep decline in colour density during fermentation on the skins, or of red juices, arises from the increasing instability of these deeply coloured molecular aggregates (copigment complexes) as the ethanol content of the surrounding medium increases. Various groups have concluded that the vertical 'π–π' stacking of pigment and pigment/copigment in such complexes is driven by hydrophobic effects (see chapter 6; Goto, 1987; Goto and Kondo, 1991; Brouillard *et al.*, 1989, 1991; Mistry *et al.*, 1991). Thus as the fermentation proceeds the increasing ethanol content of the surrounding medium, because of its increased hydrophobic character, simply acts to disrupt and destabilise anthocyanin/anthocyanin and anthocyanin/copigment complexes. In red wines the principal copigments present are the various hydroxycinnamate esters, the phenolic flavan-3-ols, (+)-catechin and (−)-epicatechin and related oligomeric procyanidins, Figure 5.16.

Young red wines have a visible absorption maximum at $\lambda \sim 520$ nm arising principally from the species contributions as indicated in Figure 5.16. Between this absorbtion maximum and the one in the ultra-violet ($\lambda \sim 280$ nm) a minimum is found at $\lambda \sim 420$ nm (Ribereau-Gayon, 1974). As the wine ages the

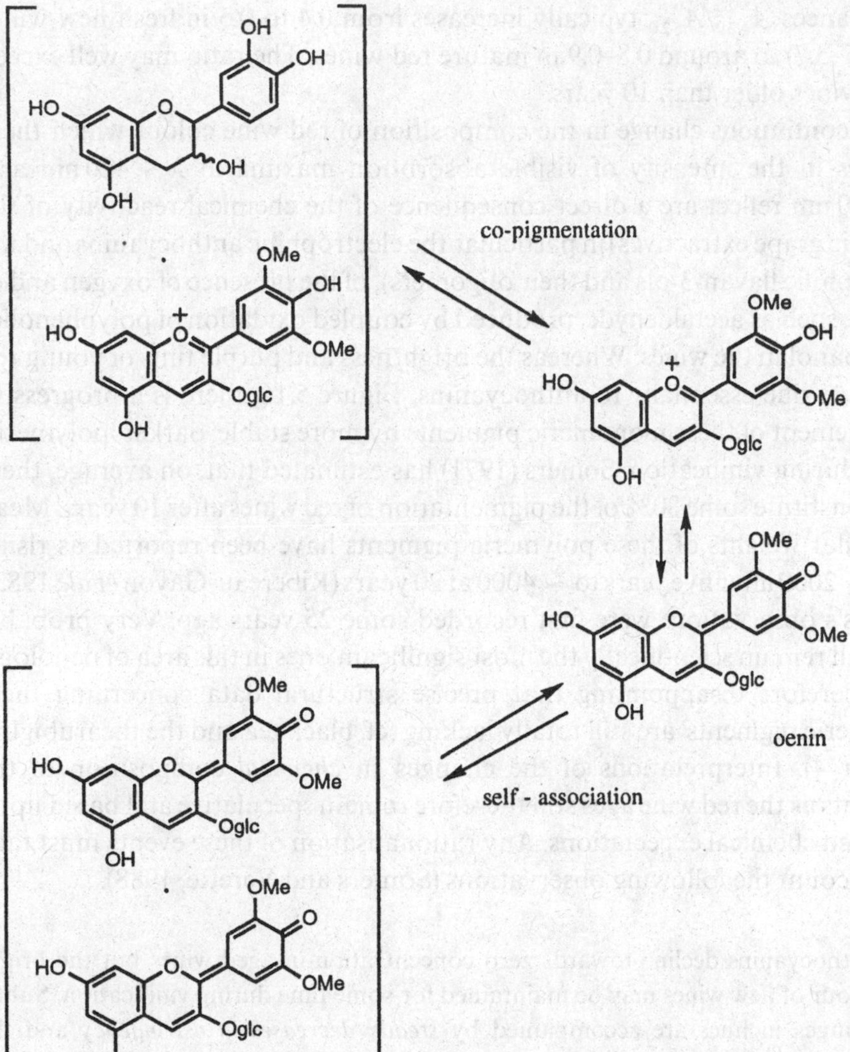

Figure 5.16. Contributions to the initial colour of young red wines: (i) free anthocyanin principally as flavylium ion and quinonoidal anhydro-base; (ii) self-association of the anthocyanin pigment (oenin); (iii) copigmentation with other phenolics present in solution, e.g. 'catechins' (shown) and hydroxycinnamate esters.

maximum at $\lambda \sim 520$ nm declines in intensity, falling to a shoulder in wines older than 10 years. This change corresponds to an increase in the yellow colour at $\lambda \sim 420$ nm and a shift, from the consumer's viewpoint, from the brightness and blue/purple tints of a young red wine to the tile-like red-orange of an older mature wine. The wine hue or tint, measured as the ratio of absorbances A_{420}/A_{520} typically increases from 0.4 to 0.5 in fresh new wines (pH 3.5–3.7) to around 0.8–0.9 in mature red wines. The ratio may well exceed 1.0 in wines older than 10 years.

The continuous change in the composition of red wine colour which these changes in the intensity of visible absorbtion maxima at $\lambda \sim 420$ nm and $\lambda \sim 520$ nm reflect are a direct consequence of the chemical reactivity of the phenolic grape extractives (in particular the electrophilic anthocyanins and the nucleophilic flavan-3-ols and their oligomers), of the presence of oxygen and of species, such as acetaldehyde, produced by coupled oxidation of polyphenolics and ethanol in the wines. Whereas the brightness and purple tints of young red wines are due essentially to anthocyanins, Figure 5.16, there is a progressive displacement of these monomeric pigments by more stable, darker 'polymeric' forms during vinification. Somers (1971) has estimated that, on average, these may constitute some 80% of the pigmentation of red wines after 10 years. Mean molecular weights of these polymeric pigments have been reported as rising from ~ 2000 after five years to ~ 4000 at 20 years (Ribereau-Gayon *et al.*, 1983). Somers's observations were first recorded some 25 years ago. Very probably they still remain scientifically the most significant ones in this area of oenology; it is therefore disappointing that precise structural data concerning these 'polymeric pigments' are still totally lacking (cf. black tea and the thearubigins, chapter 4). Interpretations of the changes in *chemical* composition of the pigments as the red wine ages still therefore remain speculative and based upon reasoned chemical expectations. Any rationalisation of these events must take into account the following observations (Somers and Verette, 1988).

 (i) Anthocyanins decline towards zero concentration in aged wines but the bright
     colour of new wines may be maintained for some time during vinification. Subtle
     changes in hues are accompanied by *steady decreases in astringency* and the
     mellowing of taste and flavour, although whether the two are *directly* linked is not
     clear.

 (ii) Initial chemical (non-enzymic) reactions of the extracted grape phenolics (princi-
      pally anthocyanin pigments and polyphenols (tannins; proanthocyanidins)) are
      fast and begin with the crushing of the berry and the breakdown of its cellular
      compartmentalisation; polymeric pigments are apparent from the very earliest
      stages (Somers and Verette, 1988; Bakker *et al.*, 1986).

(iii) The heterogeneous polymeric pigments are stable to pH and the presence of
      sulphur dioxide (Somers, 1971).

(iv) Colour stabilisation and formation of polymeric pigments occurs only in the
     presence of other wine phenolics such as polyphenols (flavan-3-ols and tannins;
     proanthocyanidins) (Ribereau-Gayon, 1982; Di Stefano and Cioafi, 1983; Glories,
     1984a,b).

Various hypotheses have been advanced, from time to time, to rationalise these experimental observations (i)–(iv) above; given the inherent complexity of red wines it seems probable that more than one process is taking place. Most of these theories are based upon the original speculations of Somers (1971) which were soundly based scientifically, sufficiently subtle to satisfy the wine cognoscenti, but laced with sound common-sense as usually befits one from the Antipodes. The susceptibility of anthocyanins to attack by nucleophilic reagents has long been realised; with carbon nucleophiles substitution at C-4 is most frequently involved (Jurd, 1967; Jurd and Waiss, 1965). Somers based his hypothesis upon this observation. The reactions whereby the colours of red wines change during maturation were formulated by Somers as shown in Figure 5.17. The first step is the nucleophilic attack at C-4 of oenin by a flavan-3-ol, (−)-epicatechin or (+)-catechin, or analogously by an oligomeric procyanidin; this is then followed by oxidation to regenerate a flavylium salt. Somers suggested that the effect of aryl-substitution at C-4 would be to stabilise the chromophore which, according to Somers, was best formulated as the quinonoidal anhydro-base, Figure 5.17. The oligomeric/polymeric nature of the pigments derives from the identity of the initial attacking nucleophile. Since phenolic compounds are known to contribute to the oxidative browning reactions of wines (Cheynier *et al.*, 1988), Somers (1971) suggested that additional 'quinone-like' structures, arising from oxidation of the polymeric pigments, may well be formed and contribute to the browner tints of elderly wines.

In a development of this work, Liao, Cai and Haslam (1992) demonstrated in model systems that reactions between anthocyanins and (−)-epicatechin or (+)-catechin in aqueous alcoholic media at ambient temperatures gave rise to water soluble orange-yellow ($\lambda_{max} \sim 440$ nm) pigmented products. Using the earlier observations of Jurd and Somers (1970) and Hradzina and Borzell (1971) these workers suggested that the products were xanthylium salts derived by cyclisation from the first formed adducts of anthocyanins and flavan-3-ols, Figure 5.18, and that their ready formation (3–6 months) probably represents a major factor in the evolution of the tile-like orange-red colour of a mature wine. The assumption that, *per se*, such products would necessarily exhibit diminished astringency in the palate has yet to be tested experimentally. However, their salt-like character would necessarily lead to an increased hydrophilic character for the procyanidin oligomers/polymers from which they were derived and hence to a decreased affinity for salivary PRPs.

The loss of astringency in red wines as they mature is also almost certainly associated with the aggregation and precipitation of procyanidins from solution. This phenomenon is facilitated by at least two processes. The first of these

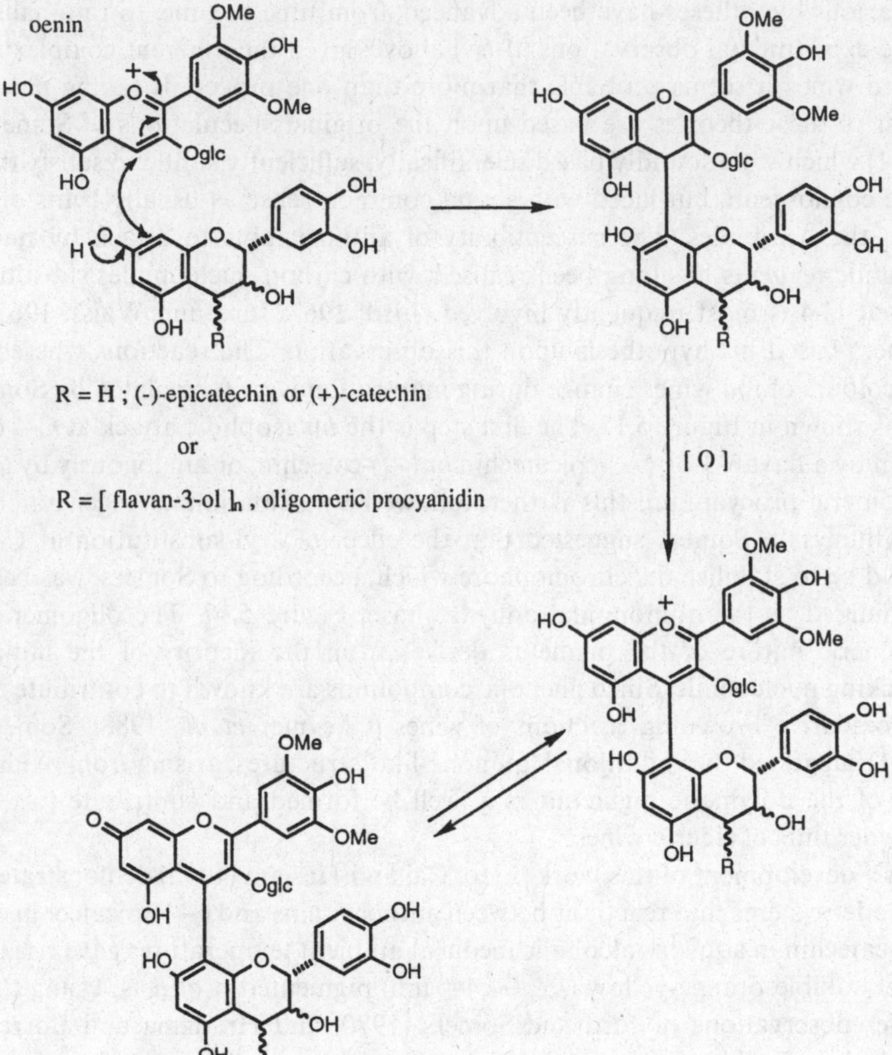

R = H ; (-)-epicatechin or (+)-catechin

or

R = [flavan-3-ol]n , oligomeric procyanidin

[O]

' polymeric ' proanthocyanidin pigments

Figure 5.17. Generation of oligomeric/polymeric pigments in red wines (Somers, 1971).

parallels observations made in the case of ripening persimmon fruit – namely
the cross-linking of proanthocyanidin oligomers by acetaldehyde (Timberlake
and Bridle, 1976; Somers and Wescombe, 1987) rendering the astringent
principles in the fruit insoluble (Figure 5.3, Tanaka *et al.*, 1994). Oxidation and
its control have been primary considerations in both wine making and ageing
for a century and more. Riberau-Gayon (1974) showed that a wine, under
barrel storage, would be expected to absorb about 30 mg of oxygen per litre per

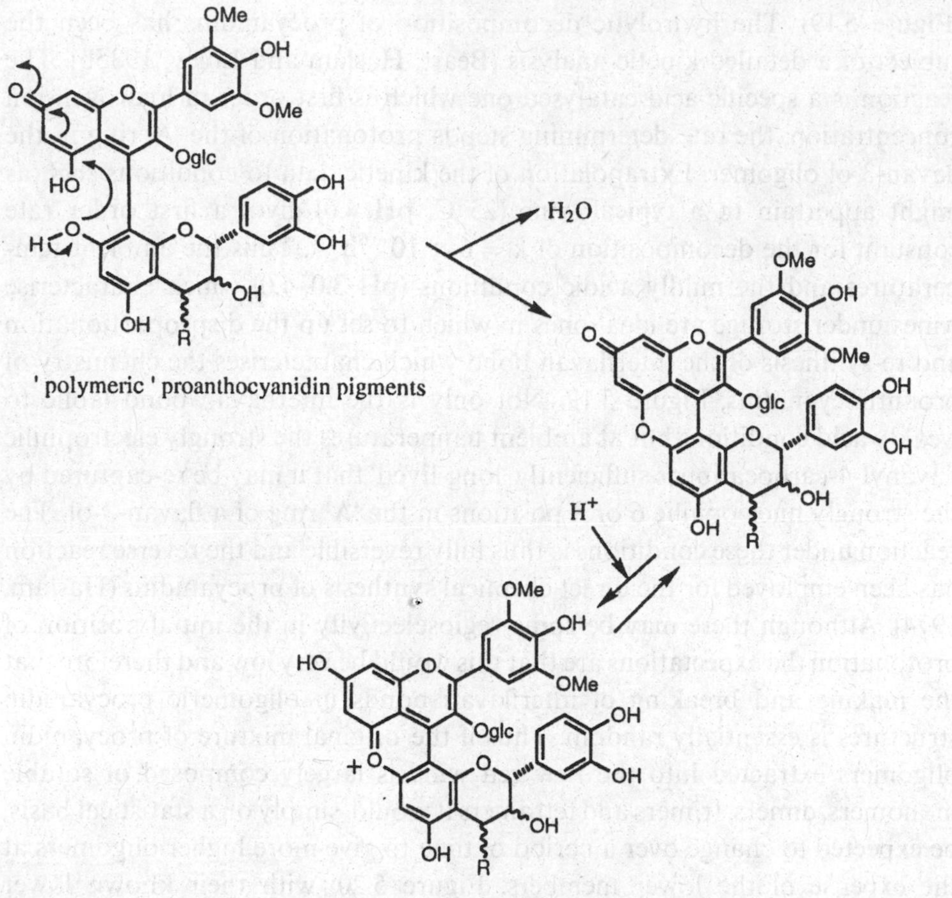

'polymeric' proanthocyanidin pigments

'polymeric' xanthylium salt pigments

Figure 5.18. Formation of water soluble xanthylium salts during red wine conservation (Liao, Cai and Haslam, 1992).

year and even in the corked bottle there is evidence for oxygen penetration. Acetaldehyde may be generated by coupled oxidation with polyphenolics in wine (Wildenrat and Singleton, 1974; Singleton, 1992), although there are wide variations in the levels of acetaldehyde because of the similarly wide variation in wine-making practices and cellar conditions. The strongly electrophilic acetaldehyde once generated would react, by processes entirely analogous to those outlined earlier, with the nucleophilic 8 and/or 6 positions of the flavan-3-ol nuclei in procyanidin oligomers, Figure 5.3, leading by cross-linking to coagulation and eventually precipitation.

Experimental work by Haslam (1980) on the role of oligomeric procyanidins in the ageing of red wines provides an alternative mechanism for the gradual polymerisation of procyanidins and/or the oligomeric procyanidin pigments

(Figure 5.19). The hydrolytic decomposition of procyanidins has been the subject of a detailed kinetic analysis (Beart, Haslam and Lilley, 1985b). The reaction is a specific acid catalysed one which is first order in hydrogen ion concentration; the rate-determining step is protonation of the 'A' ring in the flavan-3-ol oligomer. Extrapolation of the kinetic data to conditions such as might appertain to a typical wine (25°C, pH 4.0) gives a first order rate constant for the decomposition of $k = 6 \times 10^{-6} h^{-1}$. Thus the ambient temperatures and the mildly acidic conditions (pH 3.0–4.0) which characterise wines under storage are ideal ones in which to set up the disproportionation and re-synthesis of the interflavan bond which characterises the chemistry of proanthocyanidins, Figure 5.19. Not only is the interflavan bond labile to weakly acid conditions but at ambient temperatures the strongly electrophilic flavanyl-4-carbocation is sufficiently 'long-lived' that it may be re-captured by the strongly nucleophilic 6 or 8 positions in the 'A' ring of a flavan-3-ol. The reaction under these conditions is thus fully reversible and the reverse reaction has been employed for the direct chemical synthesis of procyanidins (Haslam, 1974). Although there may be some regioselectivity in the initial position of protonation the expectations are that this would be very low and therefore that the making and breaking of interflavan bonds in oligomeric procyanidin structures is essentially random. Thus if the original mixture of procyanidin oligomers extracted into the new red wine is largely composed of soluble monomers, dimers, trimers and tetramers it would, simply on a statistical basis, be expected to change over a period of time to give more higher oligomers at the expense of the lower members, Figure 5.20; with their known lower solubility in aqueous media these higher oligomers would gradually precipitate from the wine as it aged. Thus the astringency of the red wine would steadily decrease.

The polyphenolic procyanidins and the flavan-3-ols (−)-epicatechin and (+)-catechin thus play key roles in the processes associated with the maturation of red wines. These roles are manifested in several ways; the most important ones which have been identified to date are summarised in Figure 5.21. Some, if not all of these, reactions may lead to a loss in astringency of the red wine.

5.5.2   Ageing of wines and other alcoholic beverages in oak

There are barrels without taste and others that give cognac like gold. After several years one knows which barrels are which.

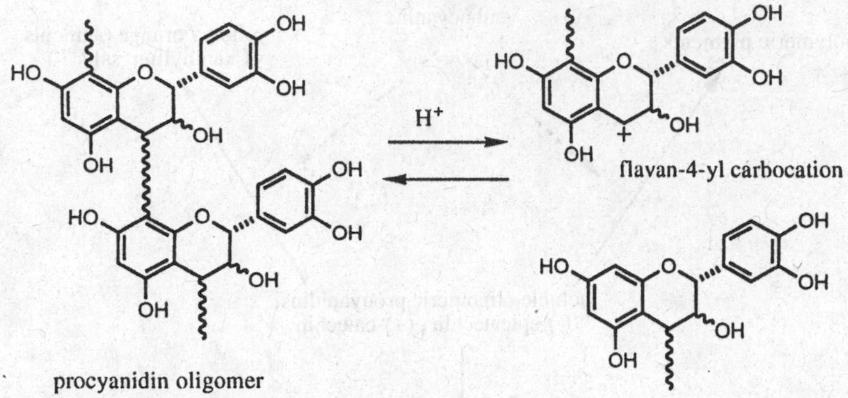

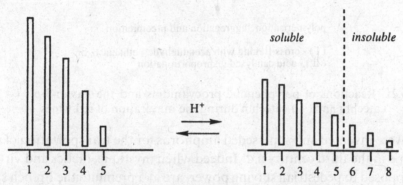

procyanidin oligomer

Figure 5.19. Reversibility of acid catalysed interflavan bond breaking and making reaction (Haslam, 1974).

Figure 5.20. Acid catalysed equilibration of procyanidin oligomers in red wines: diagrammatic representation of typical predicted changes in oligomer composition. Numbers represent oligomer composition: 1, monomers; 2, dimers; 3, trimers, etc. Bar graph gives approximate relative amounts of each oligomer: fresh wine on left; aged wine on right. Higher oligomers become progressively less soluble and precipitate from aged wine with time.

One does not need to do anything more; one must wait. The right time will come for everything. The alcohol now enters the oak, and then the wood yields everything it has. It yields sun, it yields fragrance, it yields colour.

(*A. Mosedale*, 1994)

The making of a great wine or cognac has often been said to be as much an art as a science and an undefinable romance still attaches itself to the process of ageing wines and spirits in a barrel. It is surely a revelation at the end of the twentieth century to visit an old respected Château in one of the great wine producing areas of France and to see row upon row of barrels storing the latest vintage. Changes there have been, but this is still an age old process; wood

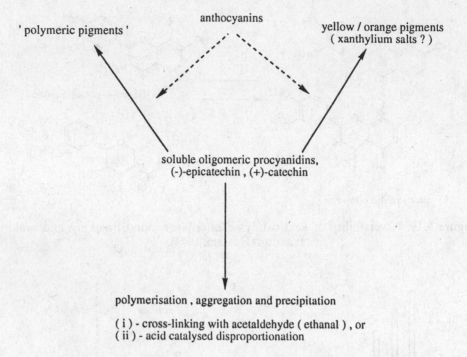

Figure 5.21. Reactions of polyphenolic procyanidins and the flavan-3-ols (−)-epi-catechin and (+)-catechin during the maturation of red wines.

barrels were invented and superseded amphoras for the transportation of wine in Europe in the third century B.C. Indeed when mystique, tactics and virtuosity, as opposed to pedestrian staying power, are at a premium the French stand head and shoulders above the rest.

Great wines are aged in oak barrels. Even in the case of less famous appelations, well managed storage in oak barrels generally improves the product (Johnson, 1989). Oak casks are also employed in the maturation of brandy/cognac and whisky. In the U.S.A. and the U.K. the legal requirements specifically prescribe their use; in the U.S.A. the raw bourbon distillate must be stored for one year in new, charred oak casks and in the U.K. the law demands that Scotch whisky be similarly stored in oak barrels for a minimum of three years. Today winemakers world-wide are increasingly seeking to add perceived, if not necessarily real, value to their products by making or ageing them in oak barrels. Thus although currently less than 10% of all the wine produced in the world today is aged in barrels, these wines attract a disproportionate attention. Oak is a hard but supple, watertight wood and is ideal for cooperage. Experiments with brandy as well as wines have demonstrated the clear superiority of air-dried over kiln-dried wood for barrel staves; indeed accelerated seasoning of oak woods sometimes lends to the final cognac or whisky undesirable 'after-tastes'. **Oak-aged** is a term which may be applied to a wine. Although it implies that the wine was matured in some form of oak barrel or cask it can also mean that it was simply exposed to chips of oak wood. Few wine producers admit to the use of this technique and prefer terms such as 'oak maturation' or 'oak influence' to describe their wines. Oak chips vary considerably in the provenance of the oak and the degree of toasting to which they have been exposed, (typically about $1 g \, l^{-1}$ of oak chips are suspended in a permeable bag). Oak extracts or essences are also available commercially, but their use is often not strictly legal.

The principal oak species used for making casks can be divided into two broad groups – the American white oak and the European oak. The former are found mainly in the Eastern part of the U.S.A. and the principal botanical species is *Quercus alba*. The European oaks comprise two major species – *Quercus robur* (pedunculate oak) and *Quercus petraea* (sessile, durmast oak). In practice many of the European oaks are hybrids of the sessile and pedunculate types and as a result oaks used for the maturation of alcoholic beverages are often referred to by geographical location rather than botanical species. Thus French casks may be sub-divided into Troncais, Gascony, Allier and Limousin. Preferences may also be noted for the maturation of particular beverages in oaks of a particular provenance – bourbon (American), cognac/brandy (French (Limousin)), armagnac (French (Gascony)), wines (French (Troncais)). During maturation a range of flavour congeners – nitrogen and sulphur compounds, carbonyl derivatives, lactones, esters and phenols – may be leached into the alcoholic beverage from the oak. American oak generally gives rise to flavours of a more vanillin type and to a greater astringency than the European counterparts. The interior of an oak barrel formed over a heat source may be 'toasted' to a greater or lesser extent by the heat of the fire. This fundamentally changes the physical and chemical composition of the layers of wood on the inner surface of the barrel. In general the less a barrel is 'toasted' the more of the natural constituents of the oak find their way into the wine or the brandy. Products resulting from the thermal degradation of lignin are more likely to be found in beverages stored in the more heavily charred or 'toasted' barrels.

Much research effort has been expended to try to understand the contribution of oak wood constituents to the organoleptic properties of wines, cognac/ brandy and whisky. Particular attention has been given in recent times to the phenols and polyphenols derived by leaching into aqueous alcohol from oak (Viriot *et al.*, 1993) since these are major solutes ($\sim$ one third of the dry extract) of spirits aged in oak casks. Their concentration increases steadily during the period of ageing. Their composition, in particular the proportions of ellagitannins and lignin oligomers, also changes over the same period of time. The

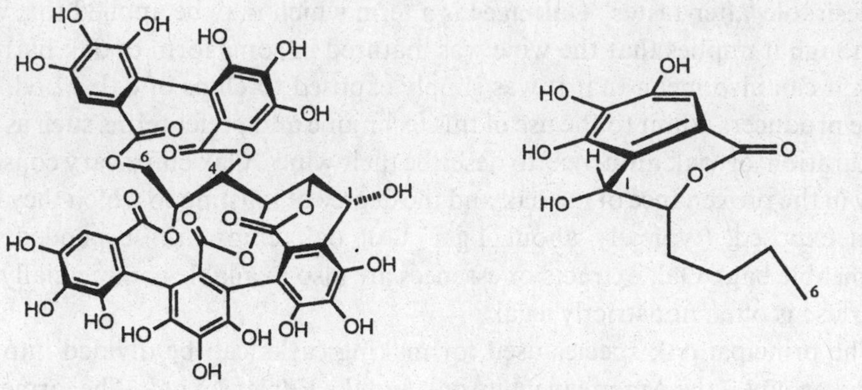

castalagin

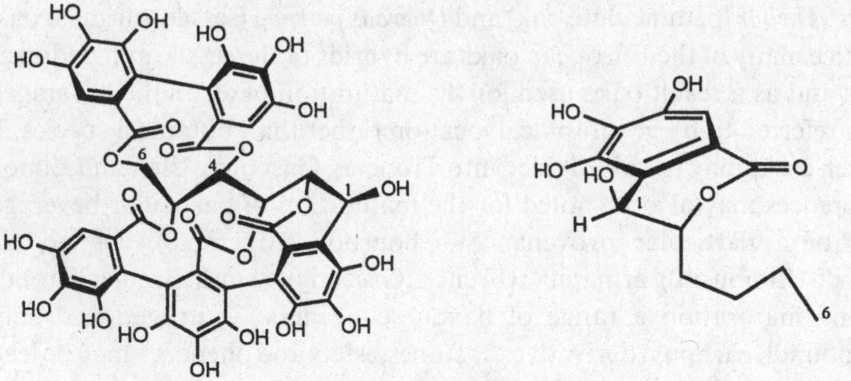

vescalagin

Figure 5.22. Ellagitannins of oak wood – castalagin and vescalagin (Mayer *et al.*, 1971a,b).

majority (although not all) ellagitannins are readily extractable (Peng *et al.*, 1991) and their maximum concentration is reached in the first years of ageing. Concomitant with their ready extraction they undergo chemical degradation. Solubilisation of fragments of the lignin structure requires cleavage of covalent bonds; it is by comparison a slow process and proceeds throughout the period of ageing. A 'young' spirit will contain a higher proportion of ellagitannins whilst in an 'old' spirit lignin fragments will predominate. Oxidative reactions are presumed to be responsible for the generation of vanillin and syringaldehyde from these lignin fragments during ageing.

Scalbert and his colleagues have given considerable attention to the ellagitannins which are extracted from oak woods during the initial phases of the ageing of spirits. They have identified the principal components, namely castalagin, vescalagin, grandinin, roburin E and the 'dimeric' roburins A–D,

(Scalbert *et al.*, 1991). Interestingly, but as predicted from earlier considerations (see chapter 4) these rigid water soluble polyphenols despite the plethora of phenolic groups which they possess do not, according to Scalbert, show significant astringency on the palate. Their main contributions to the taste and flavour of wines and spirits may therefore be derived from their roles as oxygen scavengers (antioxidants) and the generation in these reactions of quinone structures. The apparently facile degradation, and hence steady disappearance, of these ellagitannins when leached from barrels into wines and spirits is also readily rationalised not only in terms of their ease of oxidation but also in terms of their previously described chemistry. In their early seminal work on these unique polyphenolic derivatives of the 'open-chain' form of D-glucose, Mayer and his co-workers demonstrated that the 4,6-linked hexahydroxydiphenyl ester group was readily cleaved from the molecule by mild acid hydrolysis (Mayer *et al.*, 1971a,b). This would be followed by a similar acid catalysed loss of the residual nonahydroxytriphenyl ester group after cleavage of the C-1 glycosidic linkage (by a reversal of its presumed mode of biosynthesis). The relative rates at which these latter acid catalysed hydrolytic reactions occur amongst the various compounds is very probably dependent upon the configuration at this same C-1 centre and the relative ease with which the carbocation is generated at this centre; thus on this basis one would predict, under mildly acid conditions, the persistence with time of castalagin (and related derivatives; C-1 pseudo-equatorial substituent) relative to vescalagin and its analogues (C-1 pseudo-axial substituent), Figure 5.22.

References

Bakker, J., Preston, N. W. and Timberlake, C. (1986). The determination of anthocyanins in ageing red wines: comparison of HPLC and spectral methods. *Amer. J. Enol. Vitic.*, **37**, 121–6.

Beart, J. E., Haslam, E. and Lilley, T. H. (1985a). Plant polyphenols – secondary metabolism and chemical defence: some observations. *Phytochemistry*, **24**, 33–8.

Beart, J. E., Haslam, E. and Lilley, T. H. (1985b). Polyphenol interactions. Part 2. Covalent binding of procyanidins to protein during acid catalysed decomposition; observations on some polymeric proanthocyanidins. *J. Chem. Soc.* (*Perkin Trans. 2*), pp. 1439.

Brouillard, R., Mazza, G., Saad, Z., Albrecht-Gary, A. M. and Cheminat, A. (1989). The co-pigmentation reactions of anthocyanins: a micro-probe for the structural study of aqueous solutions. *J. Amer. Chem. Soc.*, **111**, 2604–10.

Brouillard, R., Wigand, M.-C., Dangles, O. and Cheminat, A. (1991). pH and solvent effects on the co-pigmentation reaction of malvin with polyphenols, purine and pyrimidine derivatives. *J. Chem. Soc.* (*Perkin Trans. 2*), pp. 1235–41.

Brown, P. J. and Wright, W. B. (1963). Interactions between milk proteins and tea polyphenols. *J. Chromatogr.*, **11**, 504–14.

Cheynier, V., Osse, C. and Rigaud, J. (1988). Oxidation of grape juice phenolics in model systems. *J. Food Sci.*, **53**, 1729–32; 1760.

Cockroft, V. (1993). Chocolate on the brain. *The Biochemist*, **15**(2), 14–16.

Dalgleish, D. A., Horne, D. S. and Law, A. J. R. (1989). Size-related differences in bovine casein micelles. *Biochim. Biophys. Acta*, **991**, 383–7.

Di Stefano, R. and Cioafi, G. (1983). Formazione di antociani polimeri in presenza di flavani ed evoluzione degli antociani monomeri durante la fementazione. *Ric. Vitic. Enol.*, **36**, 325–38.

Donnelly, W. J., McNeill, G. P., Buchheim, W. and McGann, T. C. A. (1984). A comprehensive study of the relationship between size and protein compostion in natural bovine casein micelles. *Biochim. Biophys. Acta*, **789**, 136–43.

Forsyth, W. G. C. and Roberts, J. B. (1960). Cocao polyphenolic substances. V. The structure of cacao leucocyanidin I. *Biochem. J.*, **74**, 374–8.

Fox, P. F. (1989). The milk protein system. In *Developments in Dairy Chemistry – 4: Functional Milk Proteins*, editor P. F. Fox, Elsevier: London and New York, pp. 1–53.

Glories, Y. (1984a). La couleur des vins rouge. 1. Les equilbres des anthocyanes et des tannins. *Connais Vigne Vin*, **18**, 195–217.

Glories, Y. (1984b). La couleur des vins rouge. 2. Mesure, origine et interpretation. *Connais Vigne Vin*, **18**, 253–71.

Goldstein, J. L. and Swain, T. (1963). Changes in tannins in ripening fruit. *Phytochemistry*, **2**, 371–83.

Goto, T. (1987). Structure, stability and colour variation of natural anthocyanins. *Prog. Chem. Org. Nat. Prod.*, **52**, 113–58.

Goto, T. and Kondo, T. (1991). Structure and molecular stacking of anthocyanins – flower colour variations. *Angew. Chem. Int. Ed.*, **30**, 17–33.

Griffin, M. C. A. and Anderson, M. (1983). The determination of casein micelle size distribution in skim milk by chromatography and photon-correlation spectroscopy. *Biochim. Biophys. Acta*, **748**, 453–9.

Haslam, E. (1974). Biogenetically patterned synthesis of procyanidins. *J. Chem. Soc. Chem. Commun.*, p. 594.

Haslam, E. (1980). In vino veritas. *Phytochemistry*, **19**, 2577–82.

Haslam, E. (1996). Natural polyphenols (vegetable tannins) as drugs: possible modes of action. *J. Nat. Prod.*, **59**, 205–15.

Haslam, E., Thompson, R. S., Jacques, D. and Tanner, R. J. N. (1972). Plant proanthocyanidins. Part I. Introduction; the isolation, structure and distribution in nature of plant procyanidins. *J. Chem. Soc. (Perkin Trans. I)*, pp. 1387–99.

Haslam, E. and Lilley, T. H. (1988). Natural astringency in foodstuffs – a molecular interpretation. *CRC Crit. Rev. in Food, Science and Nutrition*, **27**, 1–40.

Haslam, E., Lilley, T. H., Warminski, E., Liao, H., Cai, Y., Martin, R., Gaffney, S. H., Goulding, P. N. and Luck, G. (1992). Polyphenol complexation – a study in molecular recognition. In *Phenolic Compounds in Food and Their Effects on Health. I. Analysis, Occurrence and Chemistry*, A.C.S. Symposium Series **506**, eds. C.-T. Ho, C. Y. Lee and M.-T. Huang, American Chem. Soc.: Washington D.C., pp. 8–50.

Haslam, E., Luck, G., Liao, H., Murray, N. J., Grimmer, H. R., Warminski, E. E., Williamson, M. P. and Lilley, T. H. (1994). Polyphenols, astringency and proline-rich proteins. *Phytochemistry*, **37**, 357–71.

Hil, R. J. and Wake, R. G. (1969). Amphiphile nature of κ-casein as the basis for its micelle stabilising property. *Nature*, **221**, 635–9.

Holt, C. (1985). The size distribution of bovine casein micelles. *Food Microstructure*, **4**, 1–10.

Hradzina, G. and Borzell, A. J. (1971). Xanthylium derivatives in grape extracts. *Phytochemistry*, **10**, 2211–13.

Johnson, H. (1974). *Wine*. Mitchell Beazley: London.

Johnson, H. (1989). *The Story of Wine*. Mitchell Beazley: London.

Joslyn, M. A. and Goldstein, J. L. (1964). Astringency of fruits and fruit products in relation to phenolic content. *Adv. Food Res.*, **13**, 179–217.

Jurd, L. (1967). Catechin–flavylium salt condensation reactions. *Tetrahedron*, **23**, 1057–64.

Jurd, L. and Waiss, A. C. (1965). Anthocyanins and related compounds. VI. Flavylium salt–phloroglucinol condensation products. *Tetrahedron*, **21**, 1471–83.

Jurd, L. and Somers, T. C. (1970). The formation of xanthylium salts from proanthocyanidins. *Phytochemistry*, **9**, 419–27.

Kim, H. and Keeny, P. G. (1984). (−)-Epicatechin content in fermented and unfermented beans. *J. Food Sci.*, **49**, 1090–2.

Liao, H., Cai, Y. and Haslam, E. (1992). Polyphenol interactions. Part 6. Anthocyanins: co-pigmentation and colour changes in red wines. *J. Sci. Food Agric.*, **59**, 299–305.

Macdonald, H. (1993). Flavour development from cocoa bean to chocolate bar. *The Biochemist*, **15**(2), 3–5.

Matsuo, T. and Ito, S. (1978). The chemical structure of kaki-tannin from immature fruit of persimmon (*Diospyros kaki* L.). *Agric. Biol. Chem.*, **42**, 1637–43.

Matsuo, T. and Ito, S. (1982). A model experiment for de-astringency of persimmon fruit with high carbon dioxide treatment: *in vitro* gelation of kaki-tannin with acetaldehyde. *Agric. Biol. Chem.*, **46**, 683–9.

Mayer, W., Bilzer, W. and Schauerte, K. (1971a). Isolierung von Castalagin und Vescalagin aus Valoneagerbstoffen. *Liebig's Ann. Chem.*, **754**, 149.

Mayer, W., Seitz, H., Jochims, J. C., Schilling, G. and Schauerte, K. (1971b). Strukter des Vescalagins. *Liebig's Ann. Chem.*, **751**, 60.

Mercier, J. C., Brignon, G. and Ribadeau-Dumas, B. (1973). Structure primaire de la casein κBbovine. Sequence Complete. *Eur. J. Biochem.*, **35**, 222–35.

Mistry, T. V., Cai, Y., Lilley, T. H. and Haslam, E. (1991). Polyphenol interactions. Part 5. Anthocyanin co-pigmentation. *J. Chem. Soc.* (*Perkin Trans. 2*), pp. 1287–96.

Mosedale, A. (1994). *Variation of oak wood properties influencing the maturation of whisky*. D. Phil. Thesis, University of Oxford.

Nagel, C. W. and Wulf, L. W. (1979). Changes in anthocyanins, flavonoids and hydroxycinnamate esters during fermentation and ageing of Merlot and Cabernet Sauvignon. *Amer. J. Enol. Vitic.*, **30**, 111–16.

Norton, I. T., Goodall, D. M., Frangou, S. A., Morris, E. R. and Rees, D. A. (1984). Mechanism and dynamics of conformational ordering in xanthan polysaccharide. *J. Mol. Biol.*, **175**, 371–94.

Ozawa, T., Lilley, T. H. and Haslam, E. (1987). Polyphenol interactions: astringency and the loss of astringency in ripening fruit. *Phytochemistry*, **26**, 2937–42.

Peng, S., Scalbert, A. and Monties, B. (1991). Insoluble ellagitannins in *Castanea sativa* and *Quercus petraea* woods. *Phytochemistry*, **30**, 775–8.

Pettipher, G. L. (1986). An improved method for the extraction and quantitation of anthocyanins in cacao beans and its use as an index of the degree of fermentation. *J. Sci. Food Agric.*, **37**, 289–96.

Porter, L. J., Ma, Z. and Chan, B. (1991). Cacao procyanidins: major flavonoids and identification of some minor metabolites. *Phytochemistry*, **30**, 1657–63.

Rees, D. A. (1977). *Polysaccharide Shapes*. Outline Studies in Biology, Chapman and Hall: London.

Rees, D. A. (1981). Polysaccharide shapes and their interactions – some recent advances. *J. Pure Appl. Chem.*, **53**, 1–14.

Rees, D. A. and Welsh, E. J. (1977). Secondary and tertiary structure of polysaccharides in solution and gels. *Angew. Chem. Int. Ed.*, **16**, 214–24.

Ribereau-Gayon, P. (1974). The chemistry of red wine colour. In *The Chemistry of Winemaking*, editor A. D. Webb, Washington, D.C.: American Chemical Society, pp. 50–87.

Ribereau-Gayon, P. (1982). The anthocyanins of grapes and wines. In *Anthocyanins as Food Colourants*, editor P. Markakis, London and New York: Academic Press, pp. 209–44.

Ribereau-Gayon, P., Pontallier, P. and Glories, Y. (1983). Some interpretations of colour changes in young red wines during their conservation. *J. Sci. Food Agric.*, **34**, 505–16.

Robinson, J., editor (1994). *The Oxford Companion to Wine*, Oxford University Press: Oxford and New York.

Robinson, J. (1995). *Jancis Robinson's Wine Course*, BBC Publications: London, pp. 16–17.

Robinson, G. M. and Robinson, R. (1935). Leuco-anthocyanins and leuco-anthocyanidins. Part I. The isolation of peltogynol and its molecular structure. *J. Chem. Soc.*, pp. 744–52.

Scalbert, A., Duval, L., Peng, S., Monties, B. and Herve du Penhoat, C. (1990). Polyphenols of *Quercus robur* L. II. Preparative isolation by low-pressure and high-pressure liquid chromatography of heartwood ellagitannins. *J. Chromatogr.*, **502**, 107–19.

Scalbert, A., Michon, V. M. F., Peng, S., Viriot, C., Gage, D. and Herve du Penhoat, C. (1991). The structural elucidation of new dimeric ellagitannins from *Quercus robur* L., Roburins A–E. *J. Chem. Soc.* (*Perkin Trans. I*), p. 1653.

Singleton, V. L. (1992). Tannins and qualities of wines. In *Plant Polyphenols*: *Synthesis, Properties and Significance*, editors R. W. Hemingway and P. E. Laks, Plenum Press: New York, pp. 859–80.

Sleigh, R. W., MacKinlay, A. G. and Pope, J. M. (1983). NMR studies of the phosphoserine regions of bovine α_{s1}- and β-casein. Assignment of ^{31}P resonances to specific phosphoserines and cation binding studies by measurement of enhancement of ^{1}H relaxation rate. *Biochim. Biophys. Acta*, **742**, 175–83.

Somers, T. C. (1971). The polymeric nature of wine pigments. *Phytochemistry*, **10**, 2175–86.

Somers, T. C. (1978). Interpretation of colour composition in young red wines. *Vitis*, **25**, 161–7.

Somers, T. C. and Evans, M. C. (1979). Grape pigment phenomena: interpretation of major colour losses during vinification. *J. Sci. Food Agric.*, **30**, 623–33.

Somers, T. C. and Evans, M. C. (1986). Evolution of red wines. I. Ambient influences on colour composition during early maturation. *Vitis*, **25**, 31–9.

Somers, T. C. and Wescombe, L. G. (1987). Evolution of red wines. II. An assessment of the role of acetaldehyde. *Vitis*, **26**, 27–36.

Somers, T. C. and Verette, E. (1988). Phenolic composition of natural wine types. In

Modern Methods of Plant Analysis, New Series, Volume 6. Wine Analysis, editors H. F. Linskens and J. F. Jackson, Berlin: Springer-Verlag, pp. 219–57.

Swain, T. and Hillis, W. E. (1959). The phenolic constituents of *Prunus domestica*. *J. Sci. Food Agric.*, **10**, 135.

Taborsky, G. (1974). Phosphoproteins. *Adv. Protein Chem.*, **28**, 1–210.

Tanaka, T., Takahashi, R., Kuono, I. and Nonaka, G.-I. (1994). Chemical evidence for the de-astringency (insolubilisation of tannins) of persimmon fruit. *J. Chem. Soc. (Perkin Trans. I)*, pp. 3013–22.

Timberlake, C. F. and Bridle, P. (1976). Interactions between anthocyanins, phenolic compounds and acetaldehyde and their significance in red wines. *Amer. J. Enol.*, **27**, 97–105.

Viriot, C., Scalbert, A., Lapierre, C. and Moutounet, M. (1993). Ellagitannins and lignins in ageing of spirits in oak barrels. *J. Agric. Food Chem.*, **41**, 1872–9.

Wildenrat, H. L. and Singleton, V. L. (1974). The production of aldehydes as a result of oxidation of polyphenolic compounds and its relation to wine ageing. *Amer. J. Enol. Vitic.*, **25**, 119–26.

6

Anthocyanin copigmentation – fruit and floral pigments

6.1 Introduction

A surprisingly small number of pigments are responsible for the very wide range of fruit and floral colours observed in nature, Figure 6.1. Chlorophyll is ubiquitous, the betalains are restricted to plants of the Centrospermae, and whilst carotenoids may often contribute extensively to yellow, orange and red pigmentation, flavonoids and anthocyanins are almost universal in their distribution. Thus the distinctive colours of many ripened fruit, such as cherry, whortleberry, elderberry, black and red currants and black grapes (and hence red wines), derive from anthocyanins. These same pigments are also responsible for the exquisite shades of colour of innumerable flowering plants from the delicate salmon-pink of some roses through scarlet, magenta and violet to the deep blue of the gentian and the delphinium. Whilst the name of the Swiss chemist Paul Karrer* is intimately associated with the development of the chemistry of the carotenoids two other great figures, also themselves Nobel Laureates, Richard Willstatter** and Robert Robinson***, are inextricably linked with the anthocyanins. Their work spanned the years before and after the 1914–18 war. Willstatter thus recorded with great sadness the impact which the war had on his researches (Willstatter, 1965):

The laboratory emptied quickly. The pigment solutions spoiled ... the plantings of colourful blossoms stayed unharvested at first, and later we carried buckets of beautiful flowers to the hospitals.

and

In the post-war period I did not want to return to what I had begun. The investigation

* Paul Karrer, b. 1889, Moscow; Ph.D. (Zurich); Univ. Zurich; Nobel Prize 1937.
** Richard Willstatter, b. 1873, Karlsruhe; Ph.D. (Munich); Univ. Zurich, Berlin, Munich; Nobel Prize 1930.
*** Sir Robert Robinson, b. 1886, Chesterfield; Ph.D. (Manchester); Univ. Sydney, Liverpool, St. Andrews, University College London, Manchester, Oxford; Nobel Prize 1947.

seemed like a dismembered body to me and for years I could only think of it with pain...

The chemistry of the anthocyanins was nevertheless continued in the post-war period in Oxford by Sir Robert and Lady (Gertrude Maude) Robinson in a series of masterly researches. It was Lady Robinson's interests which doubtless also stimulated work with J. B. S. Haldane, and members of the John Innes Institute, on the chemical and genetic basis of flower colour variation. This work has been beautifully recorded by the late Rose Scott-Moncrieff, and Haldane offered the view that 'my initiation of it may have been my most important contribution to biochemistry'. Percipiently he also observed that 'at an early stage in her work Scott-Moncrieff needed great tact to collaborate simultaneously with Sir Robert Robinson and myself'. In these pioneering researches both Robinson and Willstatter drew attention to the phenomenon of *copigmentation*, although the full biological implications of these observations have only very recently been appreciated. Thus in contrast to the autonomous carotenoids and chlorophylls, the anthocyanins often require auxiliary molecules (copigments) for the full expression of colour. Some of the principal factors which it is now known are important in the development of the ultimate colour of fruit and flowers are listed in Table 6.1.

6.2 Anthocyanins

The great structural diversity present in the carotenoid floral pigments contrasts with the surprisingly small number of anthocyanidins. The properties of the anthocyanins (glycosylated anthocyanidins) and carotenoids could likewise hardly be more different; carotenoids are lipophilic whilst anthocyanins are hydrophilic and generally dissolved in the cell sap. Similarly in contrast to the carotenoids and xanthophylls, whose colours remain without the need for additional substances, anthocyanins are not generally independent and are often in need of auxiliary molecules (copigments) for the full expression of colour. This crucial observation followed from the work of Willstatter and Robinson. Thus Willstatter and Zollinger (1916) first observed that the addition of 'tannin' to a solution of oenin chloride in weakly acidic media intensified the colour and produced a change of tone, giving a much bluer-red. The Robinsons' interest arose from an examination of the fuchsia flower. The violet inner corolla gave a bluer-red acid extract than that obtained from the bluish-red petals. At first this was regarded as proof that the anthocyanins in the two tissues were different, but it was also observed that the blue-red solution from the petals became much redder on shaking with amyl alcohol. The substance

Table 6.1. *Fruit and floral pigments – colour variation*

1. Variations in pigment class (e.g. chlorophylls, carotenoids, flavonoids and anthocyanins, etc.).
2. Concentrations of individual pigments.
3. Pigment mixture composition.
4. Genetic (i.e. structural) variations within a particular pigment class.

violet / purple

scarlet / magenta

maroon

scarlet

Flower colour variation in Verbena hybrids is due to the independent 'blueing' effects of oxidation ('B' ring hydroxylation) and glycosylation at O-5 ('A' ring). Arrows indicate directions of increased 'blueing'.
5. Anthocyanins – colour is dependent upon concentration, pH, copigmentation and the presence of metal ions.

removed by the amyl alcohol was identified as a 'tannin' based on gallic acid. The phenomenon, the Robinsons proposed (Robinson and Robinson, 1931, 1932, 1939; Robinson, 1939), was the result of the formation of weak additive complexes between the 'tannin' and the anthocyanin. They suggested that it occurs in the flowers themselves and is critically important in flower colour variation. Some of their original comments are worth noting in the light of subsequent developments:

The phenomenon ... is evidently the result of the formation of weak additive complexes which are dissociated at an elevated temperature or by the action of solvents.

We have reason to believe that the formation of these complexes occurs in the flowers themselves and plays a most important part in producing variations of colour.

... we have long been of the opinion that the chief co-pigments are the tannins and

1 . Chlorophylls .

2 . Carotenoids .

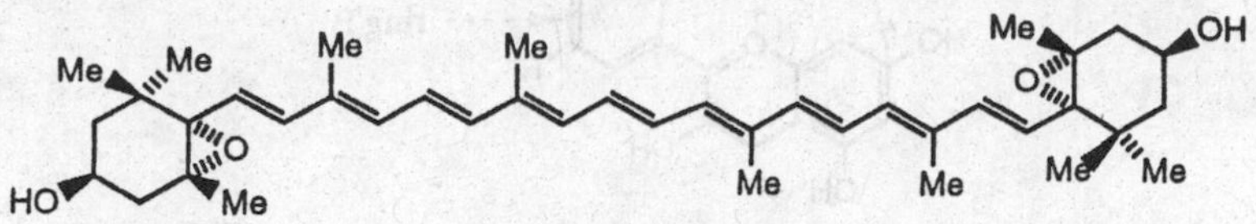

violaxanthin , orange , $\lambda_{max} = 468$ nm

3 . Betalains (Centrospermae) .

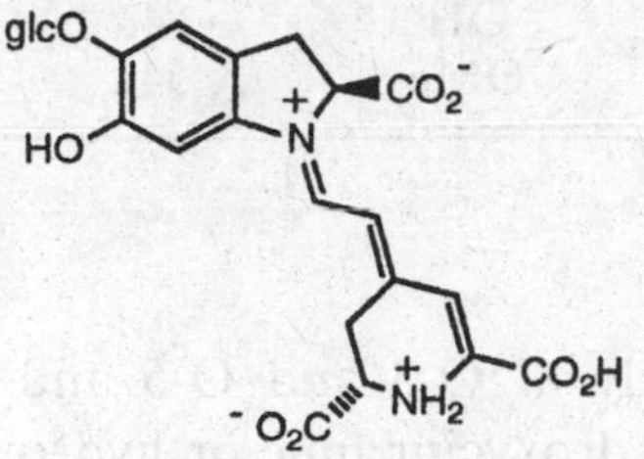

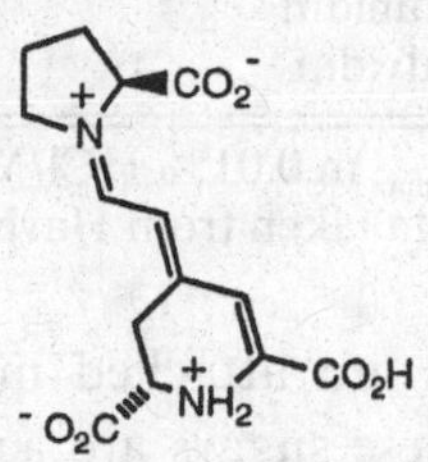

betanin
(*Beta vulgaris*)
red - violet , $\lambda_{max} = 540$ nm

indicaxanthin
(*Opuntia ficus-indica*)
yellow , $\lambda_{max} = 480$ nm

4 . Flavonoids and Anthocyanins .

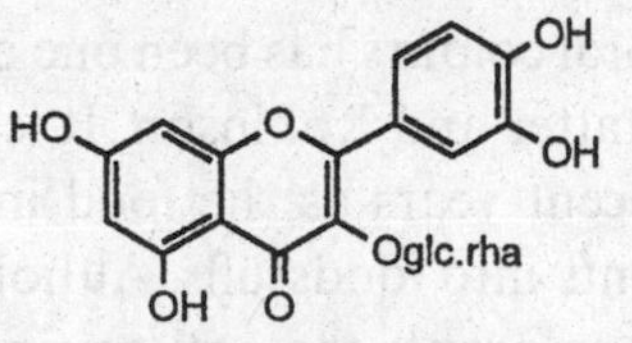

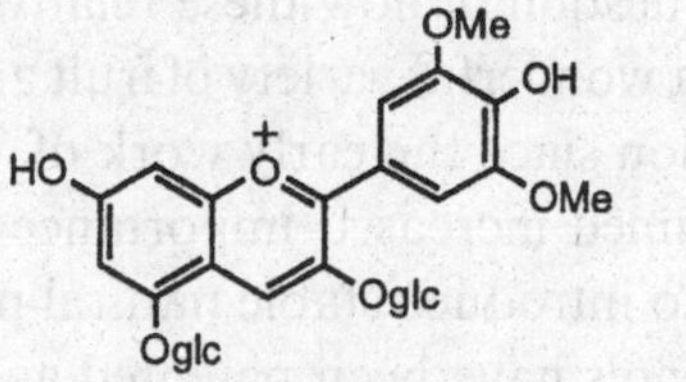

rutin
yellow , $\lambda_{max} = 359$ nm

malvin
red , $\lambda_{max} = 520$ nm
(1% TFA / H_2O)

Figure 6.1. Principal fruit and floral pigments in plants.

anthoxanthins, including the flavones and flavone glycosides… We have found the phenomenon of co-pigmentation almost universal in flower colours almost all of which are bluer than they should be at the pH obtaining in the cell sap.

Six aglycones (anthocyanidins) dominate anthocyanin structures (Table

Table 6.2. *Principal naturally occurring anthocyanidins*

Anthocyanidin	R^1	R^2	λ_{max}nm*
Pelargonidin	H	H	520
Cyanidin	OH	H	535
Paeonidin	OMe	H	532
Delphinidin	OH	OH	546
Petunidin	OMe	OH	543
Malvidin	OMe	OMe	542

* λ_{max} in 0.01% HCl/MeOH (v/v).
Data taken from Harborne (1967).

6.2). Sugars are attached most frequently at O-3 and O-5 and in many instances these sugars are acylated by hydroxycinnamic or hydroxybenzoic acids. The degree of 'blueness' of individual anthocyanins is influenced by the hydroxylation/methoxylation pattern in the anthocyanin 'B' ring (cf. Table 6.1, Verbena flower hybrids). A free hydroxyl group is however essential at one or more of the positions 4', 5, 7 in order to generate *in vivo* all of the colours responsible for fruit and floral pigmentation which are derived from anthocyanins.

The question of how these relatively few anthocyanidin structures give rise to such a wonderful variety of fruit and floral colours has been one of unending fascination since the early work of Willstatter and Robinson. It is one which has assumed increased importance in recent years as the food industry has sought to introduce stable natural pigments into foodstuffs. Although several polyphenols have been patented as food colorants the anthocyanins derived from black grape skins and red cabbage have dominated the commercial scene; as the patents indicate they all, to date, have practical limitations. In an early paper Willstatter attributed the origins of the wide range of colours associated with anthocyanins (and in particular that of blue floral pigments) to variations in the pH of the cell sap. Shibata, a Japanese botanist, and later Hayashi (see Bayer *et al.*, 1966; Hayashi *et al.*, 1958, 1959a,b) proposed an alternative 'metal-ion complex' theory according to which the blue colour due to some anthocyanins derives from complexation with metal ions such as Mg^{2+} and Al^{3+}.

In 1962, Hayashi summarised what were then seen to be the major factors (but whose relative importance was still awaiting resolution), which caused the wide range of colours, which appear in flowers as a result of the presence of anthocyanin pigments. He noted the following as the principal factors which influence flower colour:

(i) the co-existence of several anthocyanins,

(ii) variation in the cellular concentrations of anthocyanins,

(iii) the pH of the cell; Hayashi however concluded on the basis of the evidence then available that 'the cell sap is not so essential a factor as had been previously supposed,'

(iv) the phenomenon of copigmentation which had been proposed by Robinson and Robinson (1931) and observed initially by Willstatter and Zollinger (1916), according to which hypothesis the simultaneous presence of certain colourless organic compounds (tannins and flavone derivatives) in the cell sap may considerably modify the colour of the anthocyanin-containing tissues,

(v) the colloidal condition of the cell sap; an idea first postulated by Robinson and Robinson (1931). This was based upon observations that a majority of anthocyanins display a remarkable colour change when adsorbed on filter paper,

(vi) the metal-ion complex theory whereby the association of anthocyanin pigments with metals (e.g. magnesium, aluminium) gave rise to *blue* metallo-anthocyanins.

During the past three decades investigations by several groups have now thrown considerable light on these proposals. The basis for the wide variations in flower colours which arise from the presence of anthocyanins is now seen to be due to the impact of (i, ii, iii, iv and vi above) although the idea of the colloidal condition of the cell sap (v) has now seemingly been discounted as a major factor in anthocyanin pigmentation.

6.2.1 *Natural anthocyanins*

Natural anthocyanins divide essentially into two large groups: the 3-glycosides and the 3,5-bisglycosides; although anthocyanins with sugar residues at other positions of the anthocyanidin nucleus are no longer such rarities. Very frequently these sugar residues are acylated, most often by one or other of the various hydroxycinnamic acids. These patterns of acylation are invariably of considerable importance when it comes to the rationalisation of the extraordinary stability of polyacylated anthocyanins using the idea of intramolecular copigmentation. There can be little doubt that one of the outstanding (fully authenticated!) examples of the occurrence of a polyacylated anthocyanin and of the phenomenon of intramolecular copigmentation is provided by the Heavenly Blue anthocyanin from the blue petals of the Morning Glory

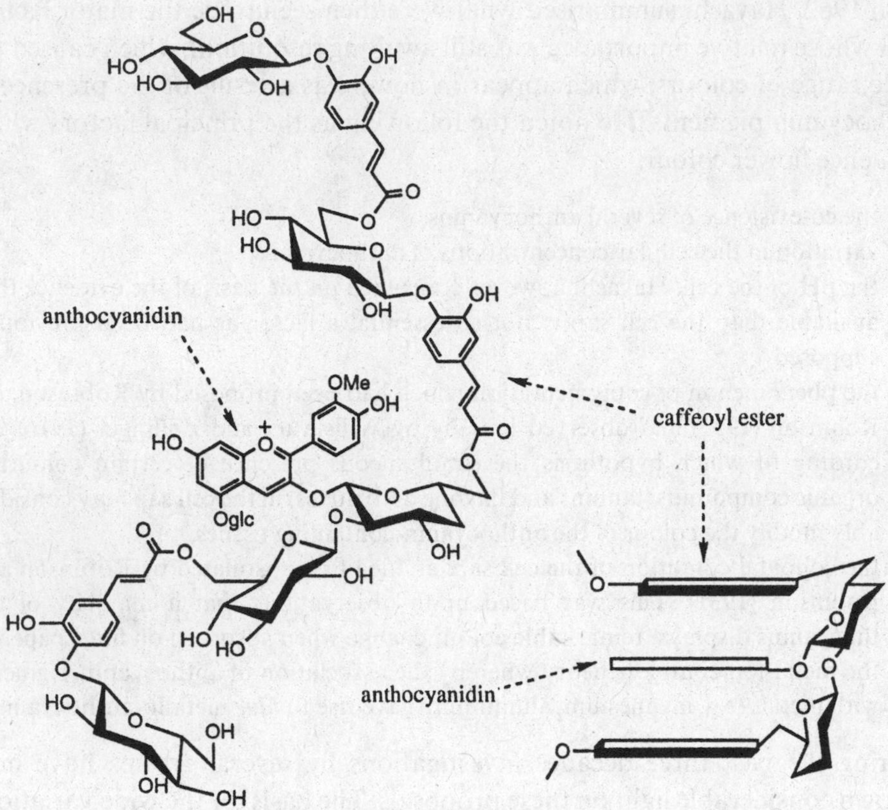

Heavenly Blue Anthocyanin

Figure 6.2. Heavenly Blue anthocyanin (*Ipomoea tricolor*): intramolecular copigmentation (Goto and Kondo, 1991; Brouillard, 1981, 1983).

(*Ipomoea tricolor*). It is the largest monomeric anthocyanin recorded to date and is composed of paeonidin with six molecules of D-glucose and three of caffeic acid; the flavylium cation has a M_R of 1759 (Goto and Kondo, 1991). The blue colour of the flower is caused primarily by the high cell sap pH of 7.5. The anthocyanidin chromophore is therefore probably present as the quinonoidal anhydro-base and/or its mono-anion. It is thought to be stabilised by at least two of the caffeoyl ester groups stacked intramolecularly with the anthocyanidin nucleus, Figure 6.2.

6.2.2 *Location of anthocyanins*

Asen, Stewart and Norris, in an elegant series of papers (Asen *et al.*, 1971, 1972, 1975a) have made several pertinent observations related to the distribution, compartmentation and analysis of anthocyanins *in vivo* in plant tissues. Microscopic examination showed a compartmentalised and sharply delimited location of pigments. In most flower petals colour due to flavonoids is enriched in the epidermal cells and adjacent sub-epidermal cells are colourless. With bi-coloured roses anthocyanins are invariably concentrated in the inner and carotenoids in the outer face of the petal. The concentration of anthocyanins in petals is of the order of 10^{-2} M whilst the pH of the pigmented cell sap lies in the region of 4.5–5.5, although extremes of 2.8 (a begonia cultivar) and 7.5 (in Morning Glory cv Heavenly Blue) were also recorded. Interestingly, in the context of the observations of Sir Robert and Lady Robinson, the pH of epidermal cells from various parts of the same flower were often quite different. Thus in the fuchsia (cv Black Knight) the pH of the pink calyx was 3.8 and that of the deep purple corolla was 5.4. Likewise increase in pH is a major factor in the colour changes (usually blueing) which ensue when flowers senesce. The optical path lengths of pigmented cells varied from 20 to 50 µm, and Asen, Stewart and Norris concluded that it was improbable that anthocyanins *alone* could contribute to the colour of those tissues whose pH is within the compass 4–6 since, in this pH range at the concentrations and optical path lengths specified, most anthocyanin solutions are virtually colourless. They concluded that the phenomenon of copigmentation offers the most logical explanation for the infinite variations of flower colour possible in a given pH range.

When the deep red/magenta coloured crystals of an anthocyanin are dissolved in aqueous media (pH ∼ 4–5) the typically deep red/violet solution which immediately ensues rapidly decolorises to leave a relatively pale red/violet solution. For a long time this decolorisation process was interpreted as damaging to the full expression of plant colour. However this first step –the loss of colour – is now seen as an essential prerequisite to the chemical mechanism which underlies floral pigmentation by anthocyanins. Brouillard (1988) has thus demonstrated that this ability of anthocyanins to exist as several colourless forms is in fact the first in a two stage **decolorisation** then **colour-stabilising** mechanism. Under very weakly acidic conditions four anthocyanin structural types exist in equilibrium (Brouillard and Lang, 1990) (illustrated for malvin, Figure 6.3): the red flavylium cation, the violet quinonoidal anhydro-base, the colourless carbinol bases (in principle four diastereoisomeric forms) and the pale yellow reversed chalcones (not shown). Equilibration between the quinonoidal anhydro-base and the colourless carbinol bases occurs exclusively via the flavylium cation, and Brouillard and Dubois (1977) quote K (carbinol base/quinonoidal anhydro-base) as 1.6×10^2 at $4°C$. The pK values which govern these structural changes fall within the 2.5–7.5 pH range. In the more acidic media the red flavylium ion predominates. At higher pH values other situations appertain and the colour shifts through violet towards blue:

(a) pH 3.5–4.5 – a mixture of the flavylium ion (H_2A^+) and the neutral quinonoidal anhydro-base (HA) is found,
(b) pH 4.5–6.0 – the concentration of the flavylium ion (H_2A^+) becomes vanishingly small and the quinonoidal anhydro-base (HA) increasingly predominates,
(c) around neutrality there is a mixture of both the neutral (HA) and the ionised (blue anionic, A^-) quinonoidal anhydro-base forms present.

Goto and his colleagues (Goto *et al.*, 1976) discovered that anthocyanins, as their quinonoidal anhydro-bases (HA) or as their flavylium cations (H_2A^+) in concentrated aqueous solutions were strongly stabilised by neutral salts such as magnesium chloride and sodium chloride. Anthocyanins when they are present in plant tissues in their quinonoidal anhydro-base forms can be extracted quantitatively by 4M magnesium chloride solutions. It has been suggested (Hoshino *et al.*, 1981a,b) that magnesium chloride stabilises anthocyanins in concentrated solutions by reducing the concentration of free water (by hydration of the magnesium ions), whilst the stabilising action of sodium chloride is probably due to its ability to promote the self-association of anthocyanins, *vide infra*.

6.2.3 Self-association of anthocyanins

Anthocyanin concentration is an important factor influencing the colour of anthocyanin containing media and hence pigmented plant tissues. Asen *et al.* (1972) thus noted that at pH 3.16 the visible λ_{max} varied from 507 to 502 nm by increasing the pigment concentration from 10^{-4} to 10^{-2} M. For this same 100-fold increase in concentration the absorbance increased 300 times. Asen and his colleagues suggested that these deviations from the Beer–Lambert law constituted *prima facie* evidence that anthocyanins (as their flavylium ion salts, Figure 6.3, H_2A^+) may undergo self-association in aqueous media. Scheffeldt and Hradzina (1978) came to similar conclusions. Asen and his group also observed (1975b) that dilution of a solution of a delphinidin glycoside at pH 6.6 (largely the quinonoidal anhydro-base form, HA, Figure 6.3) from 8×10^{-3} M to $8 \times 10^{+5}$ M gave a 'much bluer' solution; the absorbance ratio A_{624}/A_{580} changed from 0.58 to 1.10. Raising the temperature produced a similar effect

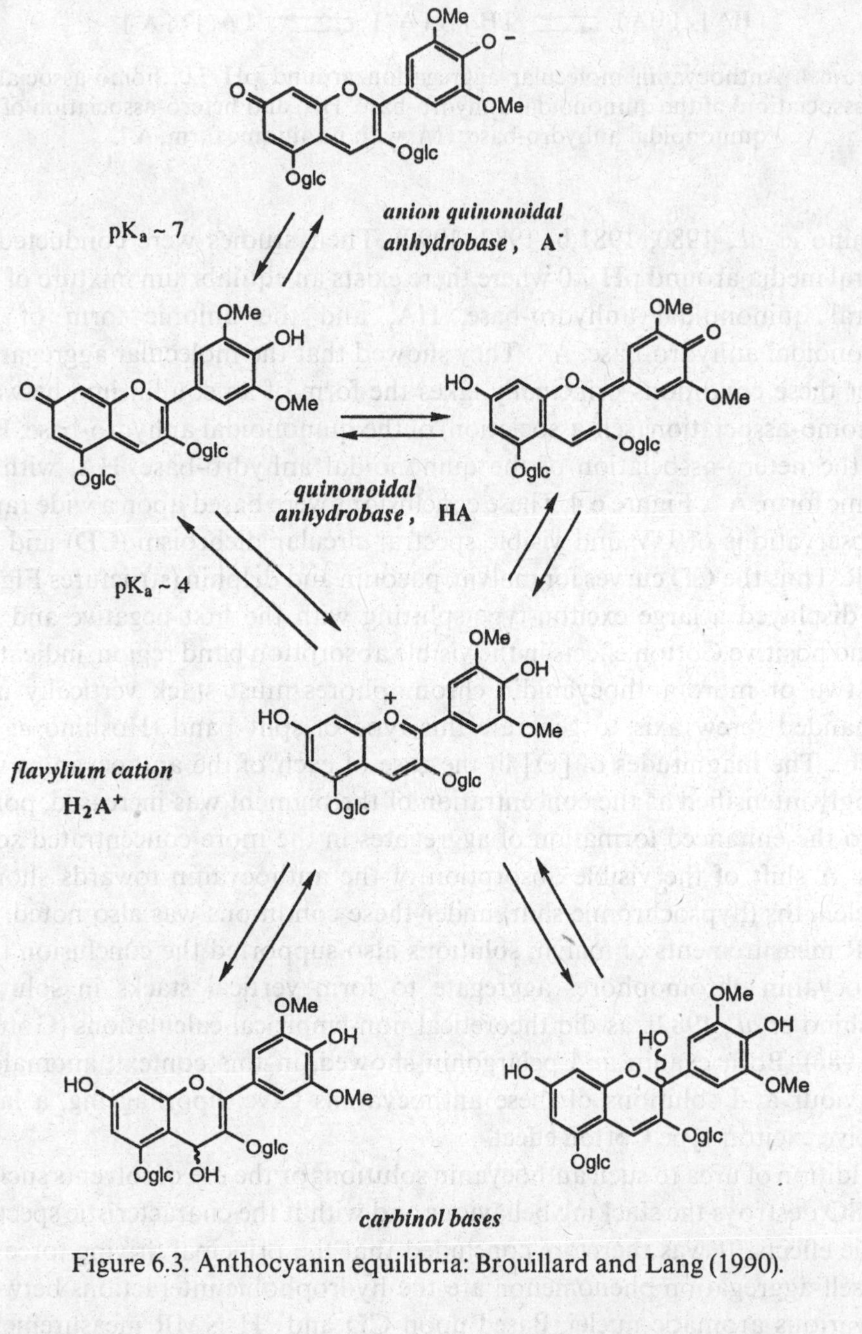

Figure 6.3. Anthocyanin equilibria: Brouillard and Lang (1990).

and in both situations they suggested that this corresponded to a transformation of the associated form of the anthocyanin to its unassociated form.

A detailed and extensive study of the self-aggregation of anthocyanin pigments has been carried out by Hoshino and his colleagues (Hoshino, 1991;

$$[\,HA\,].\,[\,HA\,] \; \rightleftharpoons \; [\,HA\,].\,[\,A^-] \; \rightleftharpoons \; [\,A^-].\,[\,A^-]$$

Figure 6.4. Anthocyanin molecular aggregation around pH 7.0: homo-association (self-association) of the quinonoidal anhydro-base, HA, and hetero-association of the quinonoidal anhydro-base, HA, with its anionic form, A^-.

Hoshino *et al.*, 1980, 1981,b, 1982, 1990). Their studies were conducted in neutral media around pH 7.0 where there exists an equilibrium mixture of the neutral quinonoidal anhydro-base, HA, and the anionic form of the quinonoidal anhydro-base, A^-. They showed that the molecular aggregation under these conditions principally takes the form of an equilibrium between the homo-association (self-association) of the quinonoidal anhydro-base, HA, and the hetero-association of the quinonoidal anhydro-base, HA, with its anionic form, A^-, Figure 6.4. These conclusions were based upon a wide range of observations of UV and visible spectra, circular dichroism (CD) and ^{1}H NMR. Thus the CD curves for malvin, paeonin and delphin (structures Figure 6.5), displayed a large exciton-type splitting with the first negative and the second positive Cotton effects in the visible absorption band region, indicating that two or more anthocyanidin chromophores must stack vertically in a left-handed screw axis to generate this type of split band (Hoshino *et al.*, 1981b). The magnitudes of $[\Theta]$ in the case of each of the anthocyanins was strongly intensified as the concentration of the pigment was increased, pointing to the enhanced formation of aggregates in the more concentrated solutions. A shift of the visible absorption of the anthocyanin towards shorter wavelengths (hypsochromic shift) under these conditions was also noted. ^{1}H NMR measurements of malvin solutions also supported the conclusion that anthocyanin chromophores aggregate to form vertical stacks in solution (Hoshino *et al.*, 1982), as did theoretical non-empirical calculations (Goto *et al.*, 1986). Both cyanin and pelargonin showed, in this context, anomalous behaviour and solutions of these anthocyanins gave, upon ageing, a large positive exciton-type Cotton effect.

Addition of urea to such anthocyanin solutions or the use of solvents such as DMSO destroys the stacking behaviour and with it the characteristic spectroscopic effects. It was therefore concluded that the principal driving force for this self-aggregation phenomenon are the hydrophobic interactions between the various aromatic nuclei. Based upon CD and ^{1}H NMR measurements, Hoshino (1991) suggested a 'head to head' stacking geometry as depicted in Figure 6.6, in which geometrical arrangement the angle between the long axes of each anthocyanin chromophore is not 0° nor 180°, and is less than 90°. Two or more anthocyanin molecules stacking in this manner would give rise to the

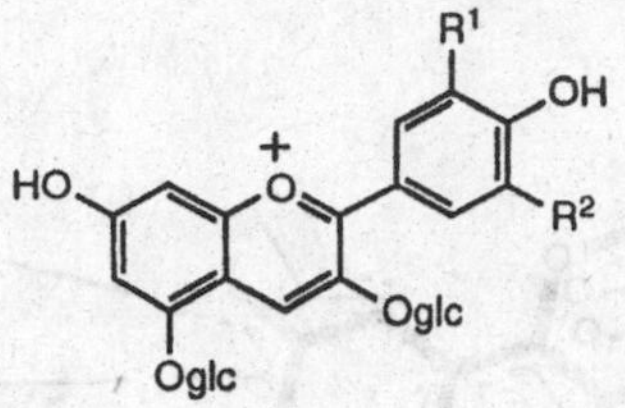

glc = β-D-glucose

Anthocyanin	R^1	R^2
Pelargonin	H	H
Cyanin	OH	H
Paeonin	OMe	H
Delphin	OH	OH
Petunin	OMe	OH
Malvin	OMe	OMe

Figure 6.5. Some common anthocyanins: anthocyanidin-3,5-bis-β-D-glucosides.

type of CD splitting patterns which are observed. The magnitude of the $[\Theta]$ value is intensified as the anthocyanin concentration is increased, thus indicating the enhanced formation of helical stacks of anthocyanin molecules (with a left-handed screw axis) in more concentrated solutions. However, Hoshino also indicated (Hoshino, 1991) that, within the compass of the experimental data, other stacking geometries were possible. Hoshino estimated typical self-association constants for anthocyanins: $\sim 18 \times 10^2$ for the $[HA]_2$ dimer and $\sim 4 \times 10^2$ for the $[A^-]_2$ dimer. Compared to self-association constants (calculated using the isodesmic model), for nucleotide base stacking, e.g. adenosine ~ 15, these values are indicative of the much stronger self-attraction of anthocyanin molecules. These values also suggest that, in the absence of other effects and at the concentrations typically found in flower petals, anthocyanin self-association may well make significant contributions to flower colours *in vivo*.

The extent of stacking of anthocyanin molecules which can occur is not yet clear but models of extended helical aggregates, as proposed by Hoshino (1991), show that the anthocyanin chromophores form an extended, essentially hydrophobic, core and the hydrophilic glucose residues surround this domain and render the supramolecular aggregate water soluble. Such extended packing would necessarily lead to the protection of both faces of the benzopyrilium ring system of the anthocyanin from attack by water and thus ensure the kinetic stabilisation against loss of colour (Hoshino, 1991).

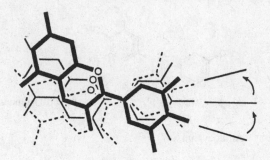

Figure 6.6. A model for the chiral stacking (self-association) of anthocyanin molecules (adapted from Hoshino, 1991; Goto, 1987). A Newman projection of the transition moments in the stacked helical aggregate gives a counterclockwise screw arrangement which is the origin of the negative chirality.

6.3 Copigmentation

Copigmentation bears many analogies to the self-association of anthocyanins; copigments may act either inter- or intra-molecularly and the thermodynamic driving forces are probably very similar in origin to those which are of importance in anthocyanin self-association.

6.3.1 Intermolecular copigmentation

The observation that the colour of isolated anthocyanins could be varied by the presence of other substances (copigments) was first made by Willstatter and Zollinger (1916) and by Robinson and Robinson (1931, 1939). Both groups noted that addition of 'tannin' induced a bathochromic shift in the visible absorption of the anthocyanin. The Robinsons observed that copigmentation was almost universal in floral colours and they concluded that it is entirely the result of the formation of weak additive complexes between the anthocyanin and the copigment which may be dissociated at elevated temperatures or by the action of solvents. Copigments have little or no visible colour by themselves but when added to an anthocyanin solution they greatly enhance the colour of that solution (Goto *et al.*, 1986; Goto, 1987; Asen *et al.*, 1972; Chen and Hradzina, 1981; Brouillard, 1983, 1988; Brouillard *et al.*, 1989; Mistry *et al.*, 1991), cf. Figure 6.7 – the copigmentation effect of quercetin-3-β-D-galactoside upon solutions of malvin chloride. Copigmentation has been shown to be dependent upon anthocyanin and copigment type, concentration, pH, temperature and the presence of metal salts. Although various groups of compounds have been identified as able to act as potential copigments, flavonoids

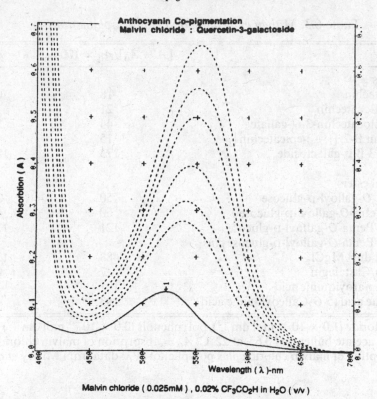

Figure 6.7. Malvin chloride (1.0×10^{-4} mol dm^{-3}) in acetate buffer (0.2 mol dm^{-3}, pH 3.65). Changes in the anthocyanin visible spectrum (———) produced by the addition of quercetin-3-β-D-galactoside (- - - - -). Final pigment to copigment ratio 1:30; from Mistry *et al.*, 1991.

(e.g. quercetin-3-β-D-galactoside) and hydroxycinnamoyl esters (e.g. chloro-genic acid) are perhaps the most commonly found in nature.

quercetin-3-β-D-galactoside

Copigmentation gives rise to two characteristic features in the UV–visible spectra of the anthocyanin: namely a positive shift in the intensity of absorb-tion in the visible range (*hyperchromism*) and a positive shift in the wavelength of the absorption maximum of the visible band (*bathochromism*), cf. Figure 6.7. It is generally assumed that non-covalent complex formation between an-thocyanin and copigment is responsible for this intensification and change in colour of the anthocyanin.

Table 6.3. *Malvin chloride–polyphenol copigmentation*

	$[A - A_0]/A_0 \times 100\%$	$\lambda - \lambda_0$, nm
Flavonoids		
(−)-Epicatechin	18	0.8
(−)-Epigallocatechin	21	1.6
(−)-Epigallocatechin-3-O-gallate	44	2.3
Procyanidin B-2: [(−)-epicatechin]$_2$	15	—
Quercetin-3-β-D-galactoside	173	18.9
Polygalloyl esters		
β-1,2,6-Tri-O-galloyl-D-glucose	50	3.2
β-1,2,4,6-Tetra-O-galloyl-D-glucose	60	4.8
β-1,2,3,4,6-Penta-O-galloyl-D-glucose	121	12.0
β-1,2,3,4,6-Penta-O-galloyl-D-glucose *plus*		
0.25 mol dm^3 MgCl$_2$	488	12.0
Vescalagin/Castalagin	7	0.5
5-O-p-Coumaroylquinic acid	101	1.6
Chlorogenic acid (5-O-Caffeoylquinic acid)	143	4.0

Malvin chloride (1.0×10^{-4} mol dm^{-3}), polyphenols (2.0×10^{-4} mol dm^{-3}) in 0.2 mol dm^{-3} acetate buffer, pH 3.65, at 22°C. A_0 = absorption of malvin chloride at λ_0, A = absorption of malvin chloride *plus* polyphenol at λ; data from Mistry *et al.*, 1991.

At pH 3.65 (0.2 mol dm^{-3} acetate buffer) and ambient temperature natural proanthocyanidins (condensed tannins) and related flavan-3-ols display very small copigmentation effects with both cyanin chloride and malvin chloride (Table 6.3). Natural galloyl esters of D-glucose (hydrolysable tannins) however give rise not only to significant bathochromic shifts in the visible absorbtion of malvin chloride (Table 6.3, $\lambda - \lambda_0$) but also to increases in absorbtivity ($A - A_0$). These effects are reduced as the temperature is increased, but are enhanced in the presence of magnesium chloride (0.25 mol dm^{-3}). Galloyl esters of flavan-3-ols similarly display significant copigmentation effects compared to the parent flavan-3-ols, and this suggests that it is the galloyl ester group which confers this property on the molecule.

The biosynthetic introduction of a hexahydroxydiphenoyl group into a metabolite in place of two vicinal galloyl ester groups by the process of dehydrogenation generally gives rise to new metabolites with a substantially reduced conformational mobility; the apotheosis of this structural effect is seen most clearly in the case of the diastereoisomeric pair, vescalagin and castalagin, metabolites of *Quercus* and *Castanea* species. Both are formally analogues of the key metabolite β-1,2,3,4,6-penta-O-galloyl-D-glucose but with six hydrogens less in the molecular structure. The presence of three diphenyl linkages and the formation of a C-glucosidic bond severely constrains their conformational mobility but conversely endows them with considerably enhanced water solubility and *hydrophilic* character. As in related studies of polyphenol complexation (e.g. peptides and proteins) these two factors give these metabolites a substantially reduced proclivity to associate with other molecules when compared to β-1,2,3,4,6-penta-*O*-galloyl-D-glucose; in this particular instance they are, unsurprisingly, poor copigments, Table 6.3. Similar observations were also made by Eugster and Nayeshiro (1989) who demonstrated that, although significant quantities of ellagitannins (hexahydroxydiphenoyl esters) are found in rose petals, they are generally poor copigments. In the context of natural copigments the flavonol glycosides, such as quercetin-3-β-D-galactoside, are however, on a comparative molar basis, clearly pre-eminent (Mistry *et al.*, 1991; Eugster and Nayeshiro, 1989).

vescalagin / castalagin

- 6H

β-1,2,3,4,6-penta-O-galloyl-D-glucose

Although it has been speculated that copigmentation is facilitated by intermolecular hydrogen bonding, Goto proposed (Goto *et al.*, 1986; Goto, 1987) that the phenomenon originates intermolecularly by the vertical 'hydrophobic stacking' of the aromatic nuclei of the anthocyanin and the phenolic copigment. Brouillard and his colleagues (Brouillard *et al.*, 1989) came to very similar conclusions and they suggested that it was unique to the aqeous environment in which copigmentation naturally takes place. In aqueous media

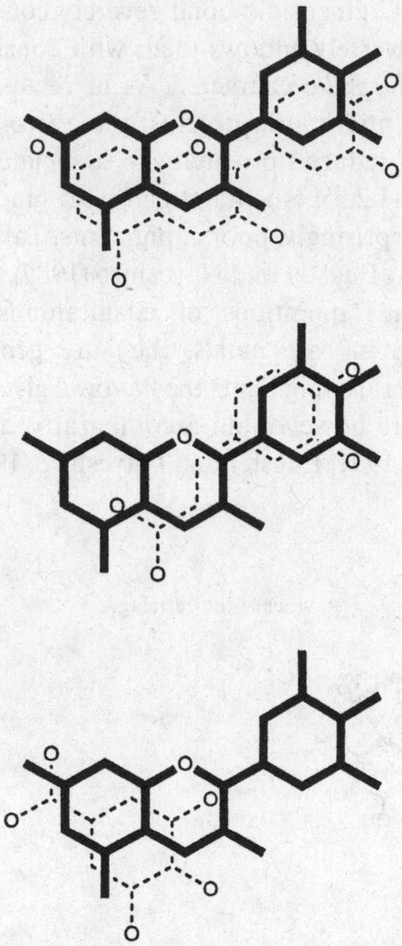

flavonols

hydroxycinnamoyl
esters

galloyl esters

Figure 6.8. Anthocyanin copigmentation: 'π–π' overlap – some comparisons.

they hypothesised that hydrophobic effects direct the anthocyanin chromophore and the copigment to form a π–π complex, which in turn protects the
anthocyanin from hydration (and hence decolorisation). This model broadly
rationalises the general observations on copigmentation since, simply on the
basis of planar spatial characteristics, the most efficient π–π overlap and
interactions would occur with a planar flavonol as compared to a hydroxycinnamoyl or galloyl ester, or the non-planar flavan-3-ols and the sterically
constrained ellagitannins such as vescalagin and castalagin, Figure 6.8.

Brouillard and his colleagues (Brouillard *et al.*, 1991) have shown that
solvent composition is an intrinsic part of the copigmentation phenomenon.
Copigmentation still takes place in binary mixtures, providing that water is the
major component of the medium. They demonstrated that the addition of

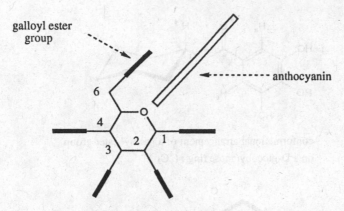

galloyl ester group

anthocyanin

6

4

2

1

3

β-1,2,3,4,6-penta-O-galloyl-D-glucose

Figure 6.9. Anthocyanin copigmentation: suggested model for the intercalation of the anthocyanin chromophore between the two galloyl ester groups at positions C-1 and C-6 in the molecule of β-1,2,3,4,6-penta-*O*-galloyl-D-glucose (Mistry *et al.*, 1991).

various cosolvents to the aqueous medium – methanol, ethanol, formamide, dimethylformamide, acetone, acetonitrile – invariably reduces the copigment effect. They suggested that the extent of copigmentation is related, not only to the polarity of the water molecule but probably, and more importantly, to the hydrogen-bonded network of the bulk water medium.

The enhanced effectiveness of many of the natural galloyl esters of D-glucose as anthocyanin copigments probably also lies in their ability to form a cleft or pocket into which the anthocyanin may intercalate, Figure 6.9. High resolution ^{1}H NMR experiments with malvin and β-1,2,3,4,6-penta-*O*-galloyl-D-glucose in D_2O (containing 1.1% CF_3CO_2D) suggested a preference for anthocyanin complexation at or near the galloyl ester groups at C-1 and C-6. A model was proposed, Figure 6.9, for the formation of a 1:1 complex in which the two galloyl ester groups at C-1 and C-6 are maintained in the optimum position to provide a sandwich conformation for hydrophobic π–π stacking with the anthocyanin chromophore. This conformation of the β-1,2,3,4,6-penta-*O*-galloyl-D-glucose molecule would be enhanced by the buttressing effect of the remaining galloyl ester groups on the D-glucopyranose ring. Parenthetically, however, it should be noted that the above suggestion is dependent on the *presumed* most favoured conformational arrangement of galloyl ester groups attached to the six-membered D-glucopyranose ring (Haslam, 1967), which was deduced from IR, X-ray and ^{1}H NMR evidence related to the preferred conformation of ester groups attached to a cyclohexane ring (Owen and Sheppard, 1963; Mathieson, 1965; Culvenor, 1966). Owen and

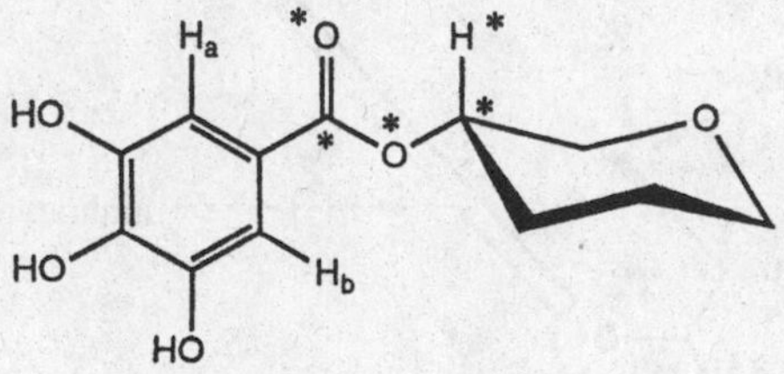

conformational arrangement of a galloyl ester group
on a D-glucopyranose ring : ($^4C_1$) .

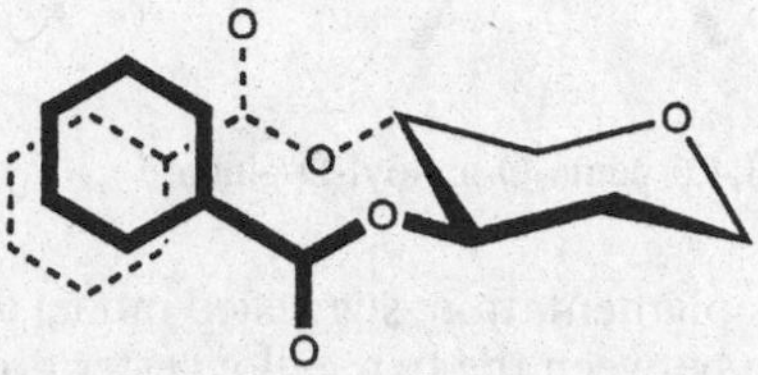

possible ' π - π ' stacking of galloyl ester groups
attached to a D-glucopyranose ring

Figure 6.10. Possible conformational arrangements of galloyl ester groups attached to
a D-glucopyranose ring (Haslam, 1967).

Sheppard indicated a conformational preference for the characteristic *cis*-
orientation of ester groups which they suggested arises from electron pair
(non-bonded or π bonds) repulsions, Figure 6.10. Mathieson and Culvenor
further argued that an ester group, attached to an equatorially disposed
oxygen atom in a six-membered ring adopts a conformation in which the
atoms indicated (*) are co-planar and in which the carbonyl oxygen atom and
the secondary hydrogen atom at the position of substitution are adjacent and
eclipsed, Figure 6.10. In the consideration given to the preferred conformational arrangement of galloyl ester groups attached to the D-glucopyranose
ring (Haslam, 1967) it was assumed that these same considerations applied
and, in order to maintain conjugation, the aromatic nucleus is also co-planar
with the carbonyl group of the ester function. If these arguments are correct
then the aroyl ester groups at C-1, C-2, C-3, C-4 in β-1,2,3,4,6-penta-*O*-
galloyl-D-glucose will lie in a plane broadly perpendicular to that of the
D-glucopyranose ring, Figures 6.9 and 6.10. In this conformational arrangement the galloyl ester groups at C-1, C-2, C-3, C-4 are positioned in such a way
that a form of 'π–π' stacking of the aromatic rings is also possible, Figure 6.10.
However, it may well be that at this stage such seductive suggestions should be
treated cautiously in the absence of any X-ray data on the conformation of
natural galloyl esters. Thus X-ray analysis of the crystal structure of methyl

gallate (Cai *et al.*, 1990) indicates that the galloyl ester groups **do not** exhibit 'π–π' stacking but prefer to adopt a typical 'herring bone' arrangement. Furthermore such a picture does not readily accommodate observations on the high resolution ^{1}H NMR spectra of compounds such as β-1,2,3,4,6-penta-*O*-galloyl-D-glucose. These show each of the five galloyl ester groups as sharp singlets and **not** as AB quartets. This indicates that at ambient temperatures there must be rapid site exchange of the two galloyl ester protons (H_a, H_b, Figure 6.10) associated with each galloyl ester group, Figure 6.10. In the absence of any other explanation the possibility of unhindered free rotation of the aromatic nuclei of each of the galloyl ester groups in β-1,2,3,4,6-penta-*O*-galloyl-D-glucose must therefore be admitted.

Although the effectiveness of a particular copigment seems to be related, at least in part, to its planarity and the potential surface area of the molecule available for hydrophobic / 'π–π' stacking, Figure 6.8, other factors are also of significance. For example Lady Robinson (1939) first noted that chlorogenic acid (5-*O*-caffeoyl quinic acid) is an important natural anthocyanin copigment in plants and its properties have been extensively investigated by Brouillard (Brouillard *et al.*, 1989). Mistry and his colleagues (Mistry *et al.*, 1991) showed that chlorogenic acid is a rather more effective copigment than its *p*-coumaroyl analogue, Table 6.3. It therefore seems probable that the electron donating capacity of the phenolic copigment may well be of additional importance in the stabilisation of the flavylium cation.

R = H ; 5-O-*p*-Coumaroyl quinic acid

R = OH ; 5-O-Caffeoyl quinic acid ,
chlorogenic acid

Some quantitative measurements of the intermolecular copigmentation effect have been reported. Brouillard's group has thus described a graphical procedure to determine the equilibrium constant (K) for the association of the flavylium cation (AH_2^+) with the copigment (CP) which also gives the number of molecules of copigment (n) complexed with the anthocyanin (Brouillard *et al.*, 1989, 1991). In the particular situation of the copigmentation of malvin with chlorogenic acid at pH 3.65, 20°C, 0.2 M ionic strength, they estimated the value of K as 390 ($\pm$ 50) $M^{-1}dm^3$ and $n = 1$. In the later study they quoted the value of 350 for the same copigmentation reaction (Brouillard *et al.*, 1991).

$$n\,CP \; + \; [\,AH_2\,]^+ \; \underset{}{\overset{K}{\rightleftarrows}} \; [\,AH_2\,]^+ \cdot n\,[\,CP\,]$$

Mistry and his colleagues (Mistry *et al.*, 1991) similarly showed that the complexation of malvin with quercetin-3-β-D-galactoside and with β-1,2,3,4,6-penta-*O*-galloyl-D-glucose involves the formation of 1:1 molecular complexes. By following the simultaneous changes in visible absorption at two wavelengths in the copigmentation experiment they estimated values of k (in H_2O containing CF_3CO_2H at 20°C) of 1686 ($\pm$ 58) $M^{-1}\,dm^3$ for quercetin-3-β-D-galactoside as copigment and 987 ($\pm$ 37) $M^{-1}\,dm^3$ for β-1,2,3,4,6-penta-*O*-galloyl-D-glucose as copigment. Using an alternative high resolution ^1H NMR procedure they determined under slightly different conditions (D_2O: CF_3CO_2H, 1.1% v/v, 45°C) a value for K [malvin]·[β-1,2,3,4,6-penta-*O*-galloyl-D-glucose] of 508 ($\pm$ 8) $M^{-1}\,dm^3$.

Asen *et al.* (1972) originally noted that anthocyanin copigmentation arose by association of the copigment with the 'red flavylium ion and the purple anhydrobase'. More recent work has shown the validity of that suggestion. The idea that anthocyanin (flavylium ion, AH_2^+) stabilisation occurs via a mechanism of 'π–π'/hydrophobic stacking of the anthocyanin chromophore with aromatic nucleic in the copigment logically led Brouillard and Haslam and their colleagues (Brouillard *et al.*, 1991; Mistry *et al.*, 1991) to studies of the effect of purine and pyrimidine bases, nucleotides and nucleic acids as anthocyanin copigments. Thus it was shown (Mistry *et al.*, 1991) that both RNA (sodium salt *ex.* calf thymus gland) and DNA (sodium salt *ex.* salmon testes) are both good copigments for the malvin flavylium ion at pH 3.42. The DNA: malvin complex was readily precipitated from solution by the addition of magnesium chloride (0.1 mol dm^{-3}). Comparisons with the disodium salt of ATP suggested that the two nucleic acids probably act in an analogous manner towards anthocyanins as galloyl ester copigments (Figure 6.9) by providing sites for intercalation between adjacent strand base pairs and subsequent 'π–π'/hydrophobic stacking. The N-methylated xanthines caffeine and theophylline also displayed distinctive copigmentation effects (Figures 6.11 and 6.12). Thus it was readily apparent that both caffeine and theophylline *preferentially* stabilise the quinonoidal anhydro-base form of malvin (AH). Indeed in the pH range 4.5–7.0 caffeine and theophylline are much more effective copigments than the various hydroxycinnamoyl and galloyl esters, giving rise to stable violet through to blue colours of the anthocyanin. These qualitative observations were given quantitative support by the work of the

Strasbourg group (Brouillard *et al.*, 1991) who measured the stability constants in water at 20°C for the complexation of malvin with chlorogenic acid and caffeine respectively, Table 6.4. The values quoted show the stability constants for complexation with malvin as the flavylium ion (AH_2^+), K_1. as the quinonoidal anhydro-base (AH), K_2, and as the anion of the quinonoidal anhydro-base (A^-), K_3.

The property of intermolecular copigmentation is most satisfactorily rationalised in terms of the existence of interactions between different 'π' systems, driven initially by strong hydrophobic effects in the aqueous media. Natural phenolic esters and flavonoids may also, to some extent, be regarded as electron rich 'π' systems able to associate with the comparatively electron deficient flavylium cation and/or the quinonoidal anhydro-base forms of the anthocyanin; similar in some ways to the analogous *p*-quinol: *p*-benzoquinone complexes alluded to elsewhere. However, although copigmentation effects for a given molecule may be predicted qualitatively and empirically, there is as yet little guidance available as to the more precise and intimate details of the complexation processes which are involved. Thus, for example, it would be very valuable to be able to explain the observation that chlorogenic acid and the various natural galloyl esters preferentially complex with the flavylium cation (AH_2^+) whilst caffeine associates more strongly with the quinonoidal anhydro-base (AH) form of the anthocyanin.

6.3.2 Intramolecular copigmentation

Although relatively few anthocyanidins, the aglycones of anthocyanins, have been found in Nature, Table 6.2, a great many anthocyanins have been isolated. The most widely distributed are anthocyanidin-3-glycosides and anthocyanidin-3,5-bisglycosides (e.g. Figure 6.5). Over the past 25 years, as extraction techniques and physical methods of analysis have been refined, many of these anthocyanidin glycosides have been shown to exist in the form of polyacylated derivatives. Hydroxycinnamic acids (*p*-coumaric, caffeic, ferulic) and *p*-hydroxybenzoic acid have thus been commonly located as acylating groups in anthocyanin structures. Typical examples are the Heavenly Blue anthocyanin from *Ipomoea tricolor*, Figure 6.2, zebrinin (*Zebrina pendula*), gentiodelphin (*Gentiana makinoi*), platyconin (*Platycodon grandiflorum*) and violdelphin (*Delphinium hybridum*); see Goto and Kondo (1991).

In 1971 Saito and his colleagues (Saito *et al.*, 1971) observed that platyconin exhibited unusual stability for an anthocyanin in neutral or weakly acidic aqueous solution. However, removal of the acylating groups rendered the anthocyanin unstable in aqueous media (*vide supra*). Both Goto and Brouillard

Table 6.4. *Stability constants in water at 20°C for the complexation of malvin with chlorogenic acid and caffeine*

Copigment	K_1, M^{-1} dm^3	K_2, M^{-1} dm^3	K_3, M^{-1} dm^3
Chlorogenic acid	350	140	40
Caffeine	150	250	110

Data from Brouillard *et al.*, 1991.

R = Me ; caffeine

R = H ; theophylline

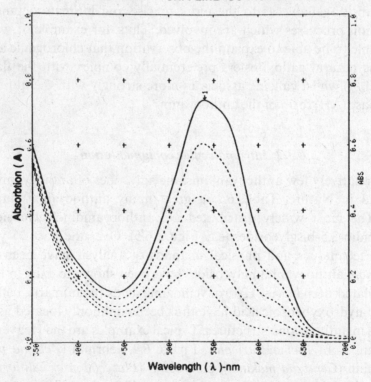

ANTHOCYANIN - CAFFEINE CO-PIGMENTATION

pH 3.42 (0.2 M acetate) , malvin chloride 2.6 x 10⁻⁴ Mole l⁻¹,
λ_0 = 523.2 nm , $\Delta\lambda$ = 16.8 nm .

Figure 6.11. Changes in visible spectrum of malvin chloride (2.6×10^{-4} mol dm^{-3}) produced by the addition of aliquots of caffeine ($\cdots$) at pH 3.42. Final pigment to copigment ratio (———, $1:150$).

ANTHOCYANIN - CAFFEINE COPIGMENTATION

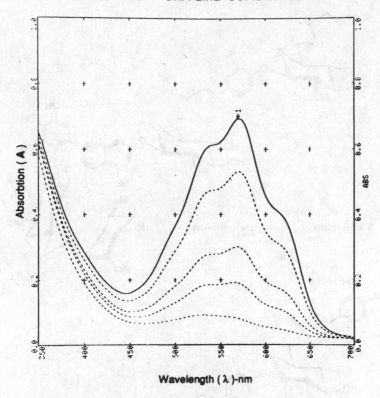

pH 4.99 (0.2M acetate buffer) , malvin chloride 2.35 x 10⁻⁴ Mole l⁻¹

Figure 6.12. Changes in visible spectrum of malvin chloride (2.3×10^{-4} mol dm^{-3}) produced by the addition of aliquots of caffeine (···) at pH 4.99. Final pigment to copigment ratio (———, 1:150).

(Goto *et al.*, 1982, 1983; Brouillard, 1981, 1983) proposed that the inherent stability of such acylated anthocyanins derived from the ability of aromatic acyl groups to form an intramolecular sandwich with the anthocyanidin as the filling, Figures 6.2 and 6.13. Such intramolecular π–π stacking arises from hydrophobic interactions between the two electron rich phenolic nuclei of the acyl groups and the electron deficient anthocyanidin (as the flavylium ion or as the quinonoidal anhydro-base) and protects the anthocyanidin from attack by water to form the pseudo-base. The driving forces for these stabilising interactions are thus entirely analogous to those suggested for intermolecular copigmentation and the self-association of anthocyanins. Evidence in favour of these proposals has been derived from high resolution ^{1}H NMR and CD methods (Goto *et al.*, 1986).

Violdelphin   (*Delphinium hybridum*)

Platyconin   (*Platycodon grandiflorum*)

6.4 Metalloanthocyanins

In an early paper Willstatter attributed the origins of the wide range of colours associated with the presence of anthocyanins in plants (and in particular the exquisite blue floral pigments such as those of the delphinium and the gentian), to variations in the pH of the cell sap. Shibata, a Japanese botanist, and later Hayashi (Hayashi *et al.*, 1958, 1959a,b) proposed a metal-ion complex theory, according to which the blue colours due to anthocyanins derive from com-

Gentiodelphin (*Gentiana makinoi*)

Zebrinin (*Zebrina pendula*)

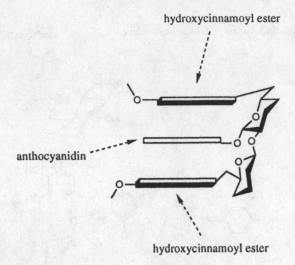

hydroxycinnamoyl ester

anthocyanidin

hydroxycinnamoyl ester

Figure 6.13. Model for the intramolecular stacking of two hydroxycinnamoyl ester groups with the anthocyanidin nucleus in a polyacylated anthocyanin (adapted from Brouillard, 1983).

plexation with metal ions, such as Mg^{2+} and Al^{3+}. Hayashi isolated a blue coloured anthocyanin, commelinin, in a crystalline form from the deep blue flower petals of *Commelina communis*. He reported that it consisted of two molecules each of an anthocyanin, awobanin, and a flavone, flavocommelin, and one atom each of potassium and magnesium. A distinctive feature of this new anthocyanin was its stability – the colour is stable and essentially remains unchanged from pH 1.0 to 7.0. Hayashi rationalised the colour and enhanced stability of the blue floral pigment in aqueous media in terms of a coordination complex of magnesium (potassium was found not to be an essential component) with two molecules of awobanin (anthocyanin) and two of flavocommelin (flavone) (Hayashi and Takeda, 1964, 1970), and gave the generic name of metalloanthocyanin to such complexes (Hayashi *et al.*, 1959a,b). However since anthocyanins generally do not (in isolation) form stable complexes with magnesium, Bayer (Bayer *et al.*, 1966) was disinclined to agree with this conclusion. Bayer himself isolated another blue metalloanthocyanin from the blue cornflower, *Centaurea cyanus*, which he named protocyanin (Bayer, 1958). Bayer indicated that the pigment had a relative molecular mass of ~ 6200 and contained $\sim 19\%$ cyanin and $\sim 80\%$ of a polysaccharide as well as ferric and aluminium ions. He proposed a structure for protocyanin in which two cyanin molecules and polygalacturonic acid were coordinatively bound by the intermediacy of trivalent metal ions. Asen and Jurd (1967) later isolated from cornflower a crystalline blue pigment which they named cyanocentaurin, but Osawa (1982) concluded that protocyanin and cyanocentaurin were in fact one and the same pigment. These two blue pigments, commelinin and protocyanin, show two very distinctive properties in common which set them apart from other anthocyanins. They are not dialysable and they migrate to the *anode* upon electrophoresis at pH 4–6, indicating that they are very probably high molecular weight pigments which bear negative charge(s). Harborne (1976) has critically reviewed much of this earlier work and the general significance of metal-ion complexation with anthocyanins and its contribution to flower colour.

Now thanks to the superb work of Professor Goto and his group in Nagoya some of these problems are nearing resolution (Kondo *et al.*, 1992, 1994; Goto and Kondo, 1991); the intimate structural details of these complex supramolecular structures and the role which metal ions play in the maintenance of their structural integrity and the stabilisation of the anthocyanin chromophore are now being revealed for the first time. Detailed structural work has thus been reported on the blue pigment, commelinin, isolated originally by Hayashi from the petals of *Commelina communis*. Commelinin is readily soluble in water and its colour is stable in concentrated solutions. When diluted, however, these solutions rapidly become colourless, suggesting that the pigment is a supramolecular complex held together by a range of non-covalent forces. Goto and his group showed that the pigment had the overall composition $[M_6F_6Mg_2]^{6-}$ in which M was the molecule malonylawobanin (an anthocyanin as its quinonoidal anhydro-base) and F a flavone, flavocommelin. The sites of metal ion chelation in the complex are the *ortho*-oxygen functionalities on ring 'B' of the anthocyanin (M). Detailed spectroscopic studies and an X-ray analysis of the cadmium analogue of commelinin have revealed the intimate structure and arrangement of the various components in the supramolecular structure (Kondo *et al.*, 1992). The gross structure has been proposed to be of the form: *cyclo*-MMFFMMFFMMFF – in which each of two M and F molecules are in an asymmetric unit correlated by three-fold crystallographic symmetry, Figure 6.14. The cyclic array is held together by hydrophobic forces and coordination to magnesium ions at the centre of the complex. Anthocyanin pairs (M_2) and flavone pairs (F_2) are stacked with the distance between aromatic nuclei ~ 3.5 Å, c.f. Figure 6.13. The three self-associated M_2 units are placed around the three-fold axis alternatively with self-associated F_2 units: each M unit is thus simultaneously copigmented with a molecule of F in a cross-parallel manner. The blue colour of *Commelina communis* is due to the anion of the quinonoidal anhydro-base (M) stabilised by coordination to magnesium.

Stabilisation of the pigment arises from (i) metal-ion complexation, (ii) self-association (M_2) and (iii) copigmentation (MF) in the supramolecular structure, Figure 6.15.

Flavocommelin (F)

Malonylawobanin (M)

More recent work has shown that protocyanin, the blue pigment of the cornflower *Centaurea cyanus* isolated by Bayer in 1958, possesses a similar supramolecular composition to commelinin (Kondo *et al.*, 1994). It has the overall composition $[S_6T_6Fe^{3+}Mg^{2+}]$ and with a molecular formula of $C_{366}H_{384}O_{228}FeMg$ a M_R of 8511! The component S is a succinyl cyanin and T is a malonylflavone and the pigment is thought to be analogous in its overall supramolecular design to that of commelinin.

6.5 The blue rose?

The imagery conjured up by the rose is richly evocative. Richard Herrick warned of the passage of time – 'gather ye rosebuds while ye may'; Robert Louis Stevenson likewise of the perils of marriage – 'marriage is like life in this – that it is a field of battle and not a bed of roses'; Robert Burns memorably compared his love to 'a red, red rose that's newly sprung in June'. The Wars of

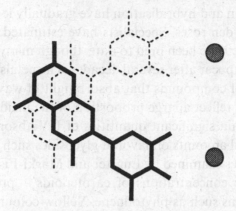

Figure 6.14. Diagrammatic view of stacking of anthocyanin units (M) in commelinin – full lines show top molecule; broken lines the bottom side molecule. The chromophores are arranged in a left-hand screw manner. Magnesium ions are shown as small spheres (Kondo *et al.*, 1992).

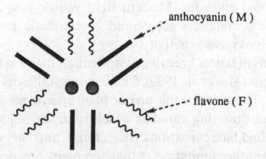

Figure 6.15. Diagrammatic representation of supramolecular structure of the anthocyanin pigment commelinin; magnesium ions at the centre of the complex are shown as small spheres (Kondo *et al.*, 1992).

the Roses began in 1455 in the streets of St. Albans and took their name from the emblems of the two rival branches of the House of Plantagenet – Lancaster (red) and York (white; though its provenance as a rose might be disputed). Today political parties use the symbol of a rose to enhance their appeal to the electorate (but roses fade and die away!). With such associations it is no surprise that the rose has been a favourite subject of plant breeders, for centuries past and to this day, in the search for novelty. Its appeal may be gauged by the fact that the rose cut-flower market is currently estimated as ~ US $5 billion per annum.

Wild roses produce only a limited range of flower colours – white, pink and red. Selection, mutation and hybridisation have gradually led, over centuries, to a wide range of garden roses. Specialists have estimated that over 50 000 varieties of garden rose have been bred to date, though many, particularly the modern ones, may disappear after several decades. The petals of the white rose are more or less free of compounds that absorb light of wavelength 400–700 nm and they therefore reflect a large proportion of the incident light. On the other hand they contain significant quantities of UV-absorbing substances, particularly substantial amounts of flavonol glycosides such as quercetin and rutin. All the white roses examined by Eugster and Marki-Fischer (1990, 1991) contained appreciable concentrations of carotenoids – predominantly the colourless hydrocarbons such as phytofluene. Yellow colours were unknown in old European garden roses and their advent in 1820 was greeted with great enthusiasm. However the emergence of varieties with an intense, highly saturated yellow did not begin until 1900 with the breeding of Soleil d'Or, a hybrid of *Rosa foetida persiana*, a red hybrid perpetual. They contain a complete carotenoid biosynthetic sequence and accumulate significant quantities of the carotenoid epoxides. Modern light yellow roses, such as Mme. Meilland, combine a complete carotenoid biosynthetic pathway with enhanced *carotenoid dioxygenase* activity.

Classical methods of flower breeding, despite their many achievements, also have their limitations (Mol *et al.*, 1989). Chief amongst these is the limited gene pool of any species. Thus for example no one plant species possesses the genetic capacity to produce flowering varieties which span the full spectrum of colours; one does not find blue carnations, roses, tulips, and chrysanthemums nor orange petunias. Another limitation is that the plant breeders are unable to alter traits in a directed manner. Thus the new combinations of genetic complements, consequent upon the crossing of two plants, may well alter flower colour but also change the delicate balance of the other factors which determine plant growth and shape. In the process, good commercial characteristics such as uniform growth and synchronous flowering may therefore be deleteriously affected. Recent advances in molecular biology, especially in gene isolation, manipulation and transfer between species are making possible the alteration of plant properties, with commercial and aesthetic value, in a highly directed fashion. Several enterprises have begun flower breeding using recombinant DNA technology; many of these are intimately associated with processes designed to change anthocyanin pigmentation.

There have been several reports in recent years of targeted gene expression in the anthocyanin biosynthetic pathway to bring about changes in floral pigmentation (Mol *et al.*, 1989; van der Krol *et al.*, 1988; Meyer *et al.*, 1987; Holton *et al.*, 1993; Holton and Tanaka, 1994; Gourtney-Gutterson *et al.*,

1994). Such a strategy is now in the course of development by those anxious to see a blue rose. Although the complex supramolecular structures of commelinin and protocyanin which lend the blue colour to flowers of *Commelina communis* and cornflower (*Centaurea cyanus*) respectively are both based upon a 'B' ring 3',4'-dihydroxylation pattern in the anthocyanin part of these structures, there is a general view that the structural requirements necessary for the formation of a blue flower colour are:

  (i) the accumulation of *delphinidin* derived anthocyanins,
 (ii) the accumulation of flavonoid copigments,
(iii) a vacuolar pH > 5.0.

Further chemical embellishment (such as acylation, methylation and glycosylation) may also be important in the formation of the final pigment structure. Anthocyanins from roses have been analysed extensively and a recent survey of 670 found only glucosides of the anthocyanidins – pelargonidin, paeonidin and cyanidin. To date none of the anthocyanins detected in roses have the typical 3',4',5'-trihydroxylation pattern of delphinidin and its derivatives (Table 6.2 and Figure 6.5). Recently the first major steps were taken towards the modification of the types of anthocyanin pigments found in roses using recombinant DNA methods (Holton *et al.*, 1993). Differences in the 'B' ring hydroxylation patterns of the various anthocyanins are controlled by the cytochrome P-450 enzymes – *flavonoid-3'-hydroxylase* and *flavonoid-3',5'-hydroxylase*. An Australian–Japanese–French consortium has very recently reported the isolation of complementary DNA clones of two different *flavonoid-3',5'-hydroxylase* genes that are expressed in petunia flowers (Holton *et al.*, 1993). The implanting of the *flavonoid-3',5'-hydroxylase* genes into cyanidin or pelargonidin producing rose cultivars should, it is thought, divert the flux of flavonoid biosynthesis towards the formation of delphinidin glucosides and hence change the flower colour towards the blue pigmentation which is desired. However, flower colour is dependent not only on the pigment but also on co-pigments and, in the case of blue colours and in the absence of metal ions, the cell pH. The observation that caffeine preferentially stabilises the violet quinonoidal base form of the anthocyanin, and its blue anion, *vide supra* clearly opens up another avenue for the development of a blue rose and other blue flowers. Transferring the genes for caffeine biosynthesis to the family Rosaceae would then truly give a hybrid tea rose.

The doctrine of the language of flowers is redolent of a bygone age; writing in 1856 from Boston in a book *Floras Interpreter and Fortuna Flora*, Sarah Jane Hale suggested that a burgundy rose meant 'you have simplicity and beauty', a white rose meant that 'I am sad' and a yellow rose said 'let us forget'. Whilst it is no doubt interesting and amusing to speculate what message a blue rose might convey midst the hustle and bustle of the late twentieth century, there is no doubt that, consonant with these times, there is a yearning in the Far East for roses that grow bright blue straight from the earth – a yearning which it is said will put a price tag of at least £50 on each rose bloom!

References

Asen, S. and Jurd, L. (1967). The constitution of a crystalline, blue cornflower pigment. *Phytochemistry*, **6**, 577–84.

Asen, S., Stewart, R. N. and Norris, K. H. (1971). Copigmentation effect of quercetin glycosides on Red Wing Azaleas. *Phytochemistry*, **10**, 171–5.

Asen, S., Stewart, R. N. and Norris, K. H. (1972). Co-pigmentation of anthocyanins in plant tissues and its effect on colour. *Phytochemistry*, **11**, 1139–44.

Asen, S., Stewart, R. N. and Norris, K. H. (1975a). Microspectrophotometric measurement of pH and pH effect on colour of petal epidermal cells. *Phytochemistry*, **14**, 937–42.

Asen, S., Stewart, R. N. and Norris, K. H. (1975b). Anthocyanin, flavonol copigments and pH responsible for Larkspur flower colour. *Phytochemistry*, **14**, 2667–82.

Bayer, E. (1958). Naturliche und Synthetische Anthocyan-Metallkomplexe. *Chem. Ber.*, **91**, 1115–23.

Bayer, E., Eggeter, H., Fink, A., Nether, K. and Wegman, K. (1966). Complex formation and flower colours. *Angew. Chem. Int. Edn.*, **5**, 791–8.

Brouillard, R., (1981). The origin of the exceptional colour stability of *Zebrina* anthocyanin. *Phytochemistry*, **20**, 143–5.

Brouillard, R. (1983). The *in vivo* expression of anthocyanin colour in plants. *Phytochemistry*, **22**, 1311–23.

Brouillard, R. (1988). Flavanoids and flower colour. In *The Flavonoids – Advances in Research*, editor J. B. Harborne, Chapman and Hall: London, pp. 525–38.

Brouillard, R. and Dubois, J.-E. (1977). Mechanism of the structural transformations of anthocyanins in acidic media. *J. Amer. Chem. Soc.*, **99**, 1359–64.

Brouillard, R., Mazza, G., Saad, Z., Albrecht-Gary, A. M. and Cheminat, A. (1989). The copigmentation reaction of anthocyanins: a micro-probe for the structural study of aqueous solutions. *J. Amer. Chem. Soc.*, **111**, 2604–10.

Brouillard, R. and Lang, J. (1990). The hemiacetal–*cis*-chalcone equilibrium of malvin, a natural anthocyanin. *Canad. J. Chem.*, **68**, 755 61.

Brouillard, R., Wigand, M.-C., Dangles, O. and Cheminat, A. (1991). pH and solvent effects on the copigmentation reaction of malvin with polyphenols, purine and pyrimidine derivatives. *J. Chem. Soc. (Perkin Trans. II)*, pp. 1235–41.

Cai, Y., Lilley, T. H., Martin, R., Spencer, C. M. and Haslam, E. (1990). The metabolism of gallic acid and hexahydroxydiphenic acid in higher plants. Part 4. Polyphenol Interactions. Part 3. Spectroscopic and physical studies of esters of gallic acid and (S)-hexahydroxydiphenic acid with D-glucopyranose ($^4C_1$). *J. Chem. Soc. (Perkin Trans. II)*, pp. 651–60.

Chen, L. J. and Hradzina, G. (1981). Structural aspects of anthocyanin–flavonoid complex formation and its role in plant colour. *Phytochemistry*, **20**, 297–303.

Culvenor, C. J. (1966). The conformation of esters and the 'acylation shift'. NMR evidence from pyrrolizidine alkaloids. *Tetrahedron Letters*, pp. 1091–9.

Eugster, C. H. and Marki-Fischer, E. (1990). Eine weitere, diesmal erfolgreiche suche nach 3'-epilutein in pflanzen. *Helv. Chim. Acta*, **73**, 1205–7, and references therein.

Eugster, C. H. and Marki-Fischer, E. (1991). The chemistry of rose pigments. *Angew. Chem. Int. Edn.*, **30**, 654–72.

Eugster, C. H. and Nayeshiro, K. (1989). Notiz uber Ellagitannine und Flavonol-glycoside aus Rosenbluten. *Helv. Chim. Acta*, **72**, 985–91.

Goto, T. (1987). Structure, stability and colour variation of natural anthocyanins. *Prog. Chem. Org. Nat. Prod.*, **52**, 113–58.

Goto, T., Hoshino, T. and Ohba, M. (1976). Stabilisation effect of neutral salts on anthocyanins, flavylium salts, anhydro bases and genuine anthocyanins. *Agric. Biol. Chem.*, **40**, 1593–6.

Goto, T., Kondo, T., Tamura, K., Imagawa, H., Iino, A. and Takeda, K. (1982). Structure of gentiodelphin, an acylated anthocyanin isolated from *Gentiana makinoi* that is stable in dilute aqueous solution. *Tetrahedron Letters*, **23**, 3695–8.

Goto, T., Kondo, T., Tamura, K. and Kawahori, K. (1983). Structure of platyconin, a diacylated anthocyanin from Chinese Bell-Flower *Platycodon grandiflorum*. *Tetrahedron Letters*, **24**, 2181–4.

Goto, T., Tamura, K., Kawai, T., Hoshino, T., Harada, N. and Kondo, T. (1986). Chemistry of metalloanthocyanins. *Ann. New York Acad. Sci.*, **471**, 155–73.

Goto, T. and Kondo, T. (1991). Structure and molecular stacking of anthocyanins – flower colour variation. *Angew. Chem. Intl. Edn.*, **30**, 17–33.

Gourtney-Gutterson, N., Napoli, C., Lemieux, C., Morgan, A., Firoozabady, E. and Robinson, K. E. P. (1994). Modification of flower colour in florist's chrysanthemum; production of a white flowering variety through molecular genetics. *Biotechnology*, **12**, 268–71.

Harborne, J. B. (1967). *Comparative Biochemistry of the Flavonoids*. Academic Press: London and New York.

Harborne, J. B. (1976). Functions of flavonoids in plants. In *Chemistry and Biochemistry of Plant Pigments*, ed. T. W. Goodwin, Academic Press: New York, pp. 736–77.

Haslam, E. (1967). Gallotannins. Part XIV. Structure of the gallotannins. *J. Chem. Soc. (C)*, pp. 1734–8.

Hayashi, K. (1962). The anthocyanins. In *The Chemistry of Flavonoid Compounds*, ed. T. A. Geissman, MacMillan: New York, pp. 248–85.

Hayashi, K., Abe, Y. and Mitsui, S. (1958). Anthocyanins. XXX. Blue anthocyanin from flowers of Commelina, the crystallisation and properties thereof. *Proc. Japan Acad.*, **34**, 373–8.

Hayashi, K., Hattori, K. and Mitsui, S. (1959a). Anthocyanins. XXXI. Commelinin a crystalline blue metallo-anthocyanin from the flowers of Commelina. *Proc. Japan Acad.*, **35**, 169–74.

Hayashi, K., Hattori, K. and Mitsui, S. (1959b). Crystallisation and properties of commelinin a blue metalloanthocyanin from Cornflower. *Bot. Mag. (Tokyo)*, **72**, 325–33.

Hayashi, K., Saito, N. and Mitsui, S. (1961). Anthocyanins. XXXIV. Metallic components in newly crystallised specimens of Bayer's protocyanin, a blue metalloanthocyanin from cornflower. *Proc. Jap. Acad.*, **37**, 393–7.

Hayashi, K. and Takeda, K. (1964). Anthocyanins. XXXVII. Violet flower colour in the pansy, *Viola tricolor*. *Proc. Jap. Acad.*, **40**, 510–14.

Hayashi, K. and Takeda, K. (1970). Studies on anthocyanins. LXII. Further purification and component analysis of commelinin showing the presence of magnesium in this blue complex molecule. *Proc. Jap. Acad.*, **46**, 535–40.

Holton, T. A., Brugliera, F., Lester, D. R., Tanaka, Y., Hyland, C. D., Menting, J. G. T., Lu, C.-Y., Farcy, E., Stevenson, T. W. and Cornish, E. C. (1993). Cloning and expression of cytochrome P-450 genes controlling flower colour. *Nature*, **366**, 276–9.

Holton, T. A. and Tanaka, Y. (1994). Blue roses – a pigment of our imagination? *Trends Biotech.*, **12**, 40–2.

Hoshino, T. (1991). An approximate estimate of self-association constants and the self-stacking conformation of malvin quinonoidal bases studied by ^{1}H NMR. *Phytochemistry*, **30**, 2049–55.

Hoshino, T. and Matsumoto, U. (1980). Evidence of the self-association of anthocyanins. I. Circular dichroism of cyanin anhydrobase. *Tetrahedron Letters*, **21**, 1751–4.

Hoshino, T., Goto, T. and Matsumoto, U. (1981a). Self association of some anthocyanins in neutral aqueous solution. *Phytochemistry*, **20**, 1971–6.

Hoshino, T., Matsumoti, U., Harada, N. and Goto, T. (1981b). Chiral exciton coupled stacking of anthocyanins: interpretation of the origin of anomalous CD induced by anthocyanin association. *Tetrahedron Letters*, **22**, 3621–4.

Hoshino, T., Matsumoto, U., Harada, N. and Goto, T. (1982). Evidence for the self-association of anthocyanins. IV. PMR spectroscopic evidence for the vertical stacking of anthocyanin molecules. *Tetrahedron Letters*, **23**, 433–6.

Hoshino, T. and Goto, T. (1990). Effects of pH and concentration on the self-association of malvin quinonoidal base – electronic and circular dichroic studies. *Tetrahedron Letters*, **31**, 1593–6.

Kondo, T. Yoshida, K., Nakagawa, A., Kawai, T., Tamura, H., and Goto, T. (1992). Structural basis of the blue colour development in flower petals from *Commelina communis*. *Nature*, **358**, 515–18.

Kondo, T., Yoshida, K., Ueda, M., Tamura, H. and Goto, T. (1994). Composition of protocyanin, a self assembled supramolecular pigment from the blue cornflower, *Centaurea cyanus*. *Angew. Chem. Int. Edn.*, **33**, 978–9.

van der Krol, A., Lenting, P. E., Veenstra, J., van der Meer, I. M., Koes, R. E., Gerats, A. G. M., Mol, J. N. M. and Stuitje, A. R. (1988). An 'anti-sense' chalcone synthase gene in transgenic plants inhibits flower pigmentation. *Nature*, **333**, 866–9.

Mathieson, A. McL. (1965). The preferred conformation of the ester group in relation to saturated ring systems. *Tetrahedron Letters*, pp. 4137–44.

Meyer, P., Heidmann, I., Forkman, G. and Saedler, H. (1987). A new petunia flower colour generated by transformation of a mutant with a maize gene. *Nature*, **330**, 677–8.

Mistry, T. V., Cai, Y., Lilley, T. H. and Haslam, E. (1991). Polyphenol interactions. Part 5. Anthocyanin copigmentation. *J. Chem. Soc.* (*Perkin Trans. II*), pp. 1287–96.

Mol, J. N. M., Stuitje, A. R. and van der Krol, A. (1989). Genetic manipulation of floral pigmentation genes. *Plant Mol. Biol.*, **13**, 287–94.

Osawa, Y. (1982). Copigmentation of anthocyanins. In *Anthocyanins as Food Colours*, ed. P. Markakis, Academic Press: New York, pp. 41–68.

Owen, N. L. and Sheppard, N. (1963). (Lone pair)–(lone pair) repulsion and molecular configurations; rotational isomerism in methyl vinyl ether, carboxylic esters and nitrites. *Proc. Chem. Soc.*, pp. 264–5.

Robinson, G. M. (1939). Notes on variable colour of flower petals. *J. Amer. Chem. Soc.*, **61**, 1606–7.

Robinson, G. M. and Robinson, R. (1931). A survey of anthocyanins. *Biochem. J.*, **25**, 1687–705.

Robinson, G. M. and Robinson, R. (1932). Developments in anthocyanin chemistry. *Nature*, **130**, 21.

Robinson, G. M. and Robinson, R. (1939). The colloid chemistry of leaf and flower pigments and the precursors of anthocyanins. *J. Amer. Chem. Soc.*, **61**, 1605–6.

Saito, N., Osawa, Y. and Hayashi, K. (1971). Platyconin, a new acylated anthocyanin in Chinese Bell-Flower *Platycodon grandiflorum. Phytochemistry*, **10**, 445–7.

Scheffeldt, P. and Hradzina, G. (1978). Copigmentation of anthocyanins under physiological conditions. *J. Food Sci.*, **43**, 517–20.

Scott-Moncrieff, R. (1981). The classical period in chemical genetics. *Notes Rec. R. Soc. Lond.*, **36**, 125–54.

Willstatter, R. (1965). *From my Life: the Memoirs of Richard Willstatter*. Translated by L. S. Hornig, Benjamin: New York.

Willstatter, R. and Zollinger, E. H. (1916). Uber die Farbstoffe der Weintraube und der Heidelbeere II. *Justus Liebig's Annalen Chem.*, **412**, 195–216.

7

Polyphenols and herbal medicines

7.1 Introduction

Natural products from plants and micro-organisms traditionally have provided the pharmaceutical industry with one of its most important sources of 'lead' compounds in the search for new drugs and medicines. Penicillins metabolised by the mould *Penicillium chrysogenum* were an early instance and their story has entered the folklore of science. People have of course always been curious about plants and other living organisms with which they share this planet. Primitive cultures exploited the toxic principals of poisonous plants for hunting, murder and euthanasia. Religious rites were often based upon the use of stimulants and hallucinogenic materials found in plants and fungi. Modern medicine has its roots in folk medicine and this has led to the development of many of the drugs, such as the penicillins, which we now take for granted. Today investigators explore the medicinal powers of sponges and algae from the deep sea bed of the Pacific and Carribean oceans; perhaps the most exciting new group of drugs to be obtained from plants are those derived from taxol, a sesquiterpene from the bark of the Pacific yew, *Taxus brevifolia*. However over the past two decades researchers have also turned to many of the traditional folk medicines themselves (Hoffman, 1987; Grieve, 1978) –invariably a 'cocktail' of natural products whose effects are most probably dependent on the synergism of several key ingredients – to uncover the scientific basis of their remedial effects; endeavours which have their roots as much in a desire to improve the efficacy as to enhance the ethics of modern medical practice. In this context particular attention has been given by various groups in the Far East, Europe and America to those traditional herbal medicines rich in polyphenols (vegetable tannins). Some typical examples of such medicines and the polyphenols which they contain are shown in Table 7.1 (Okuda *et al.*, 1989, 1991; Okuda, 1993, 1993; Haslam, 1989).

Table 7.1. *Some medicinal plants containing polyphenolic metabolites*

1. **Tree Paeony** (*Paeonia lactiflora*): outer skin of the root, used to cure disorders of the bloodstream, including high blood pressure. Principal polyphenolic metabolites – *gallotannins*.
2. **Bearberry** (*Arctostaphylos uva-ursi*): dried leaves; infusions have a soothing astringent effect which have value as diuretic, in kidney disorders and ailments of the bladder and urinary tract. Principal polyphenolic metabolites –*gallotannins, arbutin, galloyl esters of arbutin*.
3. **Agrimony** (*Agrimonia* sp.): roots and dried aerial parts of the plant: used as an astringent on the digestive system, as a diuretic and as a haemostatic agent. Principal polyphenolic metabolites – *ellagitannins*.
4. **Geranii Herba** (*Geranium maculatum, G. thunbergii*): dried rhizome and leaves; used as an astringent, anti-haemorrhagic and anti-inflammatory agent. Principal polyphenolic metabolites – *ellagitannins*.
5. **Meadowsweet** (*Filipendula ulmaria*): aerial parts of the plant – leaves and flowers used as an infusion; employed as a mild astringent, anti-rheumatic anti-inflammatory agent, and as a diuretic. Principal polyphenolic metabolites – *ellagitannins*.
6. **Raspberry** (*Rubus idaeus*): leaves and fruit; mild astringent used in disorders of the digestive system, raspberry leaf tea traditionally used during pregnancy. Principal polyphenolic metabolites – *ellagitannins*.
7. **Hawthorn** (*Crataegus* sp.): leaves and berries; used as astringent for digestive system, diuretic, cardiac tonic in treatment of high blood pressure. Principal polyphenolic metabolites – *proanthocyanidins*.
8. **Rose Bay Willow-herb** (*Epilobium angustifolium*): leaves used as substitute tea; roots and leaves have demulcent, tonic and astringent properties and are used as an intestinal astringent. It has been employed as an antispasmodic in whooping cough and asthma and in ointments for the treatment of infantile cutaneous infections. Principal polyphenolic metabolites – *gallotannins, ellagitannins*.

The ripe fruits of **hawthorn** (*Crataegus* sp.) provide one of the best tonic remedies for the heart and circulatory system. They act in a normalising way upon the heart, depending on the need, stimulating or depressing its activity. Hawthorn is a rich source of the flavan-3-ol (−)-epicatechin and proanthocyanidins related to (−)-epicatechin, e.g. epicatechin-(4β → 8)-epicatechin (procyanidin B-2), Figure 7.1.

Meadowsweet is likewise one of the best digestive remedies available. It acts to protect and soothe the mucous membranes of the digestive tract, reducing excess acidity and easing nausea. It is used in the treatment of heartburn, hyperactivity, gastritis and peptic ulceration, and relieves the pain of rheumatism. Its gentle astringency is very useful in the treatment of diarrhoea in children. Two of its principal polyphenolic constituents are tellimagrandin II and rugosin D (Haslam, 1989; Haslam and Cai, 1994); the latter polyphenol is formally derived by loss of two hydrogen atoms from the 'monomer' tellimagrandin II, Figure 7.2.

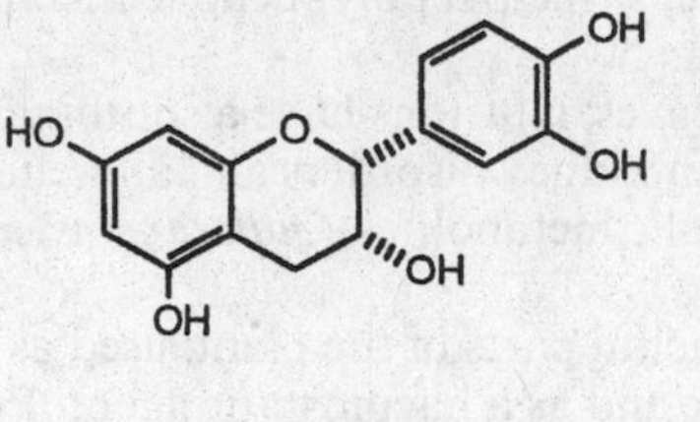

(-)-epicatechin

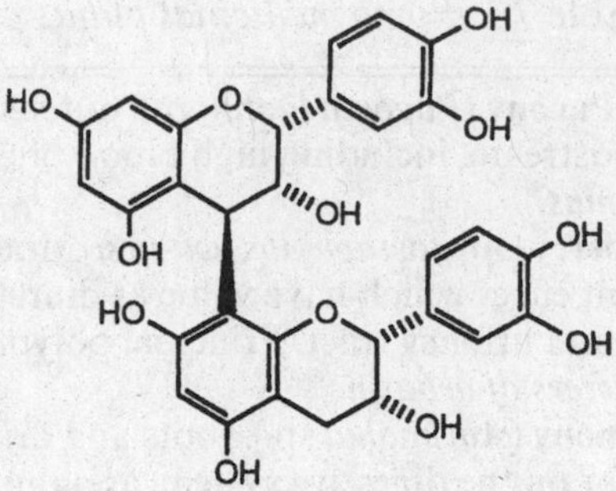

procyanidin B-2

Figure 7.1. Flavan-3-ols of *Crataegus* species.

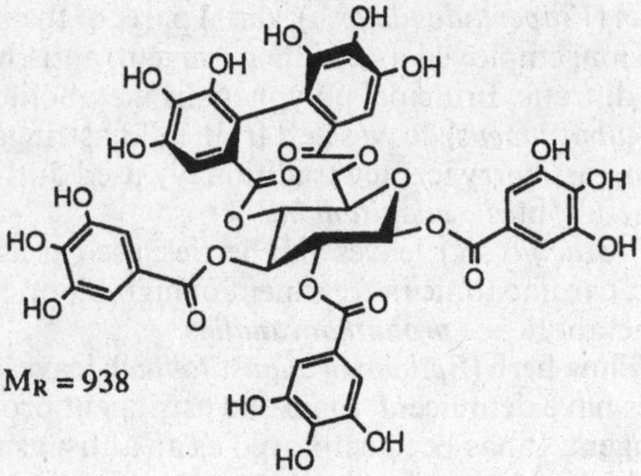

Tellimagrandin II , $M_R = 938$

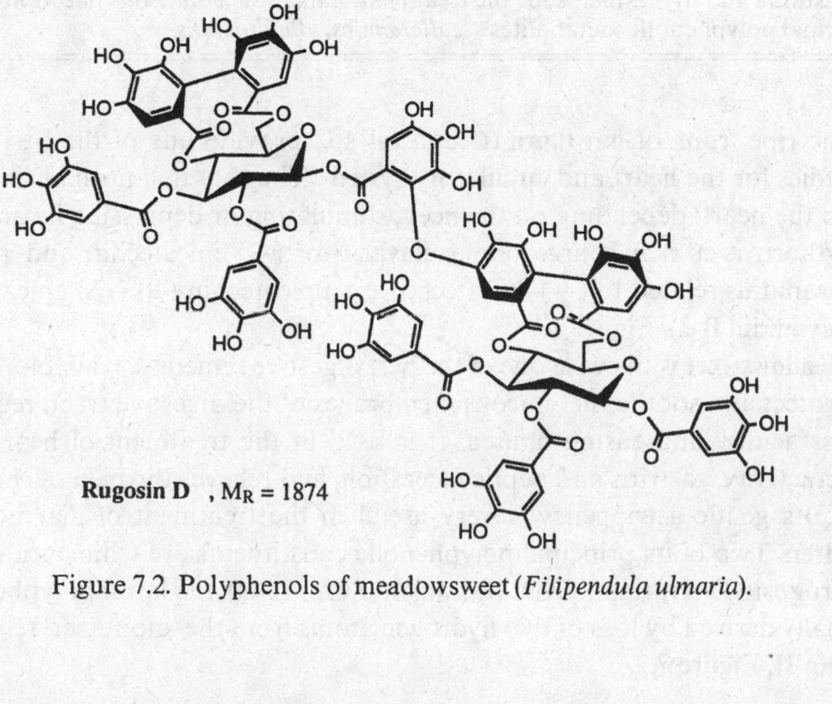

Rugosin D , $M_R = 1874$

Figure 7.2. Polyphenols of meadowsweet (*Filipendula ulmaria*).

Very recently, and related generally to this area of research, there has arisen a considerable interest in the possibility that the impact of several diseases which afflict mankind may be ameliorated, if not prevented, by the simple expedient of improving the dietary intake of nutrients with antioxidant properties. Their role has been described with varying degrees of certitude (Halliwell, 1995) as (*important* – vitamin E; *thought to be important* – vitamin C; *probably important* – β-carotene and carotenoids; *possibly important* – plant phenolics such as tannins and flavonoids). Thus although proof of a cause and effect relationship does not yet exist, substantial interest has recently been engendered by the epidemiological evidence which points to a reduced risk of certain degenerative diseases by the consumption of beverages containing polyphenols (Waterhouse, 1995), in particular green tea and red wines, both rich sources of polyphenols based on the flavan-3-ol carbon–oxygen skeleton.

7.1.1 Biological and pharmacological activity

Notable studies in this area have been made by several groups worldwide (Perchellet and Perchellet, 1989; Kinsella *et al.*, 1993a,b; Okuda *et al.*, 1989, 1991, 1992; Okuda, 1993; Scalbert, 1991; Balde *et al.*, 1990). *In vitro* testing has identified a wide range of potentially significant biological activities which are exhibited by natural polyphenols and a selection of these is shown in Table 7.2. Although these studies have revealed important differences in pharmacological activity between individual polyphenols and between classes of different polyphenol, *overall they suggest selectivity rather than high specificity towards particular biological targets*.

Thus for example considerable attention has been devoted by several groups to the inhibition of human immunodeficiency virus replication by polyphenols. In a typical study, Okuda and his collaborators (Okuda *et al.*, 1992) examined the actions of some 90 chemically defined polyphenols on the inhibition of both the cytopathic effect of human immunodeficiency virus (HIV) and the expression of HIV antigen in human lymphotropic virus type I (HTLV-1)-positive MT-4 cells. They showed that several polyphenols of the hydrolysable class based upon gallic and hexahydroxydiphenic acids (e.g. gemin D, nobotannin B, and the macrocyclic camelliin B, Figure 7.3, and trapanin B) but *not* those polyphenols of the proanthocyanidin type, and related lower molecular weight polyphenols, possessed potent anti-HIV activity. In particular these authors noted that the most potent compounds generally contain one or more hexahydroxydiphenoyl and/or valoneoyl ester groups within their molecular structures. However they also noted several examples where these same structural characteristics were present but which

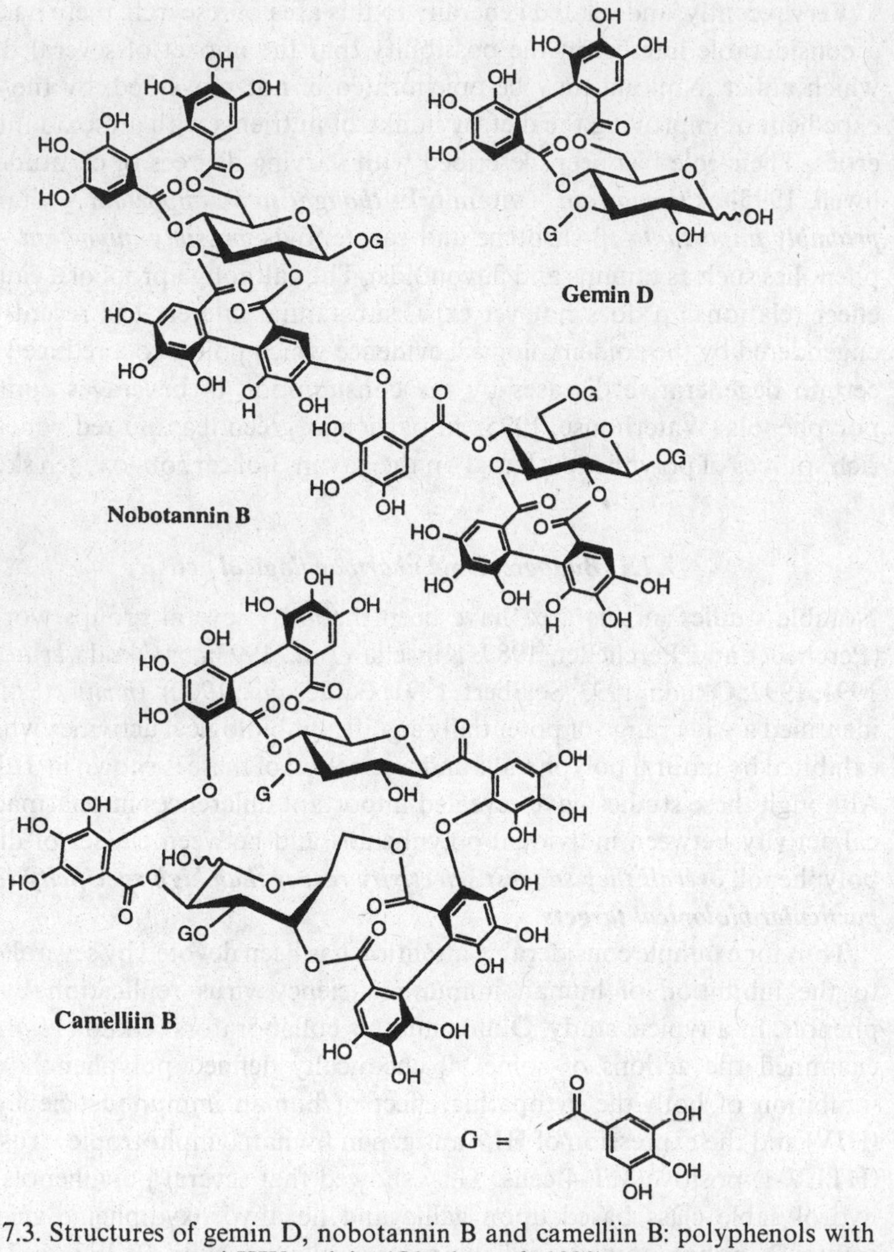

Figure 7.3. Structures of gemin D, nobotannin B and camelliin B: polyphenols with anti-HIV activity (Okuda *et al.*, 1992).

exhibited only very low anti-HIV activity, and they concluded that it is probable therefore that the anti-HIV activity is best correlated with the whole structure and stereochemistry of each individual polyphenol molecule. However they went no further to give any detailed structure/biological activity analysis. The anti-HIV activity of the various galloyl and hexahydroxydiphenoyl esters was observed to increase with increasing molecular size (generally monomers < dimers < trimers < tetramers) suggesting some link to this quantity and/or their polyanionic character *in situ*.

In a related study (Nishizawa *et al.*, 1989; Nonaka *et al.*, 1990) the inhibitory effects of polyphenols on HIV reverse transcriptase and HIV replication in H9 lymphocyte cells were studied. The fact that the reverse transcriptase (RT) plays a very important role in controlling the replication of HIV makes it one of the most attractive targets in the development of anti-AIDS (acquired immunodeficiency syndrome) drugs. Nishizawa and his colleagues demonstrated (Nishizawa *et al.*, 1989) that commercial tannic acid (chinese gallotannin) had a strong inhibitory effect against HIV RT (74% at $100 \, \mu g \, ml^{-1}$) as did three tetragalloylquinic acids containing depsidically linked galloyl ester groups (84–90% inhibition at $100 \, \mu g \, ml^{-1}$). The latter compounds also inhibited the HIV growth in cells with low cytotoxicity ($25 \, \mu g \, ml^{-1}$). Subsequently Nonaka *et al.* (1990) showed nine additional polyphenols to inhibit HIV replication in infected H9 lymphocytes with low toxicity. Three compounds were found to be particularly potent inhibitors of HIV RT with ID_{50} values of $5–20 \, \mu M$; the two most potent contain the gallagyl ester grouping, Figure 7.4, suggesting that this structure may be important for RT inhibition. However *all* the polyphenols tested appeared to inhibit virus–cell interactions and the authors concluded that, despite their anti-RT activity, the mechanism by which polyphenols inhibit HIV may not be directly associated with their action on this key enzyme. It should also be pointed out that, as yet, it is not clear in humans how or if these complex polyphenolic substrates are absorbed from the gut and this lack of precise knowledge on the fate of these compounds in the human body remains a major weakness in this area.

Thus far it has not been found possible to discuss, with any precision, the relationships between polyphenolic structure and 'biological activity'. Nevertheless, as these studies of the human immunodeficiency virus demonstrate and the generality of effects displayed by many members of the polyphenolic class of compound also show, polyphenols constitute the important active principals of many medicinal plants and medicinal plant preparations. It is now also clear that their physiological and pharmacological actions very probably derive, at least in part, from some of the chemical features which such polyphenols have in common, viz. the various physical and chemical properties which are themselves associated principally with the possession of a concatenation of phenolic nuclei within the *same* molecule. On the basis of this proposition it is therefore suggested that polyphenols exert certain of their roles in the medical treatment of diseased states by virtue of three distinctive *general* characteristics (Haslam, 1996) which they all possess to a greater or

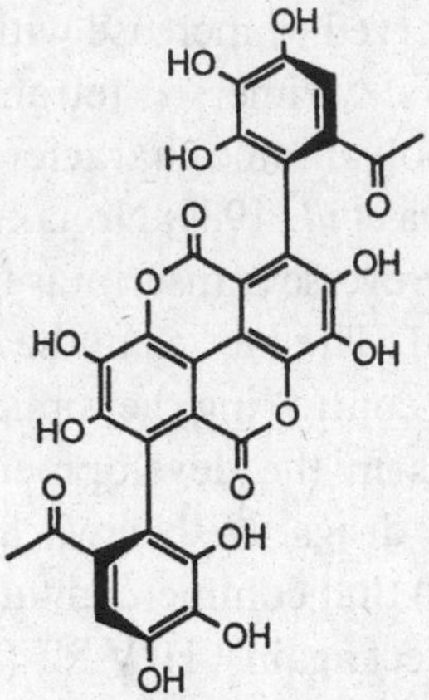

Gallagyl ester group

Figure 7.4. Gallagyl ester group: inhibition of HIV reverse transcriptase.

lesser degree and which derive in essence from the properties of the simple phenolic nucleus itself, namely:

(i) their complexation with metal ions (iron, manganese, vanadium, copper, aluminium, calcium, etc.),
(ii) their antioxidant and radical scavenging activities, and
(iii) their ability to complex with other molecules including macromolecules such as proteins and polysaccharides.

7.2 Metal-ion complexation

Pliny first described the blue-black colour produced when an aqueous infusion of oak galls is treated with iron salts and its use in the analysis of mineral waters and in the manufacture of inks was noted as early as the sixteenth century. Thereafter the reproach that 'anything that gives a blue-black colour with iron salts has been termed a tannin' has often seemed well merited. Nevertheless this property of molecules, such as natural polyphenols, with catechol and pyrogallol nuclei of forming strong complexes with metal ions such as iron, vanadium, manganese, aluminium, calcium etc., is not only a distinctive one but also an important one. In view of the importance of these metals to living systems it is logical to presume that species which form strong complexes with them, such as polyphenols, may well modify their biological activities. Insofar as the transition metals (vanadium, iron, manganese, copper and cobalt) themselves are concerned they have properties which clearly

Table 7.2. *Some physiological and pharmacological actions of polyphenols*

(i) Bacteriocidal action.
(ii) Molluscicidal action.
(iii) Anthelmintic action.
(iv) Antihepatoxic action.
(v) Stimulation of phagocytic cell iodination.
(vi) Inhibition of human immunodeficiency viral replication (HIV).
(vii) Inhibition of human simplex virus (HSV).
(viii) Inhibition of glucosyl transferases of *Streptococcus mutans* (dental caries).
(ix) Inhibition of ascorbate auto-oxidation (green tea).
(x) Inhibition of lipoxygenase dependent peroxidation; 'French paradox'.
(xi) Host-mediated antitumour activity: cytotoxic effects, inhibition of tumour promotion, inhibition of ornithine decarboxylase (ODC) response.
(xii) Inhibition of xanthine oxidase and monoamine oxidase.

distinguish them from other metals found in living systems (e.g., sodium, potassium and magnesium) namely (Frausto da Silva and Williams, 1991):

(i) they are good Lewis acids, can act as π-electron donors and are red-ox active,

(ii) biological systems have evolved in such a way that, whilst much of the particular element may remain free, a considerable fraction is bound up with particular organic ligands, e.g. Fe–haem, and

(iii) functional properties, such as red-ox potentials, are very sensitive to ligand coordination.

The case of iron may, at this juncture, be taken as representative of these various metals. Iron is common to all life and is the most abundant transition metal found in the biosphere; its involvement in biological systems is, however, complicated. It is involved not only in red-ox catalysis but has numerous other functions e.g., storage and transport of oxygen, electron transfer, hydroxylation reactions, utilisation of hydrogen peroxide, superoxide dismutation (Halliwell and Gutteridge, 1990). Because iron in its most common form is not readily available (the solubility of ferric hydroxide is 10^{-38}), many microorganisms produce siderophores – low molecular mass chelating agents that bind and solubilise iron. The one which binds iron the most strongly under physiological conditions is the siderophore enterochelin (**ent**), Figure 7.5, formed by *Escherichia coli*. The siderophore employs the three dihydroxybenzoyl rings to give a charged octahedral triscatecholate Δ-*cis* complex. The formation constant with ferric iron and the macrocycle of 10^{49} M^{-1} is the highest reported for a siderophore (Loomis and Raymond, 1990). The mechanism of capture and binding of the transition metal ion is clearly very similar to that deployed by natural polyphenols in their complexation of such ions.

The average adult male contains $\sim$ 4.5 g of iron, absorbs some 1 mg per day

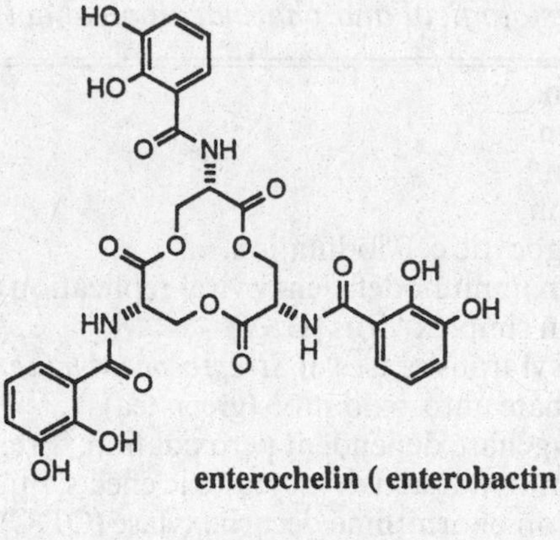

enterochelin (enterobactin)

$$Fe^{3+} + ent^{6-} = [Fe(ent)]^{3-}$$

Figure 7.5. Enterochelin.

from the diet and excretes about the same quantity when in iron balance. Plasma iron turnover accounts for some 35 mg per day and slight disturbances of iron metabolism readily lead to iron overload or iron deficiency. About two thirds of body iron is located in haemoglobin, with lesser amounts in myoglobin, various enzymes, the transport protein transferrin and a small transit pool of iron chelates (nature uncertain). Otherwise iron that is not required is stored in ferritin which consists of a protein shell surrounding an iron core which holds up to 4500 iron ions per molecule of protein.

In humans, iron is of particular medical concern because of its involvement in various red-ox reactions, its effect on infectious organisms and the diseases of iron overload and iron deficiency. Antimicrobial activity through iron depletion is, for example, well documented. Thus infection of humans by *Escherichia coli* is inhibited by the iron-chelating lactoferrin present in human milk, but restored by augmentation with additional iron. Lactoferrin is also released in septic infections by degranulation of blood circulating leucocytes. There seems, *a priori*, therefore very good reason to think that natural polyphenols have the potential (should they possess the ability to penetrate to particular sites in the human body), to modulate physiological reactions involving iron and other transition metals. Experimental developments in this area are eagerly awaited.

7.3 Antioxidant activity

Increasing attention is being given to the role of free radicals and other oxidants in the mechanism of action of many toxins and, in recent years, their involvement in the pathophysiology of major chronic diseases. Thus reactive oxygen species have been implicated in various human diseases including the processes of ageing, cancer, multiple sclerosis, Parkinson's disease, autoimmune disease, senile dementia, inflammation and arthritis and atherosclerosis. Many chronic diseases are also exacerbated by imbalances or perturbations in fatty acid and lipid metabolism. It is thought, for example, that excess dietary fatty acids are conducive to atherosclerosis and coronary arterial disease and uncontrolled lipid oxidation (enzymic or non-enzymic) is associated with arthritis, cancer and atherogenesis. During the past 400 years, whilst the average lifespan of man has increased from around 40 to 75 years, over this same period there has been no corresponding increase in the maximum lifespan: the overall effect has been an increase in the average age of the population. Two theories have been used to explain this phenomenon. The 'death gene' theory postulates that there is a genetically controlled limit to cell growth and that cell death is predestined. The homeostatic theory postulates that, for a number of plausible reasons, the body's oxidative defence mechanisms deteriorate with ageing. Whilst there is little doubt that cellular pro-oxidant states are implicated in many of the diseases noted above it is however not yet clear that they are the causative agents. Ames has argued that a deficiency of micronutrients that protect against oxidative damage to DNA is a major contributor to human cancer; on the other hand others have stated (Halliwell and Gutteridge, 1990) that increased oxidant formation is usually a *consequence* of disease.

7.3.1 Cellular pro-oxidant states

In cellular pro-oxidant states the intracellular concentration of activated forms of oxygen (reactive oxygen species) is increased, presumably because cells either overproduce these reactive substances or are deficient in their ability to destroy them (Cerutti, 1985). The ground state of the diatomic oxygen molecule (O_2) is a radical with two unpaired electrons (having parallel spins) in a π^* antibonding orbital, Figure 7.6. If oxygen is to oxidise another atom or molecule by accepting a pair of electrons from it, then both 'new' electrons must be of parallel spin to pair with the electrons in the π^* orbital. Most biomolecules are covalently bonded non-radicals and the electrons forming covalent bonds have opposite spins and occupy the same molecular orbital. As a result the reaction of ground state oxygen with biomolecules is spin restricted.

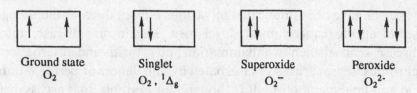

Figure 7.6. Occupancy of π^* 2p orbitals in various oxygen species.

The reactivity of oxygen can be enhanced in a number of ways.

(i) The biological role of ***transition metals*** is frequently related to their ability to participate in one-electron transfer reactions and they are often found at the active sites of oxidases and oxygenases because their ability to accept or donate an electron can overcome the spin restriction associated with reactions involving ground state oxygen.

(ii) ***Singlet oxygen.*** The reactivity of oxygen may be increased by moving one of the unpaired π^* electrons in such a way that the spin restriction is alleviated. This requires an input of energy (usually *via* a particular pigment, $h\upsilon$ and O_2) and generates the singlet states of oxygen. The pigment absorbs light, enters a higher electronic state, and transfers energy onto the oxygen molecule to give singlet oxygen. Singlet oxygen $^1\Delta_g$, the most important form of singlet oxygen in biological systems, has no unpaired electrons and is *not* a radical.

(iii) One-electron reduction of oxygen produces ***the superoxide radical O_2^-***. This species is formed in almost all aerobic cells, a major source being the 'leakage' of electrons onto oxygen from various components of the cellular electron transport chain (e.g., mitochondria). In organic solvents O_2^- is a strong base and a good nucleophile. In aqueous solution O_2^- is extensively hydrated and much less reactive, acting as a reducing agent and as a weak oxidising agent to molecules such as adrenalin and ascorbic acid. Any reaction undergone by O_2^- in aqueous media will be in competition with its spontaneous dismutation to hydrogen peroxide. Superoxide dismutase enzymes, which are specific for O_2^-, have evolved a surface arrangement of charge which accelerates the decomposition of O_2^- to hydrogen peroxide by a factor of approximately four. Protonation of O_2^- produces the hydroperoxyl radical $HO_2^·$ which is less polar than O_2^- and somewhat more reactive.

Generation of the superoxide radical in the human body can occur in several situations. Superoxide is made by 'accidents of chemistry' when certain molecules in the human body (e.g. adrenalin, dopamine, tetrahydrofolates) react directly with oxygen – an unavoidable consequence of having such molecules in a body which needs oxygen. Phagocytic cells, on the other hand, deliberately generate substantial amounts of the superoxide radical as part of the essential defence mechanism against infection by foreign organisms. It has been estimated that ~ 1–3% of the oxygen breathed is used to make superoxide radicals – some 2 kg per annum for normal humans, for those with chronic inflammation it is much more.

superoxide – generation and dismutation

$$O_2 + e \qquad \rightarrow O_2^- \text{ superoxide}$$
$$O_2^- + H^+ \quad \rightarrow 3HO_2^{\cdot} \text{ hydroperoxyl}$$
$$2\,O_2^- + 2\,H^+ \rightarrow H_2O_2 + O_2$$

(iv) *Hydrogen peroxide*. A system generating O_2^- would be expected to produce hydrogen peroxide by non-enzymic or superoxide dismutase catalysed dismutation. Pure hydrogen peroxide can cross biological membranes, but has limited reactivity. Reports of its toxicity to cells are variable.

The moderate reactivity of both O_2^- and hydrogen peroxide in aqueous media makes it unlikely that the oxidative damage done to cells can be directly attributable to these species. In general it is thought more probable that the damage incurred arises due to their conversion to more reactive radical species – notably hydroperoxyl $HO_2^{\cdot}$ and hydroxyl $HO^{\cdot}$. However the less reactive O_2^- and hydrogen peroxide may well be more damaging in the wider sphere since they can diffuse away from their site of generation and induce the formation of $HO^{\cdot}$ at remote cellular locations. *In vivo* most of the hydroxyl radicals generated come from the metal ion (e.g., Fe^{3+}) catalysed breakdown of hydrogen peroxide.

(v) Protonation of O_2^- produces the *hydroperoxyl radical* $HO_2^{\cdot}$. Although there is no clear evidence, as yet, that the hydroperoxyl radical plays a cytotoxic role, its potential arises from two factors. It is less polar than O_2^- and should cross biological membranes about as effectively as hydrogen peroxide. Secondly it is somewhat more reactive than O_2^-. Thus unlike O_2^-, the hydroperoxyl radical can attack fatty acids directly to give peroxides and can initiate peroxidation of the lipid component of low density lipoprotein.

(vi) *In vivo* the major route to the *hydroxyl radical* $HO^{\cdot}$ is the metal-ion dependent breakdown of hydrogen peroxide. The metal ion can be Ti(III) or Fe(II) and although other metal ions (or complexes) can effect this transformation *in vitro* it is probable that the Fe (II)-dependent formation of $HO^{\cdot}$ is the principal method *in vivo* under normal physiological conditions. Superoxide dismutase, by scavenging O_2^-, usually inhibits $HO^{\cdot}$ generation by iron–hydrogen peroxide systems.

hydroxyl radical generation

$$H_2O_2 + \text{Metal}^{n+} \rightarrow HO^{\cdot} + HO^- + \text{Metal}^{(n+1)+}$$

The highly reactive hydroxyl radical $HO^{\cdot}$, once formed *in vivo*, is very likely to react at or very close to its site of formation. It can for example initiate the process of lipid peroxidation by abstraction of a hydrogen atom from an unsaturated aliphatic lipid side-chain, eventually giving rise by an autocatalytic chain reaction to a *lipid hydroperoxide*, Figure 7.7. The decomposition of these lipid hydroperoxides, in the presence of transition metal ions, yields *alkoxyl* and *peroxyl radicals*

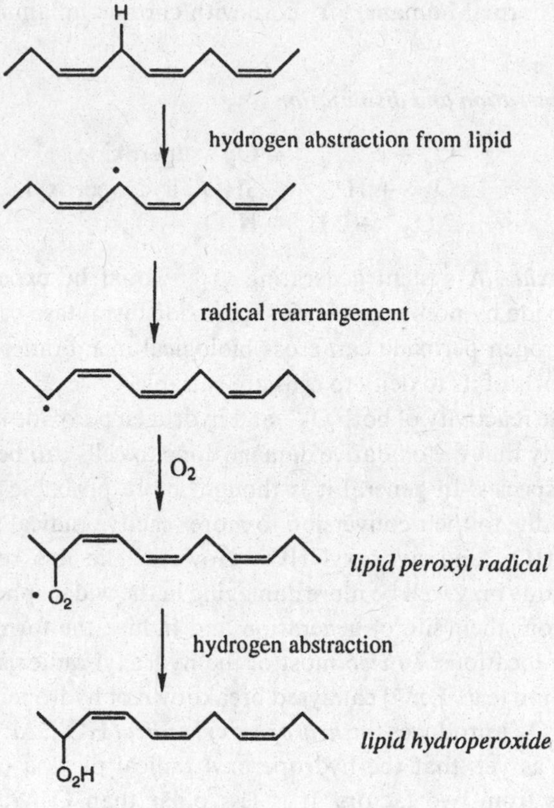

Figure 7.7. Lipid peroxidation.

which may abstract additional hydrogen atoms and contribute to chain propagation. Compared with other oxygen based radicals, lipid peroxyl radicals are more stable and capable of diffusing to cellular loci distant from their point of generation before reacting with say DNA or other macromolecules.

In cellular pro-oxidant states the intracellular concentration of reactive oxygen species (ROS: *singlet oxygen, superoxide radical, hydroperoxyl radical hydrogen peroxide, hydroxyl radical, lipid hydroperoxides, lipid peroxyl radical* etc.) is enhanced – the cells either overproduce these reactive species or excessive production of these species is beyond the capacity of the cell to destroy them; the normal balance in the cell of antioxidant and pro-oxidant properties is destroyed.

Oxidative stress represents a disturbance of the pro-oxidant/antioxidant balance of the body towards the former state and may arise from environmental or other external sources or by the endogenous production of free radicals accompanying diseased states. Cells and tissues normally possess antioxidant

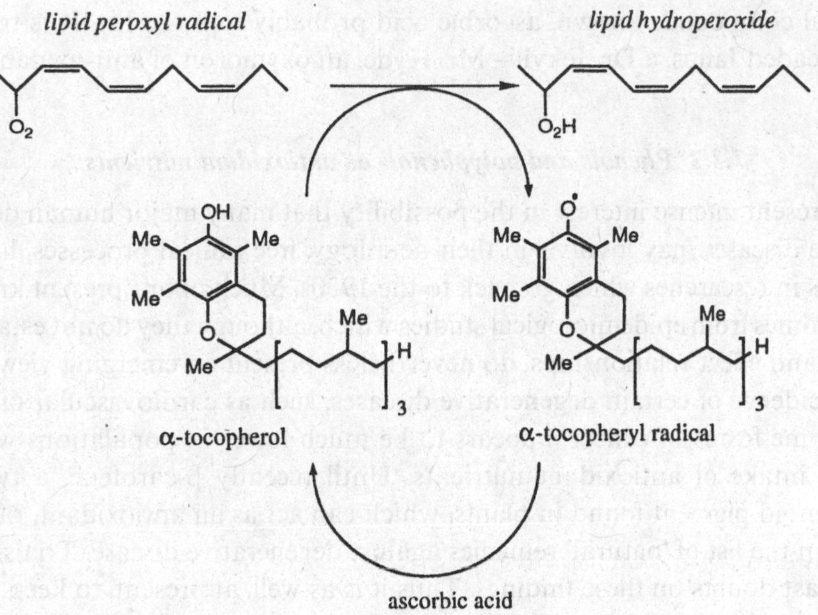

Figure 7.8. Lipid peroxidation – chain reaction blocking action of α-tocopherol.

defence mechanisms to ensure the removal of reactive oxygen species – those that are controlled endogenously (e.g., superoxide dismutase) and those (e.g., antioxidants – α-tocopherol (vitamin E), ascorbic acid (vitamin C), β-carotene) that are provided by dietary and other means. If mild oxidative stress occurs, tissues respond by raising the level of the normal antioxidant defences. However severe oxidative stress causes cell injury and death.

Vitamins E and C are known and important antioxidants. Vitamin E is the general name given to a group of lipid soluble compounds of which α-tocopherol is the most familiar. It is found in lipoproteins and membranes and acts to block the chain reaction of lipid peroxidation by scavenging the intermediate peroxyl radicals which are generated, Figure 7.8. The highly sterically hindered α-tocopheryl radical is much less reactive in attacking other fatty acid side-chains and may be converted back to the parent phenol by ascorbic acid thus breaking the chain reaction. The water soluble vitamin C (ascorbic acid) itself has many physiological roles; antioxidant activity (such as the re-cycling of vitamin E in membranes and lipoproteins) is only one of them. However it should also be noted that, *in vitro*, vitamin C is also clearly capable of pro-oxidant activity. It has long been known, for example, that the combination of ascorbate and ferrous ions generates radicals which can induce lipid peroxidation. These pro-oxidant effects of ascorbate are well known and this apparent paradox was eloquently described by Porter (1993): 'of all the paradoxical compounds known, ascorbic acid probably tops the list. It is truly a two-headed Janus, a Dr. Jekyll – Mr. Hyde, an oxymoron of anti-oxidants.'

7.3.2 *Phenols and polyphenols as antioxidant nutrients*

The present intense interest in the possibility that many major human degenerative diseases may involve, in their aetiology, free radical processes, has its origins in researches which go back to the 1930s. Much of our present knowledge comes from epidemiological studies which, although they do not establish cause and effect relationships, do nevertheless present an emerging view that the incidence of certain degenerative diseases, such as cardiovascular disease and some forms of cancer, appears to be much lower in populations with a larger intake of antioxidant nutrients. Until recently β-carotene, a typical carotenoid pigment found in plants, which can act as an antioxidant, ranked high on the list of 'natural' remedies against degenerative disease. Trials have now cast doubts on these findings. Thus it is as well, at present, to keep these ideas in perspective. Concomitant with these developments there has, notwithstanding, been a surge in the production of proprietary medicines, offered for sale by major pharmaceutical companies, which are labelled as containing vital antioxidants and often accompanied by comments such as: 'Our new antioxidant will help you stay healthy'. The title of a recent review by Halliwell (1995) puts the present position in this area with admirable clarity: 'Antioxidants: Elixirs of Life or Tonics for Tired Sheep?'.*

It is in this context of the relationship of antioxidant nutrients to the conditions of oxidative stress, cellular pro-oxidant states and lipid peroxidation that intense speculation has nevertheless very recently been generated in relation to the possible role which simple plant phenols and polyphenols, either as components of herbal medicinal extracts or as part of the regular diet, may have in the treatment/amelioration/prevention of various diseases noted earlier which are associated with the presence of pro-oxidant cellular states. According to Halliwell (1995) plant phenols and polyphenols are 'possibly important as antioxidants', and it is clear that present speculation is best replaced by the search for hard scientific evidence in this area.

Plant phenols and polyphenols are known to inhibit lipid peroxidation and lipoxygenases *in vitro* (Okuda *et al.*, 1983, 1985) and information has been accumulated over the past few years demonstrating their ability to scavenge radicals such as hydroxyl, superoxide and peroxyl, which are known to be important in cellular pro-oxidant states. Some idea of the relative reactivity of

* Attributed by Professor Halliwell to Professor T. L. Dormandy.

these reactive oxygen species towards phenolic substrates may also be gleaned from these observations. Typical data showing the radical scavenging efficiency of various phenols are shown in Table 7.3 for simple phenols and flavonoids (Bors *et al.*, 1990; Jovanovic *et al.*, 1994; Aruoma, 1993, 1994) and in Table 7.4 for the natural bis-galloyl ester – hamamelitannin (Masaki *et al.*, 1994).

The name 'superoxide' suggests an exceptional degree of reactivity for the superoxide anion (O_2^-) especially as a strong oxidant and initiator of radical reactions. However, although it contains an odd number of electrons its reactivity does not resemble that of typical organic free radicals. Superoxide is a relatively small univalent anion with an O–O bond distance intermediate between that of peroxide and dioxygen. Under strictly aprotic conditions superoxide anion (O_2^-) is a strong nucleophile but is frequently oxidatively inert. Thus it does not react with a variety of substances, such as benzaldehyde, which are normally very easy to oxidise. The aqueous solvation energy of superoxide anion (O_2^-) indicates that it is almost 'fluoride like' and that it will form exceptionally strong hydrogen bonds with water. The red-ox reactivity of superoxide anion (O_2^-) is very sensitive to this solvation. The most dominant characteristic of the superoxide anion (O_2^-), by far, is its ability to act as a strong Brønsted base; it readily removes protons from weakly acidic substrates (such as phenols) and in doing so it rapidly disproportionates to become a source of dioxygen and peroxide (Nanni *et al.*, 1980).

The propensity of superoxide anion (O_2^-) to remove protons from substrates accounts satisfactorily for its reactivity with acidic reductants and their overall oxidation. Thus combination of superoxide anion (O_2^-) with substrates such as hydroquinone, α-tocopherol, 3,5-bis-tertiarybutyl-catechol, L-(+)-ascorbic acid yields products whose formation is consistent with the apparent one-electron oxidation of the substrate and the generation of H_2O_2. The primary step is the abstraction of a proton from the substrate by the superoxide anion (O_2^-) to give the substrate anion and the disproportionation products of superoxide – dioxygen and hydrogen peroxide. In turn the substrate anion is then oxidised by dioxygen in a multi-step process to yield oxidation products and hydrogen peroxide. Significantly neither the mono-anion of 3,5-bis-tertiarybutyl-catechol nor the anion of α-tocopherol reacts with superoxide anion (O_2^-). For 3,5-bis-tertiarybutyl-catechol a plausible mechanism is represented in Figure 7.9.

Despite its name the superoxide anion is thus fairly innocuous, Tables 7.3 and 7.4; in aqueous solution (O_2^-) often acts mainly as a reducing agent (Halliwell, 1982). *In vivo* many of the deleterious effects of (O_2^-) generating systems are probably caused by other species whose formation requires (O_2^-).

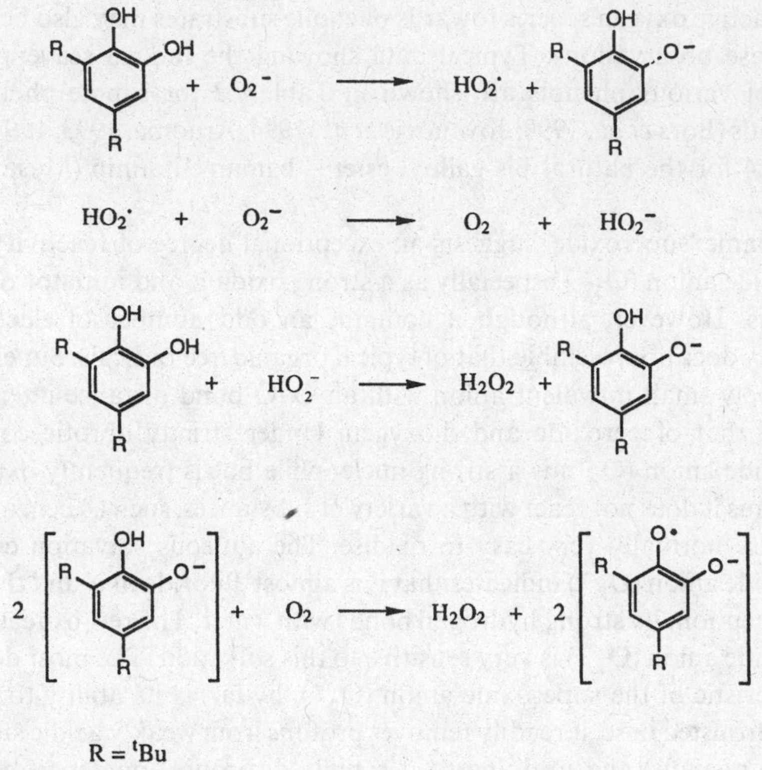

$$R = {}^tBu$$

Figure 7.9. Suggested pathway of oxidation of 3,5-bis-tertiarybutyl-catechol by superoxide anion (O_2^-) (Nanni *et al.*, 1980).

Recent evidence indicates that the hydroxyl radical HO˙ is a major candidate for such a role (Halliwell, 1982).

Evidence thus exists which supports the contention that the beneficial action of natural polyphenols in traditional medicines towards certain diseases may well derive from their ability to scavenge reactive oxygen species in cellular pro-oxidant states. However, whilst the propensity of natural plant phenols and polyphenols to act *in vitro* with reactive oxygen species in the manner predicted for natural antioxidants is clear, it should nevertheless also be pointed out that they may additionally (like ascorbate) in some circumstances show pro-oxidant characteristics (Aruoma *et al.*, 1993). Thus it has been demonstrated that the food antioxidant propyl gallate and plant phenols can in fact also react with ferrous ions in the presence of hydrogen peroxide to produce reactive oxygen species which can subsequently damage other biological molecules.

Table 7.3. *Rate constants ($M^{-1} s^{-1} \times 10^5$) for the reaction of some strongly oxidising radicals with various flavonoids and natural phenols*

Compound	O_2^-	$RCO_2^{\cdot}$	$HO^{\cdot}$
Gallic acid	—	4.5	—
Propyl gallate	—	170	—
Ascorbic acid	—	1300	—
Fisetin	0.13	4100	—
Kaempferol	0.024	—	46 000
Quercetin	0.47	390	43 000
Eriodictyol	—	—	31 000
Hesperitin	0.059	—	58 000
(+)-Catechin	0.18	61	66 000
(−)-Epicatechin	—	73	64 000

Table compiled from data derived from various sources (Bors *et al.*, 1990; Jovanovic *et al.*, 1994; Aruoma, 1993, 1994). Rates of radical reaction are directly comparable for each reactive oxygen species but are not *directly* comparable between reactive oxygen species.

Table 7.4. *Radical scavenging activities of hamameli tannin, gallic acid and propyl gallate: IC_{50} values (μM) corresponding to 50% inactivation of active oxygen species, values determined by ESR spin trapping. Data from (Masaki et al., 1994)*

Compound	Superoxide anion, O_2^-	Hydroxyl radical, $HO^{\cdot}$	Singlet oxygen
Hamamelitannin	1.31	5.46	45.5
Gallic acid	1.01	78.0	69.8
Propyl gallate	1.41	86.5	66.7
Ascorbic acid	23.3	18.8	120.4

7.3.3 Green tea and the 'French Paradox'

The present intense interest in the possibility that many major human degenerative diseases may involve, in their aetiology, free radical processes, and that antioxidant nutrients may be capable of delaying or even preventing these processes has origins which extend back to at least the 1920s and the discovery of vitamin E. Much of our present knowledge comes from epidemiological studies and indicates that the incidence of some forms of cancer and cardiovascular disease appears to be lower in populations with a larger than average

hamamelitannin
(*Hamamelis virginiana*)

propyl gallate　　　　　　　**gallic acid**

intake of antioxidant nutrients such as vitamins C and E and various carotenoids (Halliwell, 1995). Extended dietary surveys in the U.S.A. have revealed
that the calculated dietary intake of essential antioxidants is inversely related
to the risk of coronary heart disease and certain forms of cancer (Rice-Evans,
1995). Plasma concentrations of diet-derived antioxidants have revealed inverse correlations with the incidence of these diseases, although with a different
rank order of anti-oxidants.

Decreased cancer risk:
β-carotene > vitamin C > vitamin E

Decreased coronary heart disease risk:
vitamin E > β-carotene > vitamin C

It has been suggested that the following levels of intake (milligrams per day) in
the diet are likely to provide blood levels of the nutrients consonant with a low
risk of degenerative disease (Diplock, 1995; Elliott, 1995):

vitamin E (40–60 mg day^{-1}),
vitamin C (150 mg day^{-1}),
β-carotene (9–12 mg day^{-1}).

In itself this is a thought provoking observation given, for example, that in the
U.K. the estimated levels of dietary intake of these substances are generally
low. Thus for β-carotene it is, in the 16–24 age group ∼ 1.6–1.9 mg day^{-1} and
in the 50–64 age group ∼ 2.4–2.8 mg day^{-1}.

It is against this background, and the observations of Szent-Gyorgi (St.
Ruszynak and Szent-Gyorgi, 1936; Bentsath *et al.*, 1936, 1937) in the 1930s,
that the current excitements concerning the consumption of polyphenol rich

green teas and red wines should be viewed. Following his discovery of vitamin C, Szent-Gyorgi and his colleagues reported a number of findings that other substances in fruit juices probably have a synergistic effect on the actions of vitamin C. The general thrust of their observations is evident from these extracts from their papers.

Various chemical and clinical observations have led to the assumption that ascorbic acid is accompanied in the cell by a substance of similar importance and related activity. In the absence of both substances, the symptoms of the lack of ascorbic acid (*scurvy*) prevail and conceal symptoms of the deficiency of the second substance.

These results suggest that this great group of vegetable dyes, the flavons or flavanols, also play an important role in animal life, and that the dyes are of a vitamin nature.

These results suggest that experimental scurvy, as commonly known, is a deficiency disease caused by the combined lack of vitamin C and P.

The therapeutic effects observed after the administration of 'citrin' in man in septic conditions, also accompanied by polyarthritis and endocarditis, suggest that the age old beneficial effect of fruit juice is partly due to its vitamin P content.

In the event, although the story of vitamin P continued in Hungary and France well into the post-war years, substantive proof of the existence of this additional vitamin P was never forthcoming. However, it seems reasonable to assume that flavonoid compounds, such as hesperitin, hesperidin and eriodic-tyol, in the fruit juices were largely responsible for the observations, acting as antioxidants themselves or to enhance and protect the actions of vitamin C.

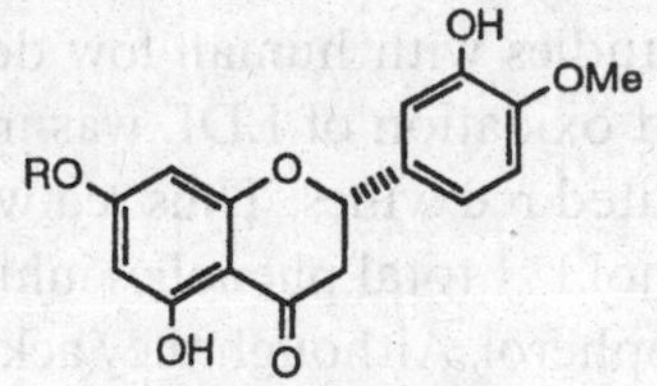

R = H , hesperitin
R = rutinose , hesperidin

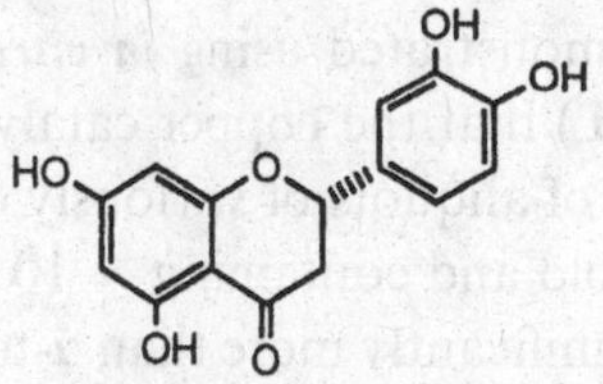

eriodictyol

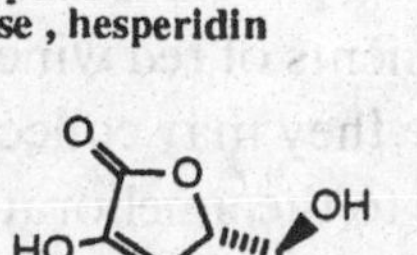

ascorbic acid , vitamin C

Contemporaneously, interest has revived in this work following observations in two other areas. Epidemiological evidence seems to indicate that consumption of polyphenol rich common items in the diet is associated with an increase in plasma antioxidant potential and this might therefore have an important role to play in the modulation of exposure to cellular oxidative stress (Kinsella *et al.*, 1993a,b; Maxwell *et al.*, 1994; Serafini *et al.*, 1994; Kondo *et al.*, 1994). Examination of WHO data shows marked differences in mortality from coronary heart disease among various countries – especially between French and U.S.A. and U.K. populations. Subjects who have similar intakes of saturated fatty acids, similar risk factors and comparable plasma cholesterol levels show a much lower incidence of death from coronary heart disease in France. The regular consumption of red wines was one of the few dietary factors that seemed to correlate with reduced coronary heart disease (CHD) mortality, although interestingly regular fruit consumption also has a good correlation with reduced CHD mortality.

Moderate alcohol consumption itself reduces CHD mortality and several possible mechanisms are now recognised – such as the ability of alcohol to alter blood lipid levels by lowering total cholesterol and raising high density lipoprotein levels (Renaud and de Logeril, 1992). The statistical data clearly show that *wine alcohol* consumption is much more strongly correlated with CHD mortality than total alcohol consumption and this has led to the view that the alcohol content of red wines may not be the sole explanation for this protection against coronary atherosclerosis. Red wine also contains significant quantities of phenolic compounds – thus a full-bodied young red wine contains up to $4.0 \, \mathrm{g} \, l^{-1}$ of phenolics – principally flavan-3-ols, oligomeric procyanidins and anthocyanin pigments (Somers and Verette, 1988), and attention has been directed, principally, but not exclusively, at the role these particular constituents and their antioxidant properties may possess in rationalising the epidemiological data and with it the so-called 'French Paradox'. Kinsella *et al.* (1993a,b) demonstrated using *in vitro* studies with human low density lipoprotein (LDL) that the copper catalysed oxidation of LDL was inhibited by the addition of aliquots of variously diluted red wines. Thus red wine diluted some 1000-fold and containing $\sim 10 \, \mu\mathrm{mol} \, l^{-1}$ total phenolics inhibited LDL oxidation significantly more than α-tocopherol. Although they acknowledged the need for more detailed work on this problem these authors suggested that if potent antioxidant phenolic components of red wine are routinely ingested by the regular consumption of red wine they may collectively reduce oxidation of lipoproteins and thereby contribute to the amelioration of atherosclerosis and morbidity and mortality from coronary heart disease. Other workers have presented data which supports these findings. Thus Japanese researchers (Kondo *et al.*, 1994) provided direct evidence that regular and long term consumption of red wine, but not ethanol, inhibited LDL oxidation *in vivo*.

However it should be very clearly pointed out as a word of caution, if not scepticism, that not all workers in this field share these enthusiasms for the view that the phenolic constituents of red wines may act as antioxidants for low-density lipoprotein (LDL) and so exert an anti-atherogenic effect. Thus de Rijke and her colleagues (de Rijke *et al.*, 1995) assessed the susceptibility of LDL to copper mediated oxidative modification by the consumption of 'alcohol-poor' red wine. They concluded, in contrast to the findings adumbrated above, that the daily intake of flavonoid-rich red wine **did not** influence the oxidisability of LDL. Halliwell similarly suggested that the enthusiasm for reports that phenolic constituents of red wine may act as antioxidants of LDL 'should be tempered by some scepticism' (Halliwell, 1993). The title of one other paper (Muller and Fugelsang, 1994) perceptively encapsulates this present uncertainty – 'Take two glasses of wine and see me in the morning'. More detailed biological observations are clearly required to substantiate the epidemiological evidence at this stage.

3*S*, (+)-catechin
3*R*, (-)-epicatechin

oligomeric procyanidins

malvidin-3-glucoside

Epidemiological evidence, principally from Japan and China, likewise strongly suggests that the habitual consumption of green tea as a beverage may protect both against cancer and the development of coronary heart disease. Attention has been similarly focused, with the presumption that polyphenols may ameliorate conditions of oxidative stress, upon the major polyphenolic component of the green tea flush (*Camellia sinensis*), namely (−)-epigallocatechin-3-*O*-gallate. Although the amounts may vary, dependent upon cultural and climatic conditions, flavanols usually constitute up to 20–30% of the dry matter of the fresh green tea leaf, of these (−)-epigallocatechin (∼ 2.5%), (−)-epigallocatechin-3-*O*-gallate (∼ 10.5%) and (−)-epicatechin-

3-*O*-gallate (~ 2.75%) overwhelmingly predominate. Serafini and his group (Serafini *et al.*, 1994) showed that black tea had an *in vitro* antioxidant activity approximately one-third of the value reported by Maxwell for red wine (Maxwell *et al.*, 1994) and that consumption of polyphenol rich common items of the diet (such as black and green tea) is associated with an increase in plasma antioxidant potential. A preliminary report (Matsumoto *et al.*, 1991) indicates that in a series of tests in rats with orally administered (−)-epigallocatechin-3-*O*-gallate, some was adsorbed in the small intestine, but most reached the large intestine where decomposition occurred. At this stage therefore the position in regard to this problem is very similar to that described for red wines − precise 'quantitative' experimental evidence in relation to the fate, after ingestion, of these phenolic compounds in humans is still awaited. Until such time the validity of the claims made, in relation to their possible effects in the amelioration of certain diseased states, for these dietary polyphenols should be treated with similar reserve.

(-)-epigallocatechin-3-O-gallate

(-)-epicatechin-3-O-gallate

In relation to the anxioxidant potential of the various catechins and gallocatechins which have been implicated in the beneficial medical effects stemming from the habitual consumption of teas and red wines, Jovanovic and his colleagues have conducted some important, substantive and quantitatively precise experiments (Jovanovic *et al.*, 1994, 1995) which are essential contributions to the fundamental background science in this area, and against which future arguments and ideas must be viewed and tested. A one-electron oxida-

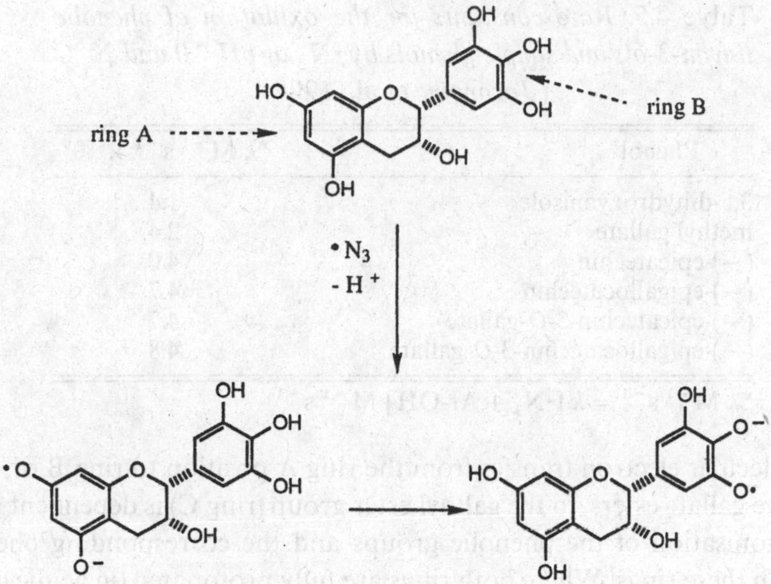

Figure 7.10. One-electron oxidation of (−)-epigallocatechin by ·N$_3$ at pH 7.0, (Jovanovic *et al.*, 1995).

tion by the azidyl radical ·N$_3$ and direct photoionisation by a 248 nm laser was utilised to generate phenoxyl radicals from the phenolic flavan-3-ol substrates. Rate constants for the reaction of the azidyl radical with some simple phenolic substrates are shown in Table 7.5. The rates are close to diffusion controlled, reflecting their favourable electron donating properties; with simple phenols both chemo-ionisation and photo-ionisation yield the same phenoxyl radical(s). In the case of the phenolic flavan-3-ols one-electron oxidation using azidyl radicals results in the formation of a short-lived and a long-lived transient, whilst photoionisation gives only the latter species. Azidyl radicals attack both the A and B rings of phenolic flavan-3-ols such as (−)-epigallocatchin at pH 7.0; the ratio of the attack on the A ring as opposed to the B ring is ∼ 1:2, Figure 7.10. In the laser flash photolysis of aqueous (−)-epigallocatechin at pH 7.0 only the B ring is photo-ionised because of its relatively lower ionisation potential. Interestingly the authors formulate the radical anion derived from the pyrogallol ring B with the radical in a *meta* position (3′ or 5′) and with the negative charge distributed between the two remaining hydroxyl groups.

Subsequent to the one-electron oxidation the phloroglucinol derived phenoxyl (ring A) transforms to the pyrogallol derived phenoxyl (ring B) via inter- and intra-molecular electron transfer. The rates of both the inter- and

Table 7.5. *Rate constants for the oxidation of phenolic flavan-3-ols and simple phenols by* $\cdot N_3$ *at pH 7.0 and 20°C (Jovanovic et al., 1995)*

Phenol	$*k\,\mathrm{M}^{-1}\,\mathrm{s}^{-1} \times 10^9$
3,5-dihydroxyanisole	1.4
methyl gallate	2.4
(−)-epicatechin	4.0
(−)-epigallocatechin	4.7
(−)-epicatechin-3-O-gallate	4.7
(−)-epigallocatechin-3-O-gallate	4.8

$*k\,\mathrm{M}^{-1}\,\mathrm{s}^{-1} = k\,[\cdot\mathrm{N}_3 + \mathrm{Ar\text{-}OH}]\,\mathrm{M}^{-1}\,\mathrm{s}^{-1}$

intra-molecular electron transfer from the ring A position to ring B or, in the case of the gallate esters, to the galloyl ester group (ring C) is dependent on the state of ionisation of the phenolic groups and the corresponding phenoxyl radicals in these rings. Where both rings are fully protonated (in acidic media) and where the oxidation produces neutral phenoxyl radicals the lowest rates appertain. Correspondingly where the products are phenoxyl radical anions, in alkaline media, the highest rates are observed. However, independent of the conditions, these secondary electron transfer processes proceed to completion leaving the ring B (or ring C) phenoxyl radicals as the final transient products. At any pH value one or two protons are exchanged subsequent to the electron transfer. Consequently, a considerable solvent reorganisation is required for intermolecular electron transfer, whereas the equivalent of a hydrogen atom is transferred in the intra-molecular process. An analysis of the activation parameters for the azidyl radical/(−)-epigallocatechin-3-O-gallate system suggests a hydrogen atom transfer rather than an electron transfer mechanism. The difference in the pK_a values of the parent phenols and the phenoxyl radicals, amounting to ~ 4 pK units, indicates as expected a high degree of delocalisation of the unpaired electron in the radicals.

Jovanovic and his colleagues (Jovanovic *et al.*, 1995) also determined the potential of the phenolic flavan-3-ols to act as antioxidants. The efficacy of these molecules to act as antioxidants in biological systems depends on their ability to intercept and inactivate potentially damaging 'foreign' free radicals or to repair damaged biomolecules or bioradicals. The electron transfer reactions of various phenolic flavan-3-ols were examined with several red-ox standards. The reduction potential of the radicals of (−)-epigallocatechin and its 3-O-gallate ester at pH 7.0 and 20°C (~ 0.42 V) is less than that of vitamin E (0.48 V) which Jovanovic and his colleagues suggested may possibly enable these substrates to participate in red-ox defence and in the recycling of vitamin

E, Figure 7.8. As a consequence the phenolic flavan-3-ols may play an important role in physiological defence of the gastrointestinal tract.

Jovanovic and his colleagues also measured the rates of reaction of various phenol derivatives with the superoxide radical, showing there was a broad correlation with the electron donating properties of the phenol. Thus the following relative rates of reaction were determined at pH 7.0 and 20°C:

3,5-dihydroxyanisole $<(+)$-catechin/$(-)$-epicatechin $<(-)$-epigallocatechin

and

$(-)$-epigallocatechin $<(-)$-epigallocatechin-3-O-gallate

$(-)$-Epigallocatechin-3-O-gallate (EGCG) is one of the most efficient scavengers of the superoxide radical: $k[EGCG + O_2^-] = 7.3 \times 10^5 M^{-1}s^{-1}$ at 20°C. The authors attributed this to the presence of the two pyrogallol derived phenolic nuclei acting as the principal antioxidant sites. They also showed that phenolic flavan-3-ols such as $(-)$-epicatechin and $(-)$-epigallocatechin and their 3-O-gallate esters rapidly quench another activated form of oxygen, namely singlet oxygen $(^1\Delta_g)$; k values were determined to be $\sim 10^8 M^{-1} s^{-1}$, comparable to that of vitamin E.

Some important general conclusions which have emerged from the work of Jovanovic and his colleagues (1996), and others, are summarised below:

(i) flavanoids, because of their abundance in vegetables and other foodstuffs, are an important class of natural antioxidants and have the ability to scavenge various peroxyl radicals,

(ii) the aromatic ring whose radical has the lower reduction potential is the principal site of antioxidant activity in any flavonoid; in the phenolic flavan-3-ols, hesperidin, rutin, dihydroquercetin and quercetin the catechol (or pyrogallol) derived ring B dominates the antioxidant activity,

(iii) all the flavonoids investigated are inferior electron donors when compared to ascorbate (vitamin C) but both quercetin and the gallocatechins possess the ability to reconstitute vitamin E from vitamin E radicals under physiological conditions,

(iv) because all the flavonoids examined are inferior electron donors when compared to vitamin C in a physiologically relevant pH range of 7–9, it therefore seems unlikely, as originally suggested by Szent-Gyorgi, that they can repair vitamin C radicals and hence 'spare' vitamin C in one-electron reactions in aqueous media.

7.4 Polyphenol complexation

Although the uses of polyphenols as medicinal agents may be summarised under several broad headings, e.g. Tables 7.1 and 7.2, many of their actions appear to devolve, either directly or indirectly, on their ability to complex with proteins and polysaccharides. They thus aid the healing of wounds, burns and inflammations. In doing so they act to produce an impervious layer (poly-phenol–protein and/or polysaccharide complex) under which the natural heal-ing processes can occur. Presumably in such instances part of their action is facilitated by their complexation with the collagen of the skin, which underlies the age-old manufacture of leather from animal skins and hides. According to the picture presented elsewhere in this text this associative process arises from the nature of the supramolecular assembly of collagen molecules in the col-lagen fibrils and fibres. The individual triple-helical collagen molecules may be regarded as stiff rod-like structures whose staggered form of organisation in the collagen fibre gives rise to a liquid crystal-like supramolecular assembly containing 'holes' or 'gap zones'. It is these latter regions which it is thought constitute the principal sites at which polyphenol molecules may interact and embed themselves within the collagen structure cross-linking fibrils and fibres (E. Haslam, unpublished observations), thereby hardening and rendering the tissue impervious to abrasion and water, and thus to additional infection. Under this surface the normal healing processes then take place. Decoctions of a number of polyphenol-rich plants were also frequently prescribed to stop internal bleeding, nose bleeds and to heal all internal wounds generally.

Similar complexation processes probably take place internally; gut secre-tions are hindered thus protecting the underlying mucosa from toxins and other irritants in the bowels. This feature very probably explains the wide-spread use of polyphenol containing plants and plant extracts in old herbal remedies for the treatment of disorders of the digestive system and the intesti-nal tract, Table 7.1. Likewise the healing action of many herbs and their ability to increase vitality has been variously ascribed to their ability to act to *purify* or to *cleanse* the bloodstream (Hoffmann, 1987). Many of these extracts contain polyphenols of one form or another, and it is tempting to suggest that the idea of a blood cleanser, whilst hinting descriptively at much but saying little of substance in medical terms, may derive from the ability of such remedies to enter the bloodstream and preferentially complex with and remove toxins and other harmful materials of a proteinaceous character.

In strictly more scientific terms this propensity to bind to proteins also presumably accounts for the fact that polyphenols inhibit virtually every enzyme that they are tested with *in vitro* (Haslam, 1989; Loomis, 1974). Assessment of the medical significance of polyphenol inhibition of particular enzymes (whose actions influence the course of development of illnesses and disease), deter-mined *in vitro*, is therefore dependent on the, as yet, unanswered questions relating to the absorption and penetration of ingested polyphenols to the desired site(s) of action *in vivo*. Where the enzymes are extracellular then these problems do not arise. Thus the mutans group of streptococci, *Streptococcus mutans* and *Streptococcus sobrinus*, are principal cariogenic organisms and their major ecological niche is the tooth surface and dental plaque. Cariogenicity is considered to be strongly associated with the ability of these organisms to synthesise extracellular water-insoluble glucans by using glucosyl transferases (GTases). The glucans are synthesised from sucrose by the co-operative action of GTases and are highly adherent to various solid surfaces, including the tooth. This results in firm, irreversible adherence of the mutans streptococci to the tooth surface and to the eventual formation of dental plaque and the development of dental caries. At least two major classes of GTase inhibitors, present in common foods and beverages, are known, namely certain mono- and oligosaccharides and polyphenolic compounds such as those found in betel nuts (*Arecha catechu*) and tea leaves (*Camellia sinensis*). Hattori and his colleagues (Hattori *et al.*, 1990) have examined the effects of tea polyphenols on the glucosyltransferase (GTase) from *Streptococcus mutans*, and they showed that whilst (−)-epicatechin, (−)-epigallocatechin and their 3-*O*-gallate esters, and their various diastereoisomers, showed modest inhibitory action against the enzyme, theaflavin and its mono- and bis-galloyl esters were potent inhibitors at concentrations of some 1–10 mM. Okuda and his group (Kakiuchi *et al.*, 1986) had earlier examined the inhibitory actions of several natural galloyl esters against the same enzyme from *Streptococcus mutans*, and they demonstrated that the level of inhibition was broadly related to the number of galloyl ester groups in the inhibitor. However, Hattori and his colleagues (Hattori *et al.*, 1990) interestingly noted in a footnote to the paper that both β-1,2,3,4,6-penta-*O*-galloyl-D-glucose and octyl gallate were both potent inhibitors of the glucosyltransferase from *Streptococcus mutans*. They commented that 'the relationship between a galloyl ester and its inhibitory potency against GTase is quite complex'. Hamada and his colleagues (Hamada *et al.*, 1992, 1993a,b) have shown that ellagic acid and various extracts of partially fermented tea (Oolong tea) inhibit glucan synthesis from sucrose by the GTase from *Streptococcus mutans* and *Streptococcus sobrinus*. These workers suggested that the active components of the Oolong tea were the polymeric fractions that were 'structurally different from those found in green and black tea', although they did not specifically characterise and identify them. Administration of polyphenolic extracts of Oolong tea into diets and drinking water led to a highly significant reduction in dental caries development and plaque accumulation in experimental animals infected with Mutans streptococci. It was suggested that such extracts (like betel nuts) may well be very useful for controlling dental caries in humans, presumably via their inhibition of the glycosyl transferases.

Polyphenols have been shown to have a broad antiviral spectrum *in vitro*, but to date their corresponding properties *in vivo* have not been well established, (Sakagami, Sakagami and Takeda, 1995). Studies on the anti-HIV activity of polyphenols and synthetic lignins have been referred to earlier (Okuda *et al.*, 1992; Nakashima *et al.*, 1992). In an early study of plant viral infection, Cadman (1960) suggested that polyphenolic extracts of the leaf of raspberry (*Rubus idaeus*) probably act on most viruses by clumping the virus particles together into complexes (by association with the protein coat and thereby facilitating cross-linking reactions between different virus particles) which are largely uninfective. In later work others have similarly deduced that viral inactivation, *in vitro*, is directly attributable to preferential binding of the polyphenol to the protein coat of the virus. In a systematic study of the anti-viral activity of a very wide range of natural products Vlietinck and his colleagues (Vanden Berghe *et al.*, 1985) concluded that polyphenols act principally by binding to the virus and/or the protein of the host cell membrane and thus arrest absorbtion of the virus. They concluded that in consequence polyphenols are probably only viricidal in nature. Sakagami and his colleagues (Sakagami *et al.*, 1995) have put forward a number of possible mechanisms whereby polyphenols may exert their anti-viral action. They suggested that the major part of the anti-viral activity due to polyphenols probably derives from their direct inactivation of the virus and/or from inhibition of the virus binding to the cells. They also noted that although polyphenols are known to inhibit viral replication enzymes (such as reverse transcriptase for HIV and RNA polymerase for influenza virus) and other enzymes (e.g. poly(ADP-ribose) glycohydrolase), these effects seem to be rather non-specific.

This affinity of natural polyphenols for proteins also extends to very simple peptides which are proline rich and/or hydrophobic in character. In this context the association of polyphenols with bio-active peptides such as angiotensin I and II and bradykinin is of particular interest. All three peptides are, for example, readily precipitated from aqueous media by polyphenols. Bradykinin ($M_R = \sim 1060$) is a nonapeptide (**Arg–Pro–Pro–Gly–Phe–Ser–Pro–Phe–Arg**) and is released from its plasma protein precursor(s) by certain snake venoms or enzymes with trypsin-like activity. Its physiological and pharmacological activities are manifold. They include stimulation of smooth muscle, inhibition of neurotransmission in the spinal cord, the release of catecholamines in the adrenal medulla, induction of acute arterial hypotension, powerful vasodilation, increased capilliary permeation, leucocyte migration and accumulation, and the initiation of pain. Bradykinin may be a mediator of conditions ranging from functional vasodilation to acute inflammation in the human body. It is well recognised that the conformation of a hormone can be related to its biological activity. In aqueous media NMR and

CD studies suggest that bradykinin is in rapid equilibrium among many conformers and does not show any persistent structural features such as β-turns or internal hydrogen bonds (Brady *et al.*, 1971; Denys *et al.*, 1982). However, in DMSO the peptide probably assumes a more rigid conformation with β-bends at both the C- and N-termini, stabilised by electrostatic interactions between the two arginine side chains which are juxtaposed as a consequence of the folding pattern (Mirmira *et al.*, 1990).

bradykinin

Bradykinin is proline rich and with two phenyl residues is hydrophobic in character; its strong association with and precipitation by natural polyphenols is therefore not unexpected. Lokman Kahn and his colleagues have conducted (Lokman Kahn *et al.*, unpublished) studies of the complexation of bradykinin with polyphenols in D_2O–d_6DMSO (20%) using high resolution NMR techniques. Association constants were determined (Table 7.6) and these show very similar trends to those predicted from earlier work with larger protein molecules, i.e. a strong dependence upon molecular size, the number and arrangement of phenolic groups and water solubility of the polyphenol.

Table 7.6. *Bradykinin – association with polyphenols; precipitation by poly-phenols*

Polyphenol	K_a, M^{-1}	*Threshold conc.
β-1,2,3,4,6-Penta-*O*-galloyl-D-glucose	33.7	< 0.2 mM
β-1,3,4,6-Tetra-*O*-galloyl-D-glucose	20.5	< 0.5 mM
β-1,3,6-Tri-*O*-galloyl-D-glucose	4.4	< 0.75 mM
Procyanidin B-2	8.2	~ 1.0 mM

Association constants (K_a) determined by titration of aliquots of polyphenol into a solution of bradykinin in D_2O–d_6DMSO at 295 K, pH 5.9, and measurement of chemical shift changes in the protons of bradykinin.
* The threshold precipitation values for polyphenols observed for the initiation of precipitation of bradykinin from a 2.0 mM aqueous solution at 295 K. Data from Lokman Kahn (unpublished) and Haslam (1996).

No evidence was obtained to show that the peptide underwent significant changes from a random coil conformation upon complexation with the poly-phenols. Although the most significant proton chemical shift changes in the bradykinin were associated with each of the three proline residues and the two phenylalanine groups, suggesting that these amino acid side-chains were participating preferentially in the complexation with polyphenol, it is not thought probable that there is a specific mode of binding between the polyphenol and peptide substrates. Rather the driving force is visualised as the relatively unselective association of the aromatic nuclei of the polyphenol with hydrophobic groups (**Pro; Phe** (π–π' interactions)) on the nonapeptide followed by secondary hydrogen bonding re-enforcing this initial complexation.

The role of the side-chain of the amino acid arginine (**Arg**) in these complexations remains unclear. However, analogues of bradykinin, which lack either of the terminal arginine groups at positions 1 and 9, were significantly inferior in their capacity to bind the polyphenol β-1,2,3,4,6-penta-*O*-galloyl-D-glucose, suggesting that these residues may also have a specific role to play in the binding of polyphenolic substrates.

It is therefore interesting to note that as a result of earlier crystallographic studies of proteins, Perutz and Levitt (1988) embarked upon an investigation to discover if the energy of conventional electrostatic interactions between amino – and benzene – groups is strong enough to speak of a hydrogen bond. Simple energy calculations showed that there was a significant interaction between a hydrogen bond donor (like –NH– and –NHC(NH)–NH_2) and the centre of a benzene ring, which acts as a hydrogen bond acceptor. Hunter and his colleagues (Hunter *et al.*, 1992) have similarly shown that the π-electrons of an aromatic ring may, in certain circumstances, represent adequate hydrogen bond acceptors. In the case of an 'electron rich' phenolic nucleus this would

Table 7.7. *Association of β-1,2,3,4,6-Penta-O-gal-loyl-D-glucose with derivatives of Bradykinin; data from Lokman Kahn (unpublished) and Haslam (1996)*

Peptide	K_a, M^{-1}
DesArginyl (*1*)Bradykinin	
Pro–Pro–Gly–Phe–Ser–Pro–Phe–Arg	20
DesArginyl (*9*)Bradykinin	
Arg–Pro–Pro–Gly–Phe–Ser–Pro–Phe	20.4
Peptide (1–5)	
Arg–Pro–Pro–Gly–Phe	~ 13.0

Association constants (K_a) determined by titration of aliquots of polyphenol into a solution of bradykinin in D_2O–d_6DMSO at 295 K, pH 5.9, and measurement of chemical shift changes in the protons of bradykinin.

doubtless possess an enhanced hydrogen bond acceptor capability and therefore such interactions may be of some importance at a secondary stage in polyphenol/bradykinin and polyphenol/salivary protein interactions.

Molecular models show that the unstructured random coil of a penta-hexapeptide is sufficient to bridge two adjacent galloyl ester groups in a substrate such as β-1,2,3,4,6-penta-*O*-galloyl-D-glucose. This minimum structural feature for bidentate binding is borne out by the relatively strong complexation of the bradykinin analogue – **peptide 1–5** – with β-1,2,3,4,6-penta-*O*-galloyl-D-glucose.

Clearly it would now be of considerable interest to ascertain if and how polyphenols may modify the *in vivo* physiological actions of bioactive peptides such as bradykinin.

Whatever finally emerges as the *in vivo* role of polyphenols in relation to their action on enzymes, viruses and other proteinaceous materials, Table 7.1 and 7.2, the mechanism of interaction of polyphenols and proteins has been a subject of detailed scrutiny over the past two decades. Present knowledge will doubtless form the fundamental basis for the future analysis of *in vivo* physiological and pharmacological studies. A discussion of the principal features of the phenomenon of protein–polyphenol interactions has been given earlier.

Insofar as the possible modes of action of natural polyphenols as drugs and medicines are concerned there is clear evidence that they have the ***potential*** to act in the three general areas specified earlier (i.e. transition metal-ion complexation, as antioxidants in cellular pro-oxidant states and by association with proteins and peptides). What is urgently required at this juncture is evidence concerning their absorption, penetration and metabolism in the

Phenylalanine residues - 'π-π' stacking

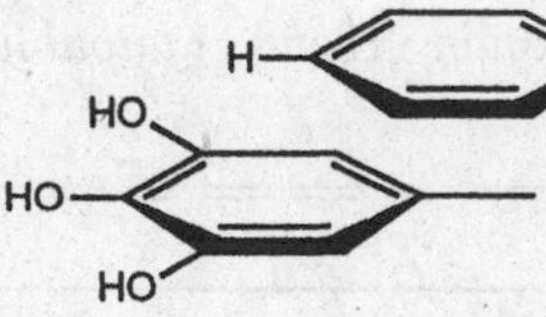

Arginyl residues - hydrogen bonding

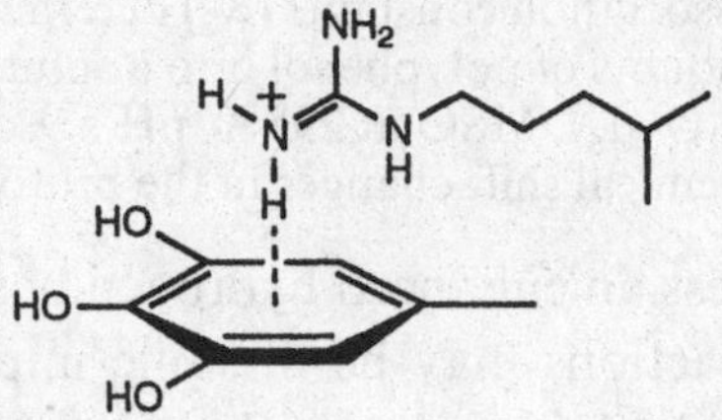

Figure 7.11. Additional modes of complexation of polyphenols with the proline rich peptide – bradykinin (Lokman Kahn, unpublished; Haslam, 1996).

human body. Until such time the evidence for their remedial effects remains based upon epidemiological evidence rather than scientific observation. Parenthetically it is worth noting that this area of medicine has been re-invigorated in the mind of the public by the impact of 'popular' scientific journalism. Thus a great deal has been written in recent years in the press concerning alternative therapies and the use of herbal medicines, and on treatments based upon the use of herbs in health and healing; there is little doubt that it has caught the imagination of many members of the general public, particularly in the so-called more affluent countries of the western world where such knowledge has largely been discarded. However 'popular' journalism in this area should perhaps also be taken with a government health warning – 'All journalists oversimplify. It is their greatest art, it is their greatest crime'.

References

Aruoma, O. I. (1993). Free radicals and food. *Chemistry in Britain*, **29**, 210–14.
Aruoma, O. I. (1994). Nutrition and health aspects of free radicals and anti-oxidants. *Food Chem. Toxicol.*, **32**, 671–83.
Aruoma, O. I., Murcia, A., Butler, J. and Halliwell, B. (1993). Evaluation of the

anti-oxidant and pro-oxidant actions of gallic acid and its derivatives. *J. Agric. Food Chem.*, **41**, 1880–5.

Balde, A. M., van Hoof, L., Pieters, L., Vanden Berghe, D. and Vlietinck, A. J. (1990). Plant anti-viral agents. VII. Anti-viral and antibacterial proanthocyanidins from the bark of *Pavetta owariensis*. *Phytotherapy Research*, **4**, 182–8.

Bentsath, A., St. Rusznyak and Szent-Gyorgi, A. (1936). Vitamin nature of flavonols. *Nature*, **138**, 798.

Bentsath, A., St. Rusznyak and Szent-Gyorgi, A. (1937). Vitamin P. *Nature*, **139**, 326–7.

Bors, W., W. Heller, W., C. Michel, C. and Saran, M. (1990). Flavanoids as anti-oxidants: determination of radical scavenging efficiencies. *Methods in Enzymology*, **186**, 343–55.

Brady, A. H., Ryan, J. W. and Stewart, J. M. (1971). Circular dichroism of bradykinin and related peptides. *Biochem. J.*, **121**, 179–84.

Cadman, C. H. (1960). Inhibition of plant virus infection by tannins. In *Phenolics in Plants in Health and Disease* ed. J. B. Pridham, Pergammon Press: Oxford and London, pp. 101–5.

Cerutti, P. A. (1985). Pro-oxidant states and tumour promotion. *Science*, **227**, 375–81.

Denys, L., Bothner-By, A. A., Fisher, G. H. and Ryan, J. W. (1982). Conformational diversity of bradykinin in aqueous solution. *Biochemistry*, **21**, 6531–6.

Diplock, A. (1995). Anti-oxidant nutrients – efficacy in disease protection and safety. *The Biochemist*, **17**, 16–18.

Elliott, R. (1995). Breaking through the language barrier – the marketers task for anti-oxidants. *The Biochemist*, **17**, 19–21.

Frausto da Silva, J. J. R. and Williams, R. J. P. (1991). *The Biological Chemistry of the Elements*, Clarendon Press: Oxford.

Grieve, M. (1978). *A Modern Herbal*. Penguin Books: Harmondsworth, England.

Halliwell, B. (1982). Superoxide-dependent formation of hydroxyl radicals in the presence of iron salts is a feasible source of hydroxyl radicals *in vivo*. *Biochem. J.*, **205**, 401–2.

Halliwell, B. (1993). Anti-oxidants in wine. *The Lancet*, **341**, 1538.

Halliwell, B. (1995). Antioxidants: elixirs of life or tonics for tired sheep? *The Biochemist*, **17**, 3–6.

Halliwell, B. and Gutteridge, M. J. C. (1990). Role of free radicals and catalytic metal ions in human disease: an overview. *Methods in Enzymology*, **186**, 1–85.

Hamada, S., Sawamura, S. and Tonosaki, Y. (1992). Inhibitory effects of ellagic acid on glucosyltransferases from Mutans streptococci. *Biosci, Biotech. Biochem.*, **56**, 766–8.

Hamada, S., Ooshima, T., Minami, T., Aono, W., Izumatani, A., Sobue, S., Fujiwara, T. and Kawabata, S. (1993a). Oolong tea polyphenols inhibit experimental dental caries in SPF rats infected with Mutans streptococci. *Caries Research*, **27**, 124–9.

Hamada, S., Nakahara, K., Kawabata, S., Ono, H., Ogua, K., Tanaka, T. and Ooshima, T. (1993b). Inhibitory effect of oolong tea polyphenols on glucosyltransferases of Mutans streptococci. *Appl. and Environmental Microbiol.*, pp. 968–73.

Haslam, E. (1989). *Plant Polyphenols: Vegetable Tannins Re-visited*, Cambridge University Press: Cambridge.

Haslam, E. (1996). Natural polyphenols (vegetable tannins) as drugs: possible modes of action. *J. Natural Prod.*, **59**, (IN PRESS).

Haslam, E. and Cai, Y. (1994). Plant polyphenols (vegetable tannins): gallic acid metabolism. *Natural Product Reports*, **11**, 41–66.

Hattori, M., Kusumoto, I. T., Namba, T., Ishigami, T. and Hara, Y. (1990). Effect of tea polyphenols on glucan synthesis by glucosyltransferase from *Streptococcus mutans*. *Chem. Pharm. Bull.*, **38**, 717–20.

Hoffman, D. (1987). *The Herb Users Guide – The Basic Skills of Medical Herbalism*. Thorson Press: Vermont, U.S.A.

Hunter, C. A., Hanton, L. R. and Purvis, D. H. (1992). Structural consequences of a molecular assembly that is deficient in hydrogen-bond acceptors. *J. Chem. Soc. Chemical Communications*, pp. 1134–6.

Jovanovic, S. V., Steenkens, S., Tosic, M., Marjanovic, B. and Simic, M. G. (1994). Flavanoids as anti-oxidants. *J. Amer. Chem. Soc.*, **116**, 4846–51.

Jovanovic, S. V., Steenken, S., Hara, Y. and Simic, M. G. (1995). Anti-oxidant potential of gallocatechins. A pulse radiolysis and laser photolysis study. *J. Amer. Chem. Soc.*, **117**, 9881–8.

Jovanovic, S. V., Steenken, S., Hara, Y. and Simic, M. G. (1996). Reduction potential of flavonoid and model phenoxyl radicals. Which ring in flavonoids is responsible for anti-oxidant activity? *J. Chem. Soc. (Perkin Trans. 2)*, pp. 2497–504.

Kakiuchi, N., Hattori, M., Nishizawa, T., Yamagishi, T., Okuda, T. and Namba, T. (1986). Studies on dental caries prevention by traditional medicines. VIII. Inhibitory effects of various tannins on glucan synthesis by glucosyltransferase from *Streptococcus mutans*. *Chem. Pharm. Bull.*, **34**, 720–5.

Kashiwada, Y., Nonaka, G.-I., Nishioka, I., Chang, J.-J. and Lee, K. H. (1992). Tannins and related compounds as selective cytotoxic agents. *J. Nat. Prod.*, **55**, 1033–43.

Kinsella, J. E., Frankel, E. N., German, J. B. and Kanner, J. (1993a). Possible mechanisms for the protective role of anti-oxidants in wine and plant foods. *Food Technology*, pp. 85–9.

Kinsella, J. E., Frankel, E. N., German, J. B., Parks, E. and Kanner, J. (1993b). Inhibition of oxidation of human low-density lipoprotein by phenolic substances in red wine. *The Lancet*, **341**, 454–7.

Kondo, K., Matsumoto, A., Kurata, H., Tanahashi, H., Koda, H., Amachi, T. and Itakura, H. (1994). Inhibition of oxidation of low-density lipoprotein with red wine. *The Lancet*, **344**, 1152.

Lokman Kahn, Md., Lilley, T. H., Williamson, M. P. and Haslam, E. (1995) unpublished observations.

Loomis, L. D. and Raymond, K. N. (1991). Solution equilibria of enterobactin and metal–enterobactin complexes. *Inorg. Chem.*, **30**, 906–11.

Loomis, W. D. (1974). Overcoming problems of phenolics and quinones in the isolation of plant enzymes and organelles. *Methods in Enzymology*, **31**, 528–44.

Masaki, H., Atsumi, T. and Sakurai, H. (1994). Hammameli tannin as a new potent active oxygen scavenger. *Phytochemistry*, **37**, 337–43.

Matsumoto, N., Tono-oka, F., Ishigaki, A., Okushio, K. and Hara, Y. (1991). The fate of epigallocatechin gallate (EGCg) in the digestive tract of the rat. *International Symposium on Tea Science*, Shizuoka, Japan, Abstract 42.

Maxwell, S., Cruickshank, A. and Thorpe, G. (1994). Red wine and anti-oxidant activity in serum. *The Lancet*, **344**, 193–4.

Mirmira, S. M., Durani, S., Srivastava, S. and Phadke, R. S. (1990). Occurrence of β-bends in bradykinin dissolved in DMSO d_6. *Mag. Res. Chem.*, **28**, 587–93.

Muller, C. J. and Fugelsang, K. C. (1994). Take two glasses of wine and see me in the morning. *The Lancet*, **343**, 1429–30.

Nakashima, H., Murakami, T., Yamamoto, N., Naoe, T., Kawazoe, Y., Konno, K. and Sakagami, H. (1992). Lignified materials as medicinal resources. V. Anti-HIV (human immunodeficiency virus) activity of some synthetic lignins. *Chem. Pharm. Bull.*, **40**, 2102–5.

Nanni, E. J., Stallings, M. D. and Sawyer, D. T. (1980). Does superoxide ion oxidise catechol, α-tocopherol and ascorbic acid by direct electron transfer? *J. Amer. Chem. Soc.*, **102**, 4481–5.

Nishizawa, M., Yamagishi, T., Dutschman, G. E., Parker, W. B., Bodner, A. J., Kilkuskie, R. E., Cheng, Y-C. and Lee, K-L. (1989). Anti-Aids agents. I. Isolation and characterisation of four new tetragalloylquinic acids as a new class of HIV reverse transcriptase inhibitors from tannic acid. *J. Natural Prod.*, **52**, 762–8.

Nonaka, G-I., Nishioka, I. Nishizawa, M., Yamagishi, T., Dutschman, G. E., Kashiwada, Y., Bodner, A. J., Kilkuskie, R. E., Cheng, Y-C. and Lee, K-L. (1990). Anti-Aids agents. 2. Inhibitory effects of tannins on HIV reverse transcriptase and HIV replication in H9 lymphocyte cells. *J. Natural Prod.*, **53**, 587–95.

Okuda, T. (1993). Natural polyphenols as anti-oxidants and their potential use in cancer prevention. In *Polyphenolic Phenomena*, editor A. Scalbert, Paris, INRA publications, pp. 221–35.

Okuda, T., Kimura, Y., Yoshida, T. Hatano, H. and Arichi, S. (1983). Studies on the activity of tannins and related compounds from medicinal plants and drugs. I. Inhibitory effects on lipid peroxidation in mitochondria and microsomes of liver. *Chem. Pharm. Bull.*, **31**, 1625–31.

Okuda, T., Mori, K. and Hatano, H. (1985). Relationship of the structure of tannins to the binding activities with haemoglobin and methylene blue. *Chem. Pharm. Bull.*, **33**, 1424–33.

Okuda, T., Yoshida, T. and Hatano, T. (1989). Ellagitannins as active constituents of medicinal plants. *Planta Medica*, **55**, 117–22.

Okuda, T., Yoshida, T. and Hatano, T. (1991). Chemistry and biological activity of tannins in medicinal plants. *Economic and Medical Plant Research*, **5**, 129–65.

Okuda, T., Nakashima, H., Murakami, T., Yamamoto, N., Sakagami, H., Tanuma, S., Hatano, T. and Yoshida, Y. (1992). Inhibition of human immunodeficiency virus replication by tannins and related compounds. *Antiviral Research*, **18**, 91–103.

Perchellet, J. P. and Perchellet, E. M. (1989). Anti-oxidants and multistage carcinogenesis in mouse skins. *Free Radicals in Medicine and Biology*, **7**, 377–408.

Perutz, M. F. and Levitt, M. (1988). Aromatic rings act as hydrogen bond acceptors. *J. Mol. Biol.*, **201**, 751–4.

Porter, W. L. (1993). Paradoxical behaviour of anti-oxidants in food and biological systems. *Toxicol. Indust. Health*, **9**, 93–122.

Renaud, S. and deLogeril, M. (1992). Wine, alcohol, platelets and the French paradox for coronary heart disease. *The Lancet*, **339**, 1523–6.

Rice-Evans, C. (1995). Anti-oxidant nutrients in protection against coronary heart disease and cancer. *The Biochemist*, **17**, 8–11.

de Rijke, Y. B., Demacker, P. N. M., Assen, N. A., Sloots, L. M., Katan, M. B. and Stalenhoef, A. F. H. (1995). Red wine consumption and the oxidation of low-density lipoproteins. *The Lancet*, **345**, 325–6.

Sakagami, H., Sakagami, T. and Takeda, M. (1995). Antiviral properties of polyphenols. *Polyphenol Actualites*, **12**, 30–2.

Sawyer, D. T. and Valentine, J. S. (1981). How Super is Superoxide? *Acc. Chem. Res.*, **14**, 393–400.

Serafini, M., Ghiselli, A. and Ferro-Luzzi, A. (1994). Red wine, tea and antioxidants. *The Lancet*, **344**, 626.

Scalbert, A. (1991). Antimicrobial properties of tannins. *Phytochemistry*, **30**, 3875–83.

St. Rusznyak, A. and Szent-Gyorgi, A. (1936). Vitamin P: flavonols as vitamins. *Nature*, **138**, 27.

Somers, T. C. and Verette, E. (1988). Phenolic composition of natural wine types. In *Modern Methods of Plant Analysis*, New Series, Vol. 6. Wine Analysis, eds. H. F. Linskens and J. F. Jackson, Springer Verlag: Berlin, pp. 219–57.

Vanden Berghe, D. A., Vlietinck, A. J. and Van Hoof, L. (1985). In *Advances in Medical Plant Research*, eds. A. J. Vlietinck and R. A. Dommisse, Wissenschaftliche Verlagsgesellschafte mbh, Stuttgart, p. 47.

Waterhouse, A. L. (1995). Wine and heart disease. *Chem. & Ind.*, pp. 338–41.

White, T. (1957). Tannins – their occurrence and significance. *J. Sci. Food Agric.*, **8**, 377–85.

8

Quinone tanning and oxidative polymerisation

8.1 Introduction

Most phenolic metabolites, under normal circumstances of growth and development, are found safely sequestered in the vacuole of a plant cell. However if the cell structure is degraded adventitiously, by a predator, as the tissue senesces or as part of a technical process, then the phenols are liberated into the general milieu of the cell cytoplasm and here they have the opportunity to display their chemical virility and promiscuous reactivity with a host of other cytoplasmic constituents, and in particular the ubiquitous proteins. Although the initial steps may well be promoted by reversible association of phenolic metabolites with another substrate (e.g. protein), the influence of external agencies (oxygen, enzymes, metal ions, etc.) invariably then brings about subsequent chemical transformations which may simply involve the phenols themselves, but more often changes which are characterised by irreversible covalent bond formation between the phenol and the co-substrate (e.g. protein), Figure 8.1. In either situation new products are formed, often in a random, unregulated manner. In some instances the transformation of the phenolic constituents which results when the plant is harvested or deliberately damaged is not only desirable but is crucial to the manufacture of certain foodstuffs – notably teas, red wines and cocoa. These 'secondary' reactions then imbue the products with their characteristic and recognisable tastes, flavour and appearance. In other instances these same reactions undoubtedly cause serious problems for the producer. The most familiar situation is that of the browning of fruit and vegetables – general browning reactions caused by accidental damage during harvesting, storage and transport of the fruit and vegetable crop. The visual appearance and frequent loss of palatability which accompanies such changes – invariably grouped under the general heading of 'enzymic browning' – often present serious problems for the producer. It is safe

335

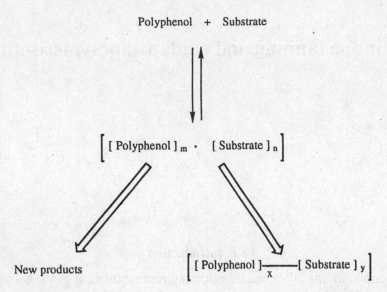

Figure 8.1. Irreversible polyphenol association: (i) the formation of new products, and
(ii) the irreversible covalent binding of substrate to polyphenol(s).

to say that these reactions, despite their great significance and importance,
typify the classic fathomless 'black hole' of polyphenol chemistry; indeed
except in special cases extraordinarily little is known about any of these
reactions. Knowledge is most often in inverse ratio to importance; hard factual
evidence still runs a distant second to informed speculation.

The most commonly encountered situations which fall within the compass
of the general picture depicted in Figure 8.1 are those in which irreversible
chemical transformations, initiated by polyphenol oxidation, take place. Polyphenols, particularly those which contain one or more *o*-di- or tri-hydroxy
phenyl groups, may be oxidised in the presence of metal ions (e.g. iron,
manganese, vanadium) and autocatalytically at pH values around 7.0 and
above. However, the most commonly encountered situations are those in
which polyphenol oxidation is mediated by enzymes, variously described as
polyphenoloxidases; two principal groups of which are recognised – catechol
oxidases and laccases (Mason, 1956; Mayer and Harel, 1979). Tissues rich in
polyphenols are often good sources of such enzymes, but despite their ubiquity
in plants there is little agreement concerning their physiological function. A
biosynthetic role in the synthesis of *o*-di- or tri-hydroxy phenyl groups in
natural phenolics, functions in electron transport, plant growth, the healing of
damaged tissues and disease resistance have all been advocated. *In vitro* the
action of these enzymes upon substrates containing catechol and/or pyrogallol
nuclei is, however, to generate *ortho*-quinones whose extraordinary reactivity

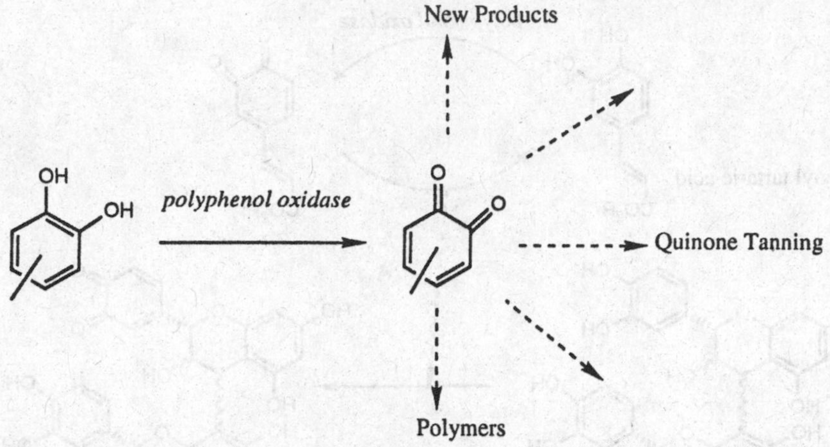

New Products

polyphenol oxidase

Quinone Tanning

Polymers

Figure 8.2. *In vivo* oxidation of catechol (pyrogallol) nuclei by polyphenoloxidase: generation of chemically reactive *ortho*-quinones and their *in vitro* transformations.

then leads to a wide range of products by subsequent chemically directed reactions, Figure 8.2. These chemical reactions may or may not involve other substrates.

The oxidation of several complex polyphenolic substrates has also been shown to occur by enzyme (polyphenol oxidase) catalysed *coupled* oxidations in which a simpler phenolic substrate acts as the initial substrate and is first oxidised to a quinone. The species, once generated, then takes part in a coupled redox system to oxidise the more complex polyphenol and itself is reduced back to the free phenolic state (Haslam *et al.*, 1992). Caffeoyl tartaric acid is thus believed to act in this way in the enzyme coupled oxidation of red wine polyphenolics, Figure 8.3 (Cheynier *et al.*, 1988).

The action of polyphenoloxidases is of crucial importance in the fermentation of the young tea flush (*Camellia sinensis*) and cocoa beans (*Theobroma cacoa*) in the formation of black tea and cocoa respectively. The case of the tea flush is of particular interest as the most obvious case in which polyphenols are transformed (without the intervention of other substrates) to new products – the theaflavins (and thearubigins), Figure 8.1. This distinction probably arises because the young tea flush is such a rich source of phenolic flavan-3-ol substrates (principally (−)-epigallocatechin, (−)-epicatechin and their derivatives, ~ 20% dry weight). As a result the initial action of the tea polyphenoloxidase enzyme is to give a sufficiently high concentration of quinones to allow chemical reactions not only with other cellular constituents (such as proteins) but also between these oxidised phenolic substrates alone (to give in this case the theaflavins, thearubigins, and related compounds). In other

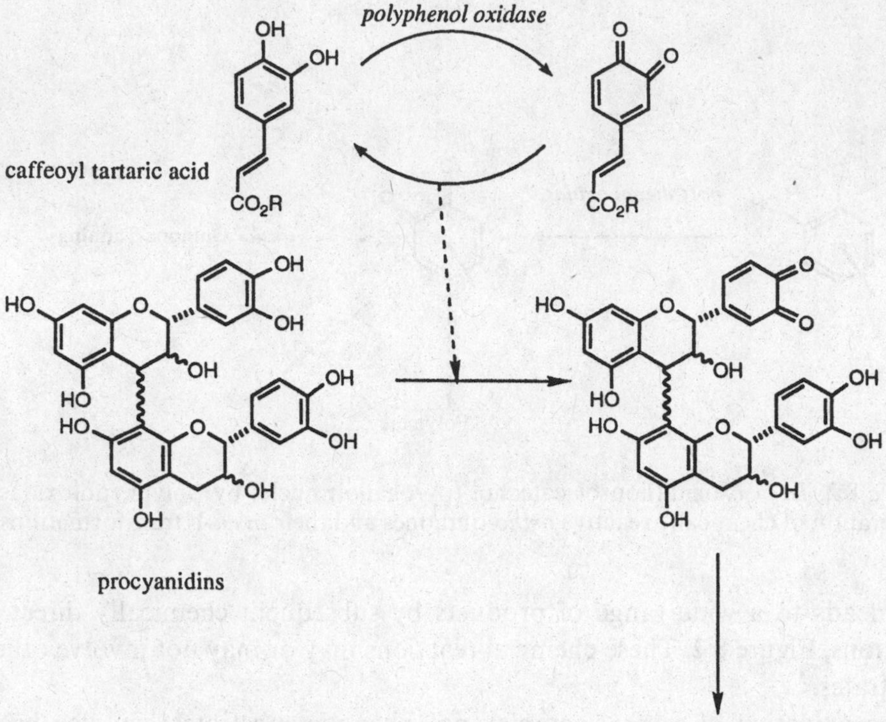

Figure 8.3. Enzyme coupled oxidation of procyanidins in red wines (Cheynier *et al.*, 1988).

circumstances where the initial concentration of phenolic substrates in the plant tissue is much lower, such quinone–quinone, quinone–phenol reactions are, if at all, correspondingly harder to detect.

There is little doubt that the oxidatively mediated reactions of polyphenolic metabolites are of immense significance in the plant kingdom and particularly as a result of the ways in which man treats plants as raw materials in the food and beverages industry and in the production of leather from animal skins. However, the very nature of the reactions involved and their speed means that in most cases all that is presently possible is to point to the general as opposed to the specific chemical processes which are likely to be taking place. Paradoxically despite their importance, but because of their undue complexity, systematic studies in this crucial area still remain scientifically 'unfashionable'. Irreversible polyphenol complexation reactions are widely encountered. Some typical examples – from the plant *and animal* kingdoms – are shown in Table 8.1.

Table 8.1. *Polyphenol complexes: chemical and enzymic transformation*

(i) Enzymic and non-enzymic browning of fruits and fruit juices; preparation of protein leaf concentrates.
(ii) Fermentation processes in the manufacture of black teas (generation of thearubigins and theaflavins) and cocoa.
(iii) The ageing of red wines: changes in pigmentation and moderation of astringency.
(iv) Formation of permanent hazes in beers and lagers.
(v) 'Fixed tannage' – leather manufacture.
(vi) Necrosis and the protection of plant tissues against infective agents, predators and parasites.
(vii) *Sclerotisation and the development of the hard exoskeleton in insects.*
(viii) *The manufacture of adhesives by marine organisms (mussels).*
(ix) Humic acid formation as organic matter is degraded in the formation of humus in soils.

(*Italics* – animal kingdom.)

8.1.1 Quinone intermediates

Whatever pathway is formulated for the oxidative process one of the first formed and most readily identified intermediates in the oxidation of an *ortho*- or *para*-dihydroxyphenol is the corresponding *ortho*- or *para*-quinone. The generic term quinone derives from the fact that the simplest and commonest member of this class of compounds, *para*-benzoquinone, was first recognised in Liebig's laboratory as a product of oxidation of quinic acid. A familiar characteristic of quinones is their colour; *para*-quinones are often yellow, whilst the majority of *ortho*-quinones are orange or red. Another distinctive feature is their chemical reactivity. Quinones are analogous to open chain α,β-unsaturated ketones, but they are considerably more reactive, since nucleophilic addition to the quinone usually regenerates an aromatic ring system, Figure 8.4. The legendary Emil Fischer and Schrader (1910) for example showed that *para*-benzoquinone readily reacts with glycine ethyl ester to give, finally, a bis adduct of the original quinone. The reaction (which bears remarkable analogies to those involved in quinone tanning and provides a very simple and

quinic acid   *para*-benzoquinone

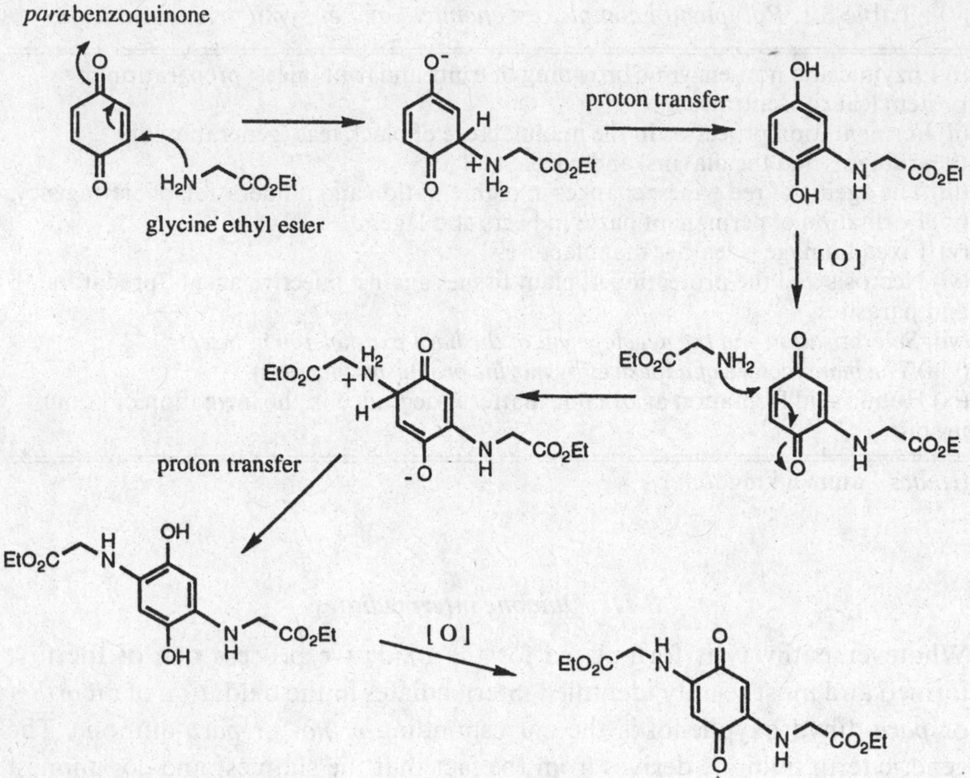

Figure 8.4. Nucleophilic substitution of *para*-benzoquinone with glycine ethyl ester (Fischer and Schrader, 1910; Cranwell and Haworth, 1971).

elegant model for this process) was carried out in air and the reaction can be formulated as a double sequence of nucleophilic addition, enolisation and finally oxidation to regenerate the substituted quinone system, Figure 8.4. Presumably the mildly basic conditions appertaining to the reaction facilitated the aerial oxidations (Cranwell and Haworth, 1971).

The reduction of *para*-benzoquinone occurs in two one-electron steps. The product of the first step is a radical anion that can be detected in dilute solution by electron spin resonance (ESR) spectroscopy. The same radical anion is produced by one-electron oxidation of the hydroquinone dianion, and one of the most important aspects of this red-ox system is that it is electrochemically reversible, Figure 8.5. Standard reduction potentials ($E^\ominus$) may be measured for a range of quinones. The more positive the value of $E^\ominus$, the more ready the reduction of the quinone. Thus the more electron-rich a species is, the easier is its oxidation and conversely the more difficult is the reduction of the quinone. Thus electron attracting substituents (e.g. Cl, Br) cause the quinone to be

Table 8.2. *Reduction potentials of quinones (25°C, volts)*

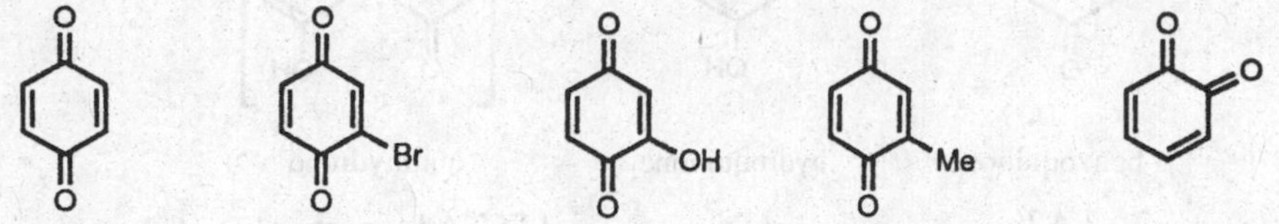

| 0.699 V | 0.715 V | 0.59 V | 0.645 V | 0.78 V |

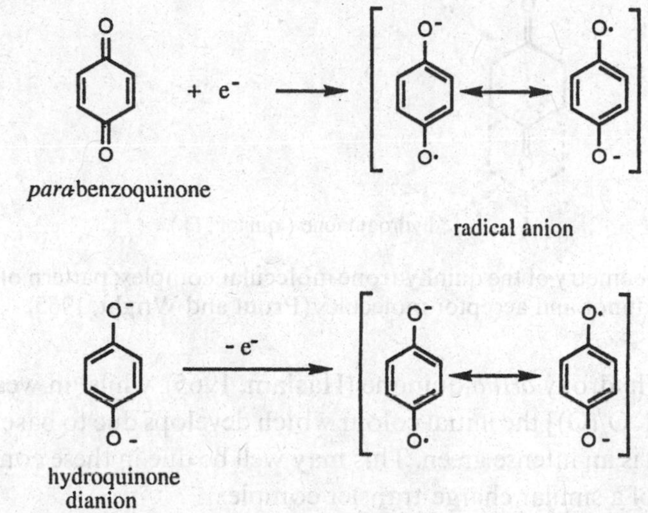

para-benzoquinone

radical anion

hydroquinone
dianion

Figure 8.5. Red-ox reactions of *para*-benzoquinone and the hydroquinone dianion.

reduced more easily. Electron donating substituents (OH, Me) similarly cause the quinone to be reduced less easily, Table 8.2.

An equimolar mixture of the hydroquinone and *para*-benzoquinone forms a dark green crystalline molecular complex 'quinhydrone' (melting point of 171 °C) which is normally referred to as a charge-transfer complex (see chapter 2), Figure 8.6. Such complexes are often highly coloured and are characterised by the presence of one component that is electron rich (donor, D) and another that is electron deficient (acceptor, A); they are best formulated as a hybrid of two resonance structures, Figure 8.6. The second resonance structure – the charge-transfer structure – makes only a very small contribution to the overall hybrid structure.

In this context it is interesting to note that oxidation of esters of gallic acid in weakly acidic media (KIO_3 /H_2O) gives the transient rose-cerise colour of the

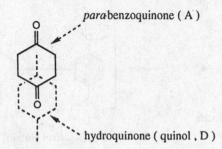

p-benzoquinone   hydroquinone   quinhydrone

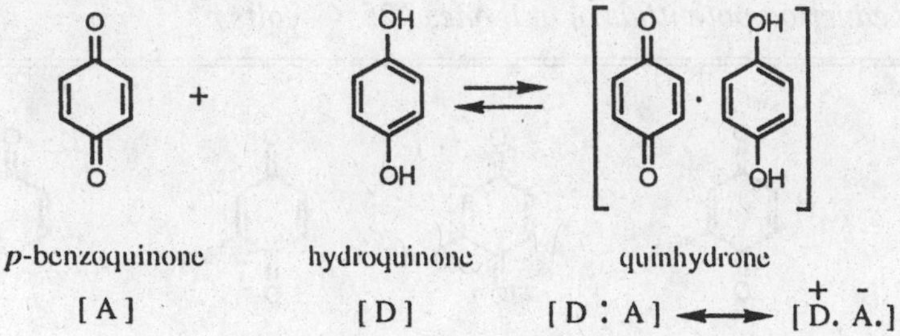

[A]   [D]   [D : A] $\longleftrightarrow$ [D. A.]

Figure 8.6. Quinhydrone: charge transfer complex.

para-benzoquinone (A)

hydroquinone (quinol , D)

Figure 8.7. The geometry of the quinhydrone molecular complex; pattern of overlap of the donor and acceptor molecules (Prout and Wright, 1965).

corresponding hydroxy *ortho*-quinone (Haslam, 1965), whilst in weakly basic media [$NH_3/H_2O/(O)$] the initial colour which develops due to base catalysed auto-oxidation is an intense green. This may well be due in these conditions to the formation of a similar charge-transfer complex.

The structure of the quinhydrone complex consists of approximate plane to plane stacks of alternate donor and acceptor molecules, with an interplanar distance of 3.16 Å. In the crystal these are held together by hydrogen bonds (2.71 Å) to form chains in which the kinds of molecules similarly alternate. The relative orientation of the donor (hydroquinone) and acceptor (*para*-benzoquinone) molecules in the stacks is as shown in Figure 8.7 with the highly polar carbonyl group lying roughly over the centre of the aromatic ring of the donor (hydroquinone) molecule.

In so far as the *in vitro* self-polymerisation of *ortho*-benzoquinones and the formation of new products from these same quinones is concerned, Figure 8.2, then the generation of 'quinhydrone type' π–π complexes as described may well be a key initial step in these processes. Such considerations are, for example, helpful in understanding the mode of generation of the theaflavins in tea fermentation. The generally accepted mechanism for this reaction is as depicted in Figure 8.8; whilst this 'quinone dimerisation' mechanism very satisfactorily rationalises the formation of the benzotropolone structure, it

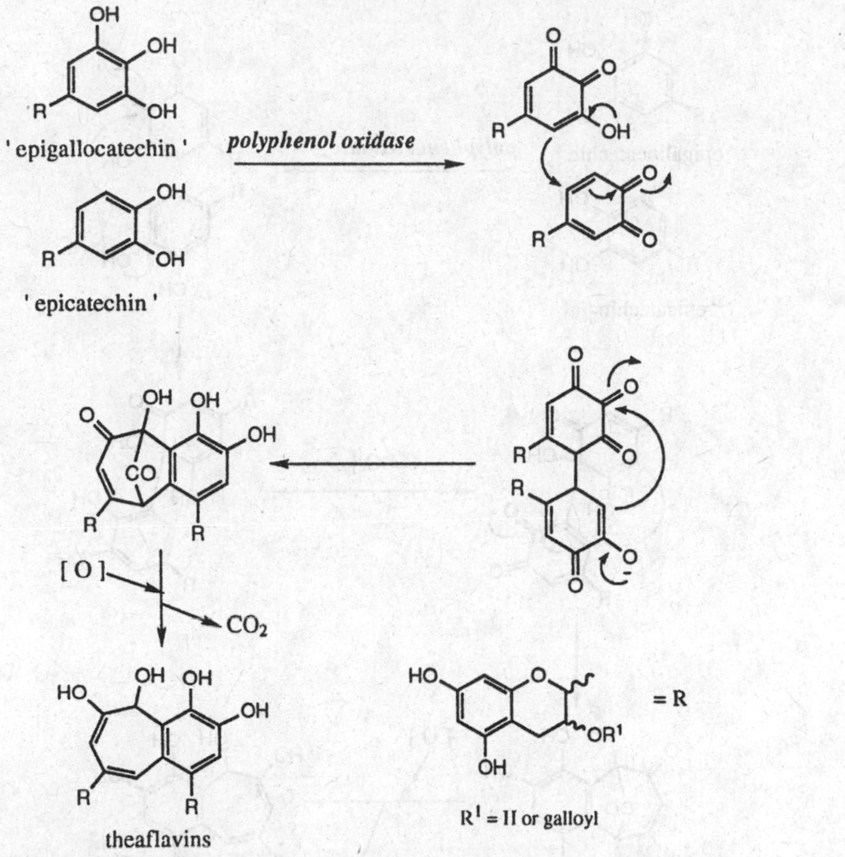

Figure 8.8. Theaflavin(s) formation: 'quinone dimerisation' mechanism.

provides no insight into the reasons for the unusual regioselectivity of the initial 'dimerisation' to give the suggested tricyclic intermediate.

However, the unusual mode of dimerisation that is invoked to rationalise the formation of the benzotropolone ring system of the theaflavins suggests some form of pre-organisation of the two aromatic nuclei which then directs the cyclisations in the particular manner depicted. The alternative mechanism, shown in Figures 8.9 and 8.10, presumes the initial formation of a 'quinhydrone type' of complex between the hydroxy *ortho*-quinone (the oxidised ring B of the epigallocatechin derivative) and the catechol ring B of the epicatechin derivative, Figure 8.10. Intermolecular hydrogen bonding would presumably stabilise the intermolecular π–π complex in the configuration shown and would thereby bring the two centres required for the two intermolecular reactions into precisely the correct orientation and closely juxtaposed. In this mechanism it is the initial formation of such an intermolecular complex that determines the subsequent reaction pathway.

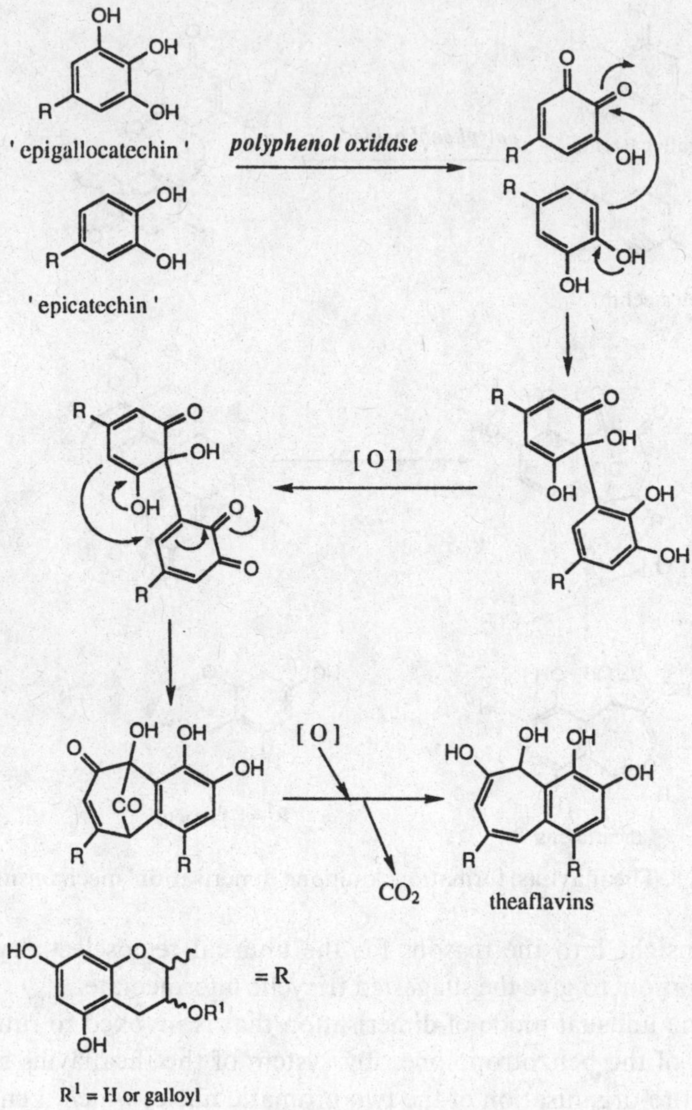

Figure 8.9. Alternative reaction mechanism for the formation of theaflavins in tea fermentation; initial generation of an intermolecular π–π complex between the oxidised 'epigallocatechin' and the un-oxidised 'epicatechin' derivatives.

8.1.2 Black tea pigments – thearubigins – thoughts and speculations

The colouring matter of black tea infusions, it is agreed, consists of two major groups of compounds – theaflavins and thearubigins – which are formed in the fermentation process by endogenous enzymic oxidation of the green tea leaf polyphenols – very largely phenolic flavan-3-ols, based on (−)-epigal-

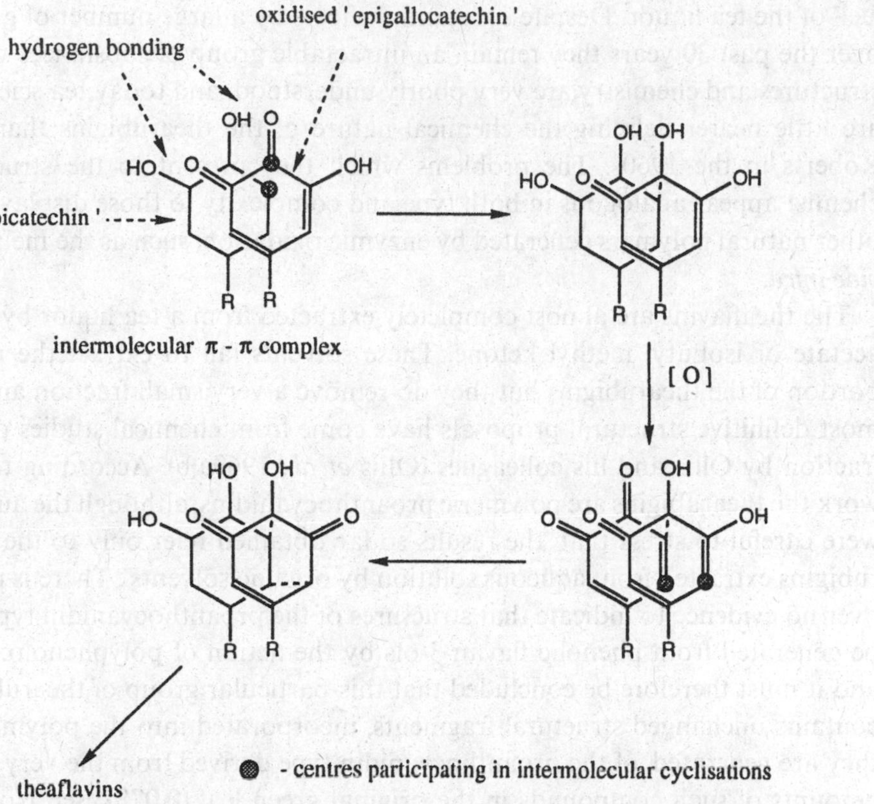

Figure 8.10. Alternative reaction mechanism for the formation of theaflavins in tea fermentation; pictorial representation of the way in which the geometry of the initially formed intermolecular π–π complex between the un-oxidised 'epicatechin' and the oxidised 'epigallocatechin' derivatives directs the chemical reaction (R as in Figure 8.9).

locatechin and (−)-epicatechin, and their gallate esters (Roberts, 1962; Sanderson, 1972). The chemical nature of the theaflavins as phenolic derivatives with a benzotropolone structure, and their probable mode of formation is now firmly established (see above and Figure 4.8). They constitute up to 2% of the dry weight of black tea and up to 6% of the solids extracted in the brewing of a cup of tea. They give a bright reddish colour in solution and an important role has been attributed to their presence in teas to give the qualities of 'brightness' and 'strength'. However, the major part of the colour in a tea liquor is due to a very ill-defined heterogeneous mixture of compounds called thearubigins. Accurate analysis of their concentrations in teas is difficult but they are believed to constitute up to 20% of the black tea leaf and up to 60% of the solids in a black tea infusion. Besides being responsible for a great deal of the colour they also make significant contributions to the 'strength' and 'mouthfeel' of the tea liquor. Despite substantial efforts by a large number of groups over the past 30 years they remain an intractable group of substances whose structures and chemistry are very poorly understood, and today tea scientists are little nearer defining the chemical nature of the thearubigins than was Roberts in the 1960s. The problems which they present to the structural chemist appear analogous in both type and complexity to those displayed by other natural polymers generated by enzymic oxidation, such as the melanins, *vide infra*.

The theaflavins are almost completely extracted from a tea liquor by ethyl acetate or isobutyl methyl ketone. These solvents fail to extract the major portion of the thearubigins but they do remove a very small fraction and the most definitive structural proposals have come from chemical studies of this fraction by Ollis and his colleagues (Ollis *et al.*, 1969a,b). According to this work the thearubigins are polymeric proanthocyanidins, although the authors were careful to stress that 'the results so far obtained refer only to the thearubigins extracted from aqueous solution by organic solvents'. There is moreover no evidence to indicate that structures of the proanthocyanidin type can be generated from phenolic flavan-3-ols by the action of polyphenoloxidase and it must therefore be concluded that this particular group of thearubigins contains unchanged structural fragments, incorporated into the polymers as they are generated, of the proanthocyanidin type derived from the very small amounts of such compounds in the original green leaf (0.07%; see Nonaka, Kawahara and Nishioka, 1983; Figure 4.7).

It is generally assumed that the thearubigins are complex heterogeneous polymers formed by the oxidation (predominantly) of phenolic flavan-3-ols in the green tea leaf, all of which contain a 'catechol' or 'pyrogallol' 'B' ring and may in addition contain a gallate ester group. The first step in this enzymic oxidation is the conversion of some or all of these catechol and pyrogallol based nuclei to the respective *ortho*-quinones. Thereafter these intermediates partition to give either the theaflavins or the thearubigins, Figure 8.11. The routes to the various theaflavins have been established and follow clearly defined chemical pathways. The routes to the thearubigins are ambiguous but it is normally assumed that these involve oxidatively induced oligomerisation and very probably further oxidation steps. The colour of the thearubigins may thus derive from the presence of unreacted *ortho*-quinone groups in the final polymeric structures. Although model tea fermentation systems indicate that thearubigin type substances are formed by oxidation of any one of the green tea leaf phenolic flavan-3-ols, or a combination thereof, it seems highly probable that the principal substrates are (−)-epigallocatechin-3-*O*-gallate and, to a lesser extent, its parent flavan-3-ol. Such a conclusion seems inescapable for,

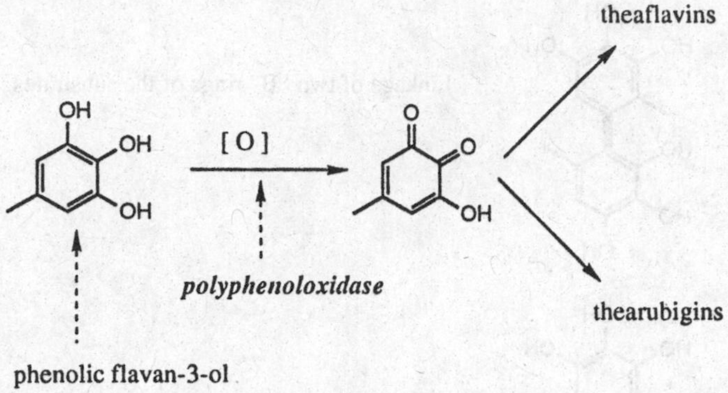

theaflavins

[O]

polyphenoloxidase

thearubigins

phenolic flavan-3-ol

Figure 8.11. Principal pathways for the oxidative conversion of phenolic flavan-3-ols to theaflavins and thearubigins in tea fermentations: partitioning of the *ortho*-quinone intermediate.

as Roberts first pointed out, there are no other oxidisable substrates present in sufficient quantity to yield the amounts of thearubigins formed. Minor pathways doubtless are in operation utilising not only the other green tea leaf phenolic flavan-3-ols but also the theaflavins (as they are formed) as substrates in direct or coupled oxidations.

Given this simplification, one possible way forward would be to treat this problem in an analogous way to that taken much earlier by Freudenberg and others with the structure of the phenolic polymer lignin. Thus one might consider what other types of reaction the postulated key intermediate *ortho*-quinones, Figure 8.11, derived from (−)-epigallocatechin and its 3-*O*-gallate ester, might be expected to undergo on the basis of reasoned chemical expectations. Assuming that at least one of these processes is that of oligomerisation, then at least three types of oligomerisation reaction may be plausibly envisaged, namely C–C, C–O bond formation and cycloaddition. There is good chemical precedent in the area of the chemistry of phenolic flavan-3-ols to assume that oligomerisation *via* C–C bond formation might be expected to occur by linkage of two 'B' rings of the substrates (cf. the theasinensins, Figure 4.7); by linkage of ring 'A' to ring 'B' of the substrates (cf. the oxidation of (+)-catechin, Figure 4.10); and by linkage of galloyl ester groups to form hexahydroxydiphenoyl and nonahydroxytriphenoyl groups (see chapter 1). Thus on this basis structural fragments produced by oligomerisation might be predicted to have one or more of the forms shown in Figure 8.12. However, unless free rotation about the diphenyl linkages in the hexahydroxydiphenoyl and flavogallonyl ester groups were restricted then these ester groups would not be expected to survive in the fermentation. Facile intramolecular cyclisation (Cai and Haslam, 1994) would lead to the formation of ellagic acid and its

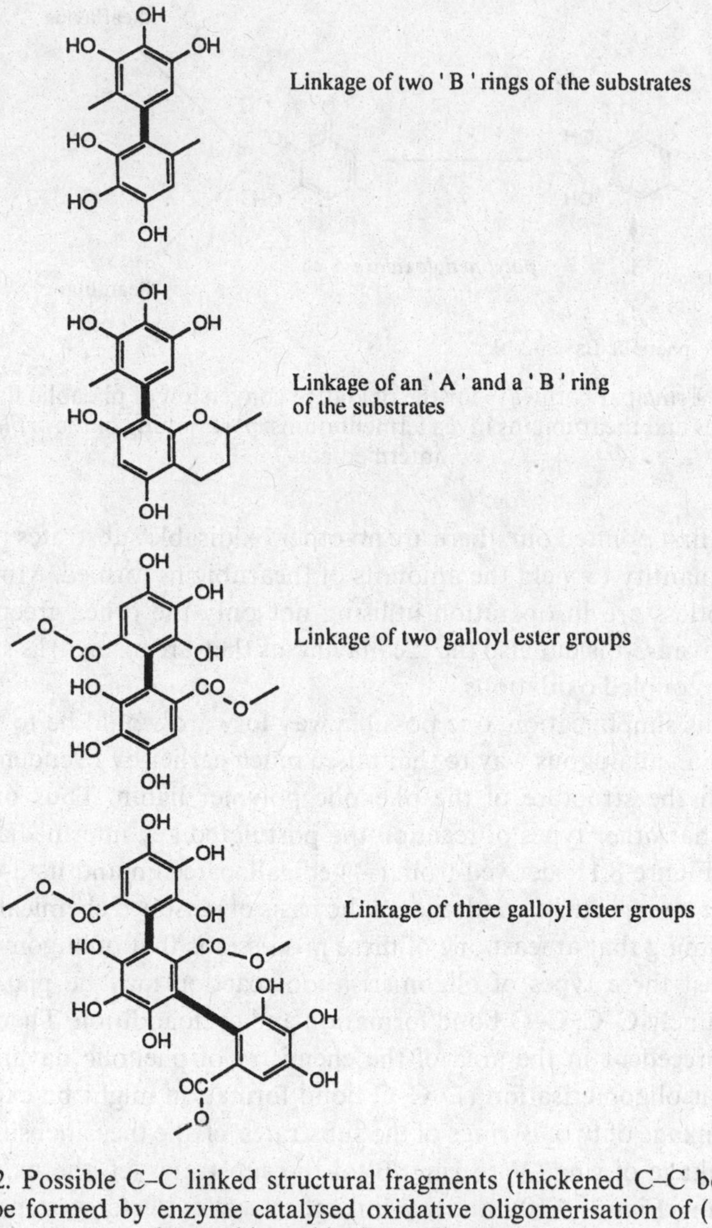

Figure 8.12. Possible C–C linked structural fragments (thickened C–C bonds) predicted to be formed by enzyme catalysed oxidative oligomerisation of (−)-epigallocatechin and its 3-*O*-gallate ester.

triphenoyl analogue respectively and thus loss of the oligomer linkages, Figure 8.13. An important corollary of this would be that oxidative oligomerisation – assuming it to be an essentially random process with little discrimination between the different types of 'pyrogallol' nuclei – would lead to the loss of gallic acid equivalents as the polymer is elaborated.

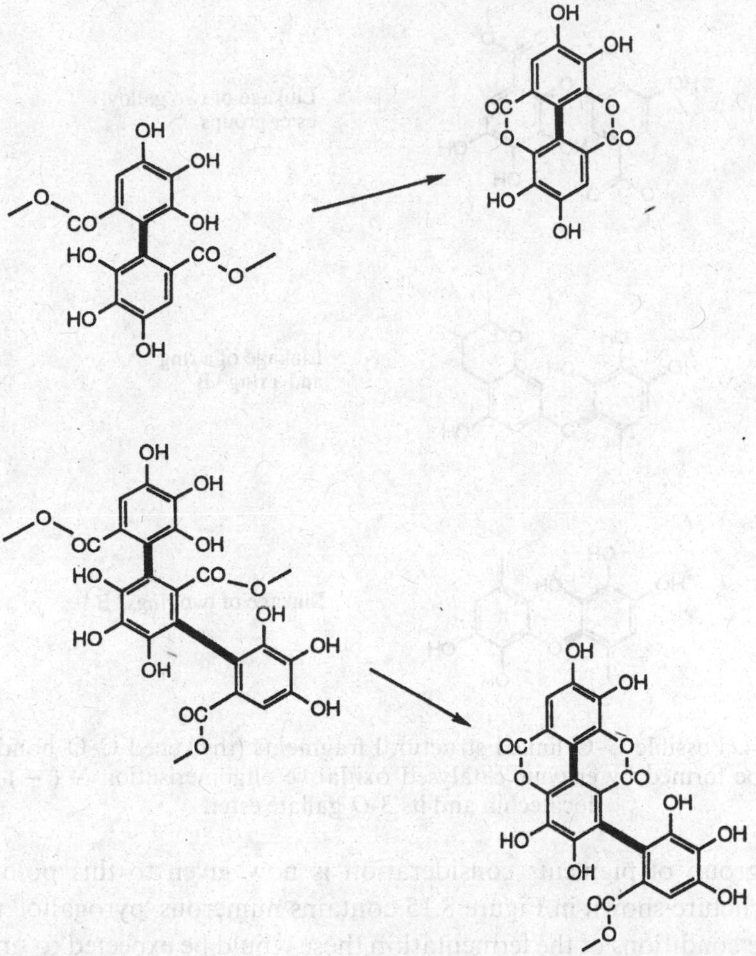

Figure 8.13. Loss of intermolecular C–C oligomer linkages by spontaneous intra-molecular cyclisation of hexahydroxydiphenoyl and flavogallonyl ester groups.

Oxidation might also produce, in principle, C–O oligomer linkages by analogous processes. Although no examples are known in the case of phenolic flavan-3-ols there is ample precedent in the field of the chemistry of the galloyl ester derivatives (see chapter 1) for the formation of C–O linkages. Several of these possibilities are shown in Figure 8.14.

Using these ideas a typical oligomeric structure formed by the enzymic oxidation of (−)-epigallocatechin and its 3-*O*-gallate ester might therefore be predicted to contain structural fragments of the type shown in Figure 8.15.

As Roberts first noted, the thearubigins display a marked acidity and in this theoretical discussion of the possible general structural features of this hetero-

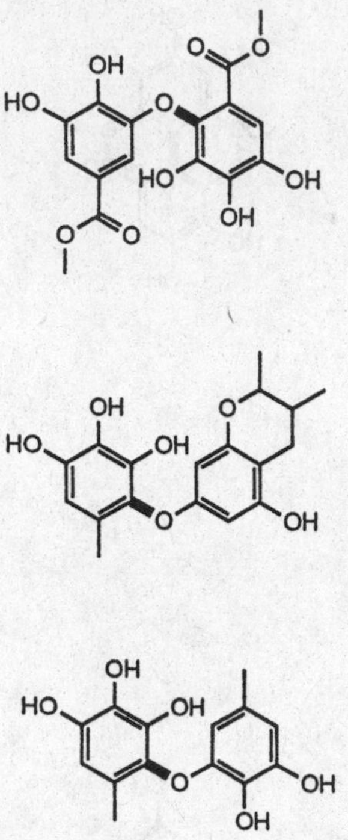

Linkage of two galloyl ester groups

Linkage of a ring ' A ' and a ring ' B '

Linkage of two rings ' B '

Figure 8.14. Possible C–O linked structural fragments (thickened C–O bonds) predicted to be formed by enzyme catalysed oxidative oligomerisation of (−)-epigallocatechin and its 3-*O*-gallate ester.

geneous group of pigments consideration is now given to this point. The partial structure shown in Figure 8.15 contains numerous 'pyrogallol' nuclei. Under the conditions of the fermentation these would be expected to undergo further direct or coupled enzymic oxidation to give the corresponding *ortho*-quinones. Whilst certain of these would remain to give colour to the thearubigins, it is postulated that a fraction might undergo hydrolytic ring opening to give carboxylic acid groups. Possible pathways are outlined in Figure 8.16.

These proposals are finally incorporated into a proposal for the structural patterns which may be encountered in the structure of the complex phenolic pigments – the thearubigins of black tea – and are shown in Figure 8.17.

These structural proposals are not definitive, rather they should be taken as a starting point for further discussion and experimentation. Nevertheless they suggest explanations for a number of the characteristics of the thearubigins. The polymers are heterogeneous and this would result from the largely random modes of oxidative polymerisation. The phenolic properties would derive generally from the residual rings 'A' of the (−)-epigallocatechin-3-*O*-gallate precursor; acidic functionality would result from the oxidatively promoted

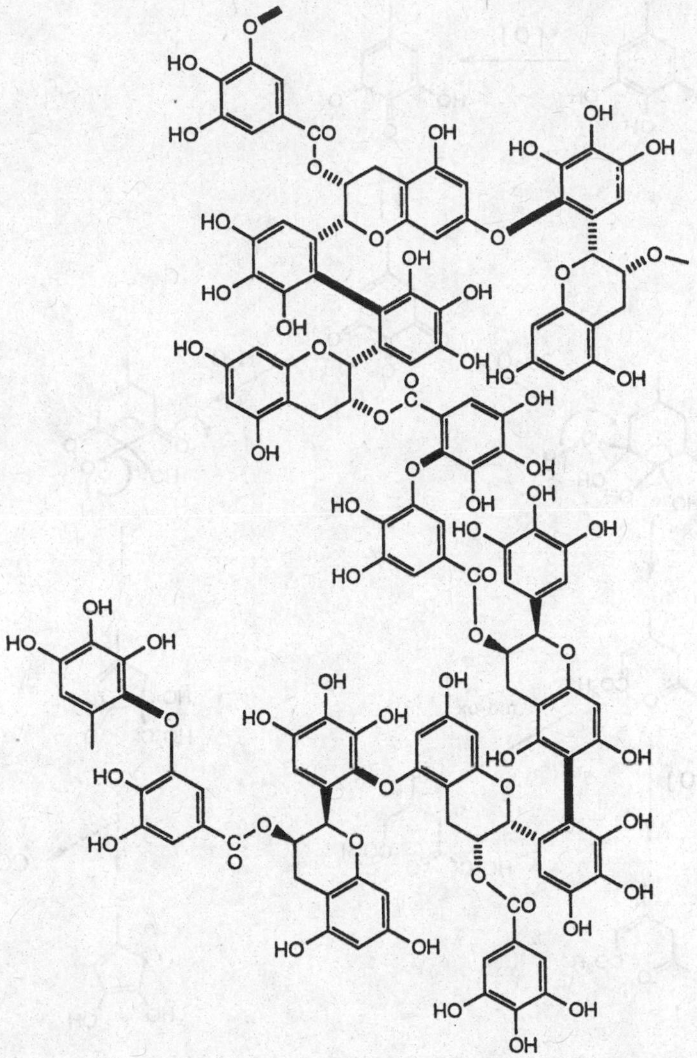

Figure 8.15. Hypothetical oligomeric partial structure derived by enzymic oxidation of (−)-epigallocatechin and its 3-*O*-gallate ester: new C–C and C–O interflavan bonds indicated by thickened bonds.

ring fission of the various 'pyrogallol' nuclei; colour likewise is provided by the presence of residual quinone structures within the polymers. Finally it is worthy of note that Roberts's analyses of tea creams consistently showed that caffeine preferentially precipitated the theaflavins relative to the thearubigins (62% of the theaflavins as against 28% of the thearubigins). This observation is entirely in accord with the structural proposal shown in Figure 8.17. Thus it is well established that the presence of galloyl ester groups in a phenolic flavan-3-

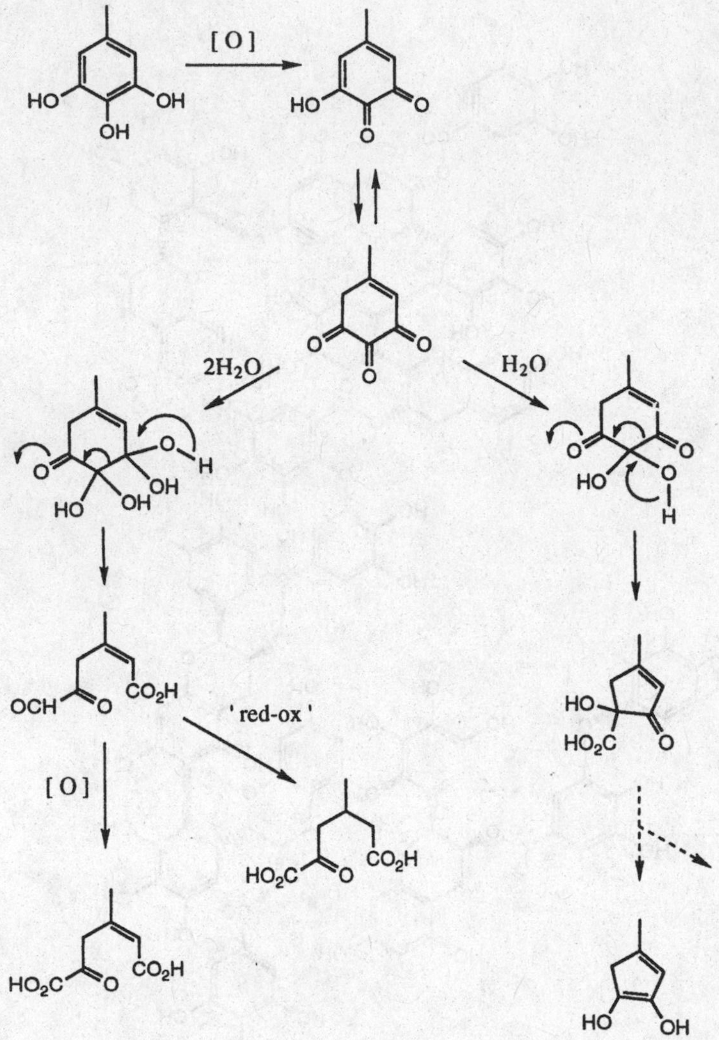

Figure 8.16. Oxidation of 'pyrogallol' nuclei and hydrolytic ring opening to give carboxylic acid functions.

ol substrate enhances the ability to complex with caffeine; these groups are the principal sites for non-covalent complexation with caffeine. According to the arguments above, these groups would be substantially depleted in the thearubigins compared to their presumed precursor(s) and the theaflavins.

Support for some of these proposals has been provided most recently by Bailey, Nursten and McDowell (1992) who isolated a fraction of the polymeric thearubigins by chromatography on Solka-Floc. They named this group of thearubigins the *theafulvin* fraction, and showed that it was free of nitrogen. X-ray analysis in an electron microscope indicated the presence of potassium,

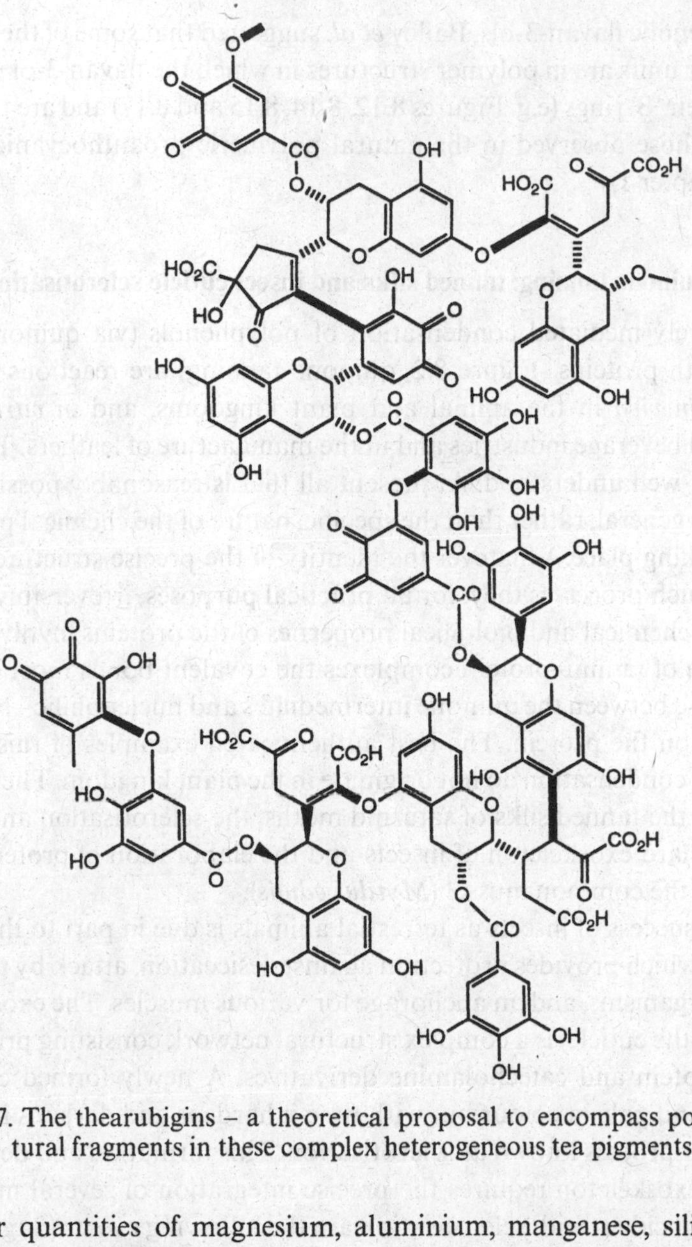

Figure 8.17. The thearubigins – a theoretical proposal to encompass possible structural fragments in these complex heterogeneous tea pigments.

and lesser quantities of magnesium, aluminium, manganese, silicon, phosphorus and sulphur. Infra-red spectroscopy showed that it was a phenolic polymer. The ^{13}C NMR spectrum of the theafulvin fraction showed, not unexpectedly, a poor signal-to-noise ratio and line broadening which the authors attributed to the presence of the paramagnetic manganese and to the heterogeneous nature of the polymer fraction. However, the spectrum showed a good general similarity to that obtained from a mixture of green tea phenolic flavan-3-ols. On the basis of comparisons with this ^{13}C NMR spectrum of

green tea phenolic flavan-3-ols, Bailey *et al.* suggested that some of the theafulvin monomer units are in polymer structures in which the flavan-3-ol units are
linked via their 'B' rings (e.g. Figures 8.12, 8.14, 8.15 and 8.17) and are therefore
different to those observed in the natural polymeric proanthocyanidins discussed in chapter 1.

8.2 Quinone tanning: tanned silks and insect cuticle sclerotisation

The oxidatively mediated condensation of polyphenols (via quinone intermediates) with proteins (Figure 8.2, quinone tanning) are reactions of great
importance *in vivo* in the animal and plant kingdoms, and *in vitro* in the
foodstuff and beverage industries and in the manufacture of leathers. They are,
however, not well understood. At present all that is reasonably possible is to
point out the general, rather than the specific, nature of the chemical processes
which are taking place. Whatever the identity of the precise structures which
result from such processes they, for all practical purposes, irreversibly change
the physical, chemical and biological properties of the proteins involved.

In the case of tannin–protein complexes the covalent bonds most likely to
form are those between the quinone intermediates and nucleophilic $-NH_2$ and
$-SH$ groups on the protein. The best authenticated examples of this type of
cross-linking condensation do not originate in the plant kingdom. They are the
formation of the tanned silks of saturnid moths, the sclerotisation and generation of the hard exoskeleton of insects and the elaboration of protein-based
adhesives by the common mussel (*Mytilus edulis*).

The great success of insects as terrestial animals is due in part to their hard
exoskeleton which provides protection against desiccation, attack by parasites
and micro-organisms, and an anchorage for various muscles. The exoskeleton
of the insect (the cuticle) is a complex structural network consisting principally
of chitin, protein and catecholamine derivatives. A newly formed cuticle is
usually soft and pale in colour but with time it hardens and darkens in colour
by means of a process known as sclerotisation. This formation and deposition
of the hard exoskeleton requires the precise integration of several metabolic
activities. Coincident with sclerotisation are the following events (Sugumaran,
1987):

(i) the mobilisation of catecholamine derivatives to the cuticle,
(ii) the decrease in solubility of structural proteins and an increase in their resistance
to attack by enzymatic and chemical degradation,
(iii) extrusion of water from the cuticle,
(iv) formation of chitin–protein and protein–protein cross-links,
(v) development of pigmentation.

As early as 1940, Pryor (1940, 1962) characterised quinones as one type of sclerotising-agent in studies of the hardening and darkening of cockroach egg capsules. His studies suggested that 3,4-dihydroxybenzoic acid is first oxidised to its *ortho*-quinone derivative by cuticular phenoloxidase and is then incorporated into the structural proteins of the egg capsule.

3,4-dihydroxybenzoic acid

Structural proteins ⟶

Cross-linked proteins

Later workers have developed these idea and two general schemes – quinone tanning and β-sclerotisation – have been put forward to rationalise the events which take place during the darkening and hardening of the insect cuticle. According to the quinone tanning hypothesis catechol derivatives (such as *N*-acetyldopamine or *N*-β-alanyldopamine) are secreted into the cuticle, oxidised to the corresponding *ortho*-quinones, which then react with available nucleophilic groups on structural proteins (e.g. the ε-amino groups of lysine and the thiol groups of cysteine). In concert with further oxidation (cf. scheme above) this fundamental reaction brings about cross-linking (hardening) and pigmentation of the exoskeleton of the insect, Figure 8.18. The reaction sequence is, in many respects analogous to the model reaction shown in Figure 8.4, except that the nucleophilic species is the surface of a multifunctional protein.

Sugumaran (1987) has elaborated an alternative method of protein cross-linking – namely the β-sclerotisation process (Andersen, 1971, 1974) in which the key intermediates are quinone methides derived from various catecholamines, principally *N*-acetyldopamine and/or *N*-β-alanyldopamine. Whereas in the quinone tanning hypothesis it is the aromatic ring of the quinone precursor which provides the points of attachment for the proteins, Figure 8.18, in the β-sclerotisation process it is suggested that it is the aliphatic side-chain of the catechol amine which acts to generate points of anchorage for the protein, Figure 8.19. This is achieved by the formation from the catecholamine of quinone methide intermediates in the oxidation; cross-linking of different protein skeletal fragments requires bis-substitution of the quinone methide at the β-position of the side-chain, Figure 8.20. From a stereochemical

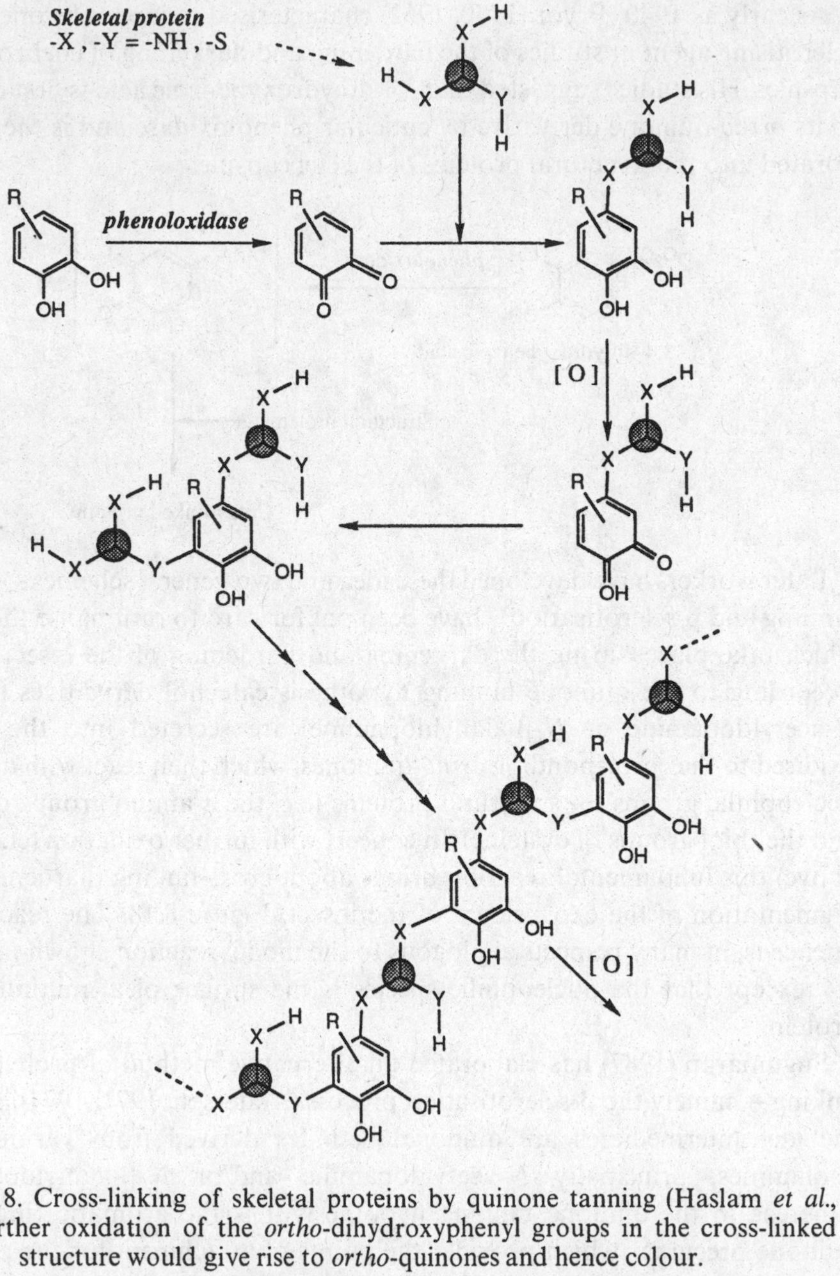

Figure 8.18. Cross-linking of skeletal proteins by quinone tanning (Haslam *et al.*, 1992). Further oxidation of the *ortho*-dihydroxyphenyl groups in the cross-linked structure would give rise to *ortho*-quinones and hence colour.

viewpoint this looks a more difficult process to develop for protein cross-linking than that envisaged in quinone tanning, Figure 8.18.

Together the quinone tanning and β-sclerotisation reactions account very satisfactorily for many of the physical and chemical features contingent upon the sclerotisation of the insect cuticle. Whilst they are amongst the most familiar and best authenticated examples of the irreversible chemical reactions

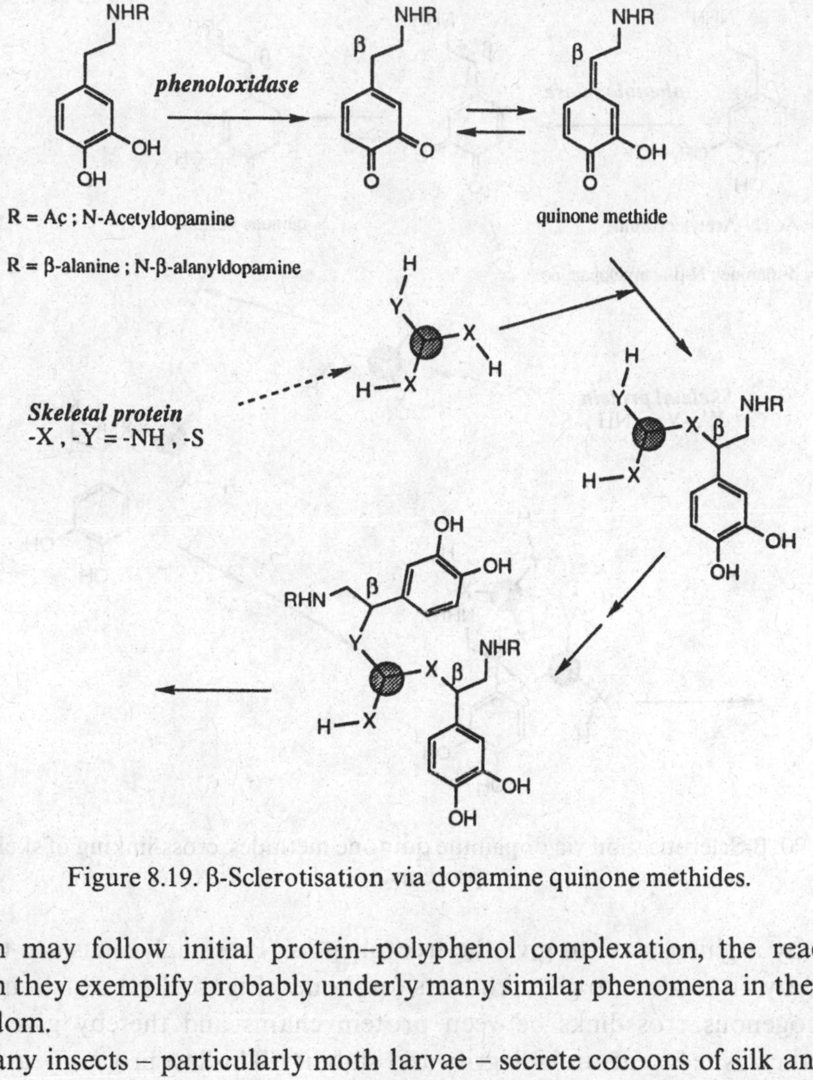

Figure 8.19. β-Sclerotisation via dopamine quinone methides.

which may follow initial protein–polyphenol complexation, the reactions which they exemplify probably underly many similar phenomena in the plant kingdom.

Many insects – particularly moth larvae – secrete cocoons of silk and it is tacitly assumed that, in some way, these are 'protective'. The overall process draws heavily on the resources of the insect and some 50% of the proteins of the larva may go into the construction of the cocoon. Silk, though fluid when secreted, dries out very quickly. It is pure white and if artificially kept dry then it remains white. If it becomes moist then it darkens. A typical silkworm's cocoon is constructed from a continuous thread of the protein fibroin cemented by a gum provided by the protein sericin which endows the otherwise flexible fibroin with a degree of mechanical stability. The silks of the large cocoons of saturnid moths range in colour from pale to dark brown and they consist not only of the proteins fibroin and sericin, but also enzymes and phenols. The presence of moisture allows these to interact. The phenols of the silk gland are in the form of glucosides which, in the presence of moisture, are

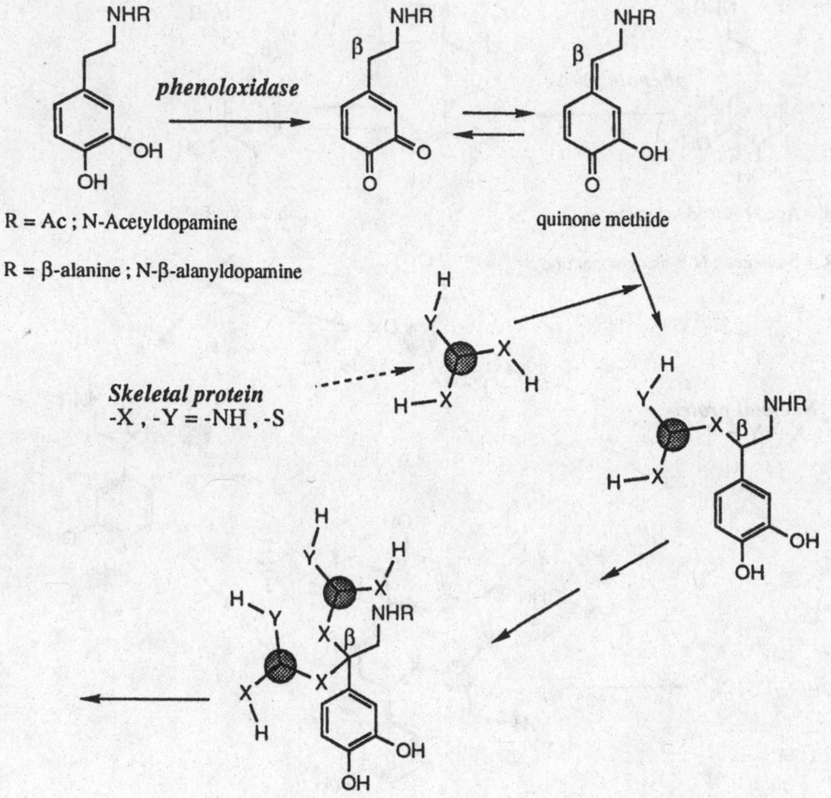

Figure 8.20. β-Sclerotisation via dopamine quinone methides: cross-linking of skeletal proteins.

cleaved by a glucosidase to give the free phenols. These aglycones are then oxidised by an oxidase to give products (quinones, Figure 8.21) which introduce exogenous cross-links between protein chains and thereby give the cocoon the desired structural integrity and stability. This last process bears all the hallmarks of quinone tanning described above, Figure 8.18. Brunet and Coles identified (Brunet and Coles, 1974) two of the phenolic glucosides from *Bombyx* moths as the *O*-glucoside of 3-hydroxyanthranilic acid (most probably a degradation product of L-tryptophan) and the 5-*O*-glucoside of gentisic acid (probably arising as a conjugate from the food-plant), Figure 8.21.

8.3 Melanins

Melanin, it has been facetiously remarked, is a pigment of the imagination: this statement simply reflects the difficulty that has been experienced, over many decades, in the definition of its chemical structure and the precise routes to its

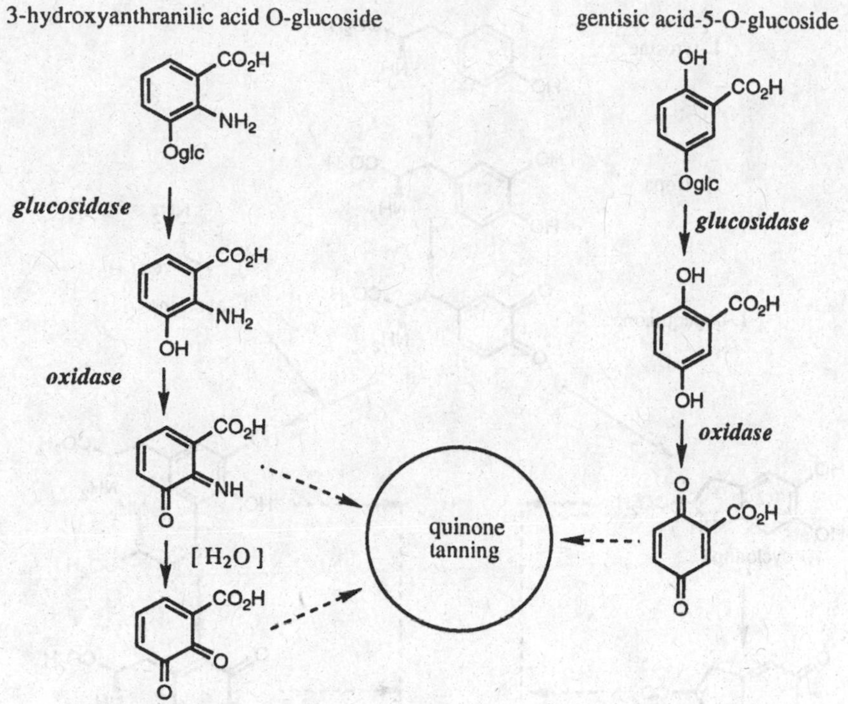

Figure 8.21. Post-secretory generation of quinone intermediates from *O*-glucosides in the tanning of silk by the silkworm (Brunet and Coles, 1974).

formation. Although melanogenesis may seem some distance scientifically from the processes of quinone tanning and oxidative polymerisation of plant polyphenols, all these processes involve quinone substrates and generate, by their polymerisation and reaction with other substrates, dark brown pigmentation of ill-defined structure. For these reasons brief mention is made of the melanin pigments at this juncture.

The melanins are of high molecular mass and are insoluble in virtually all solvents. They have significant ion-exchange capability, can undergo reversible red-ox processes, show persistent ESR signals of free radical origin, and are capable of scavenging free radical species. Human skin colour is determined largely by the size, type and distribution of cytoplasmic organelles, melanosomes, in which melanin pigments are synthesised and deposited. The whole process of melanosome formation and pigment synthesis is stimulated by UV light and current interest in this area has been re-awakened because of the significant increase in the last 30 years of the incidence of malignant melanoma, possibly through the increased exposure to UV radiation (Benasson *et al.*, 1993; Prota, 1988). However, present views on the structure and

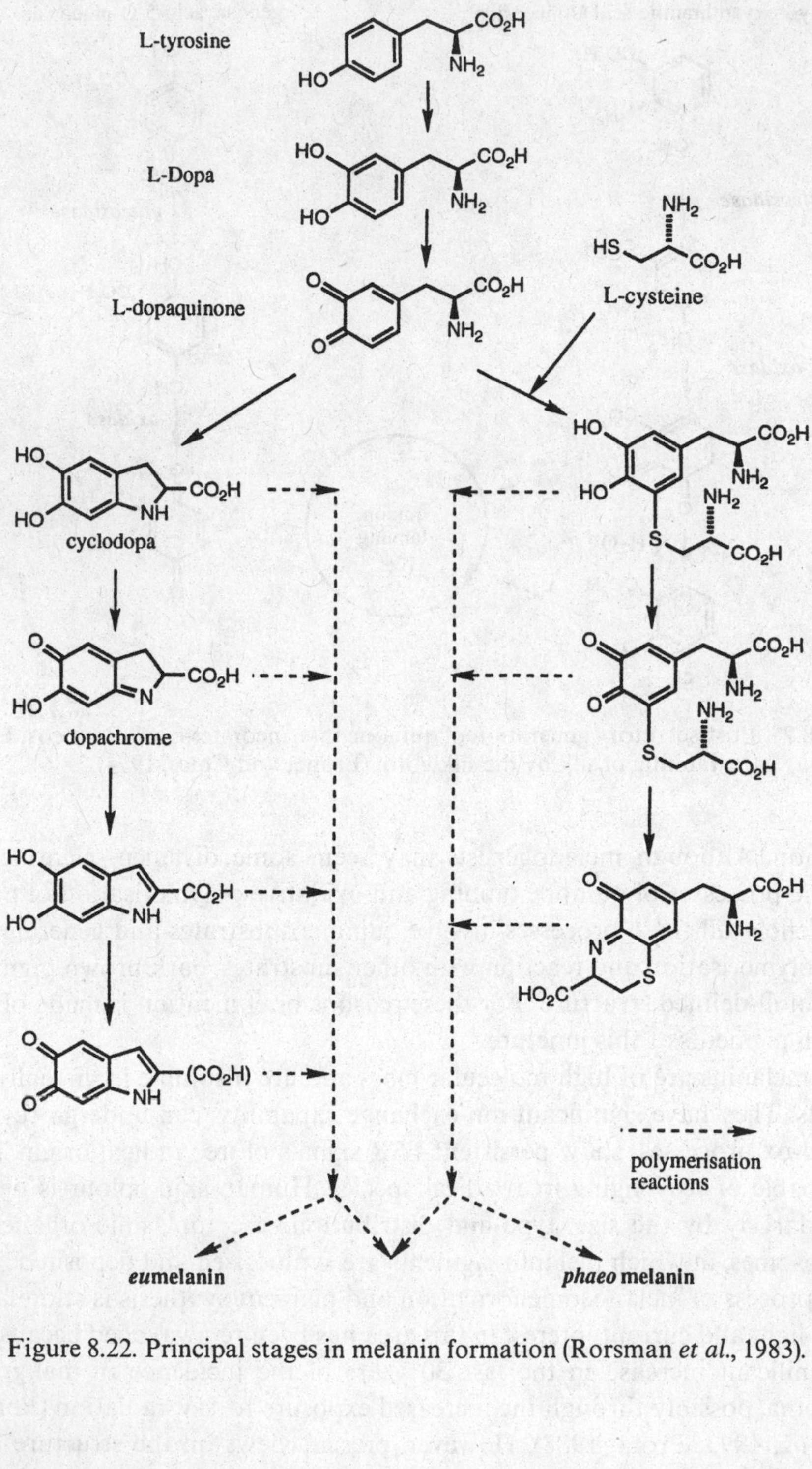

Figure 8.22. Principal stages in melanin formation (Rorsman *et al.*, 1983).

generation of melanin rest principally on work conducted half a century ago. Human epidermal pigment was originally believed to be composed of an intimate mixture of two chemically distinct polymers – black or brown *eu*melanin and red–brown *phaeo*melanin. Photochemically the pigments differ; *phaeo*melanin is much more photosensitive than *eu*melanin. More recent evidence suggests that the two melanins show greater similarities than originally thought; they both thus contain sulphur derived from cysteine.

The classic Raper–Mason scheme (Raper, 1928; Mason, 1948) for melanogenesis is shown in Figure 8.22. In this proposal the conversion of L-tyrosine to melanin is suggested to occur by a series of oxidation steps. The two initial steps, the tyrosinase-catalysed oxidation of L-tyrosine to L-DOPA and the latter's conversion to the highly reactive L-dopaquinone are believed to be under enzymic control. The subsequent reactions may then proceed spontaneously under purely chemical control and several of the suggested intermediates have not been characterised. Both types of melanin are polymers in which all intermediates may take part in the copolymerisation processes.

8.4 The mussel byssus: an underwater adhesive

Seemingly equally distant from the subject of quinone tanning by plant polyphenols is the common mussel (*Mytilus edulis*). However, the mussel byssus or beard, which mediates strong and durable adhesion to surfaces under water, is a technological facet of the life of the mussel that 'many entrepreneurs would give their right arm to know' (Waite, 1983, 1990, 1991). Amongst other features it appears to involve enzymes with catechol oxidase-like activity and subsequent quinone tanning reactions.

A common limitation in every day use of adhesives is the presence of moisture. Water forms a fine boundary between a glue and its intended bonding surface and thus subverts adhesion. Mussels (*Mytilus edulis*) have evolved a strategy whereby the mussel byssus or beard brings about strong and durable adhesion to surfaces *under water*. The mussel byssus is a bundle of silky threads that is proximally connected to the mussel by a root-like mechanism and distally to an appropriate surface by adhesive plaques. Each byssal thread consists of collagen/elastic protein gradients predominating in collagen at the distal or plaque end and proximally in elastic fibre. By drawing itself up on these threads the mussel can control the tension of its attachment. The insulation consists of a tough durable varnish derived from stoicheiometric mixtures of a polyphenolic protein and an enzyme with catechol oxidase activity, secreted by an accessory gland at the foot of the thread. Curing occurs

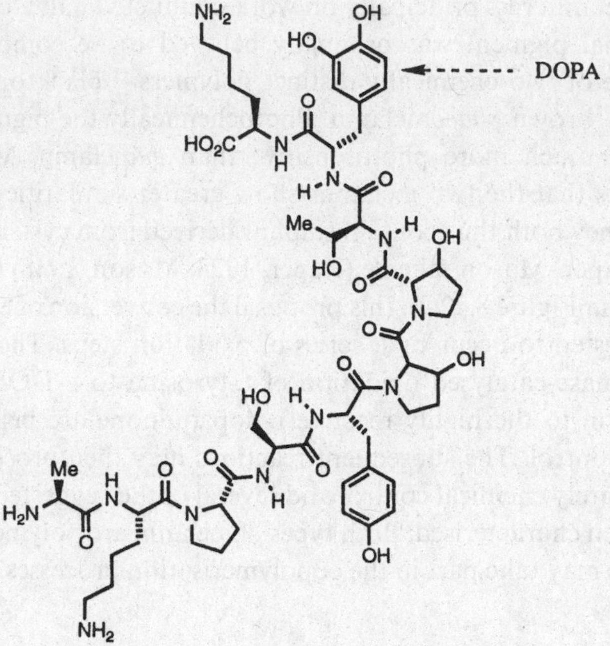

[ala. lys. pro*. ser. tyr*. hyp. hyp. thr. DOPA. lys]₇₀

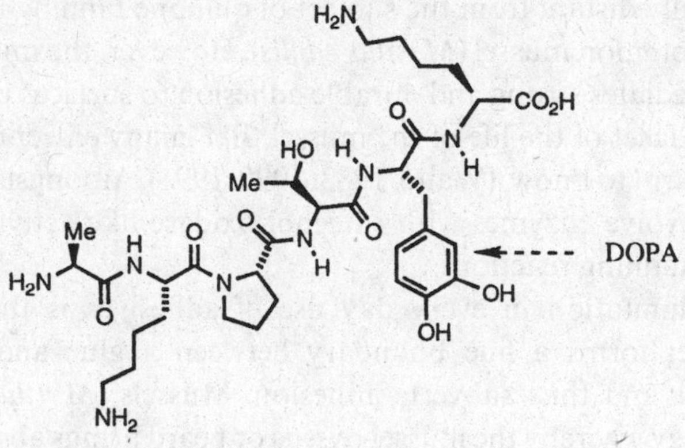

[ala. lys. pro*. thr. tyr*. DOPA. lys]₁₃

Figure 8.23. Repeat peptide motifs in the polyphenolic pre-polymer in the mussel byssus (Waite *et al.*, 1985). Variations in the peptide repeats occur at the positions shown (*): **pro** to **hypro** and **tyr** to **DOPA**.

during coalescence and proteolytic activation of the catechol oxidase; evidence suggests that this enzyme has roles as both a copolymer and a catalyst.

The adhesive plaques of the mussel contain significant levels of a polyphenolic protein. Earlier immunological observations, which suggested that

tyrosyl group   DOPA group

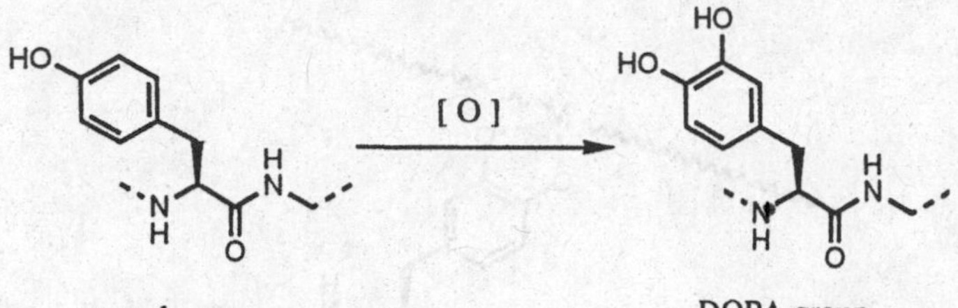

Figure 8.24. Post-translational oxidative conversion of a tyrosyl residue in the pre-polymer to a DOPA residue (Marumo and Waite, 1986).

this protein is located at the interface between the plaque and the substratum, await confirmation. Various features of the chemical composition and properties of the polyphenolic protein have now been delineated (Waite *et al.*, 1985; Williams *et al.*, 1989; Marumo and Waite, 1986). Noteworthy features of this strongly basic pre-polymer are its high molecular mass ($M_R \sim 125\,000$), its random coil conformation and apparent absence of secondary structure, and the presence of the unusual protein α-amino acid, L-dihydroxyphenylalanine (L-DOPA). Eight amino acids (lysine – lys, alanine – ala, serine – ser, threonine – thr, proline – pro, hydroxyproline – hyp, tyrosine – tyr, and dihydroxyphenylalanine – DOPA) account for more than 90% of the amino acid content of the protein. It is a protein which has a highly repetitive primary structure. Typical repeat motifs found in the protein, with the approximate numbers of these repeats in the polyphenolic pre-polymer, are shown in Figure 8.23. Both the hydroxyproline and the DOPA residues in the polyphenolic pre-polymer are believed to be derived by post-translational hydroxylation of proline and tyrosine residues respectively, Figure 8.24, hence the earlier analogy to melanin.

Although much remains to be learnt about the structure and function of DOPA containing proteins in the mussel byssus the peptidyl–DOPA is a pendant group which probably acts in a dual capacity – as a surface coupler and, in oxidatively mediated transformations, as a cross-linking functionality between disparate protein chains (quinone tanning). The *ortho*-dihydroxyphenyl group of DOPA has a well recognised ability to form complexes with metals such as iron and aluminium, and thus admirably fulfils the role of surface coupling agent in the adhesive, Figure 8.25. Similarly the phenolic pre-protein has a multiplicity of lysyl residues in its structure and thus it is entirely reasonable to expect that oxidation of the DOPA residues to create DOPA quinone groups will be followed by multivariant cross-linking of the peptide chains by quinone tanning, Figure 8.26, cf. Figure 8.18.

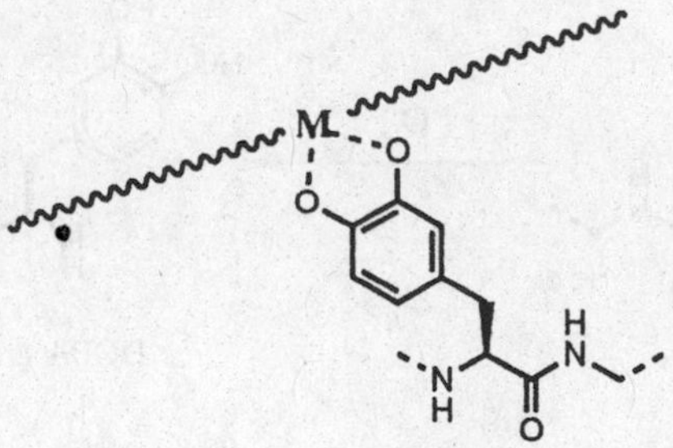

DOPA group

Figure 8.25. Anchorage of the polyphenolic protein adhesive of the mussel to a metal surface by means of metal-ion complexation with the pendant DOPA groups of the protein.

Whilst the role of the pendant catechol nuclei of the L-DOPA residues in the protein pre-polymer is of undoubted significance in the context of the phenomenon of quinone tanning it is also of interest in the light of the enhanced affinity which polyphenols demonstrably have for proline rich proteins (PRPs) generally (see chapter 3). The proline plus hydroxyproline content of the protein pre-polymer is ~ 20–25% and these residues are very probably primarily responsible for the flexible, unstructured conformation of the protein. In addition it also seems very reasonable to suggest that they may facilitate a loose pre-organisation of the peptide chains which favours rapid oxidative cross-linking and hence polymerisation. This would result from the selective inter- and intra-molecular complexation of the DOPA groups and the prolyl residues in different regions of the peptide structures *prior* to oxidative quinone tanning, Figure 8.27. It is therefore interesting to note that the site N-terminal to the proline group in for example the heptapeptide repeat, Figure 8.23, and the one in this scenario which would be best positioned to participate in the subsequent oxidative reactions is occupied by the lysyl residue. If all of these roles are actually realised in practice then it is clear that nature is responsible for a particularly neat and effective design of adhesive/varnish which can be employed *under water*.

Despite its evident importance and widespread occurrence in living systems the phenomenon of quinone tanning remains something of a chemical enigma, shrouded in uncertainties and lacking in chemical definition. One reason is that quinone-tanned structures are amongst the most intractable known to chemists; the science is not attractive and in today's sleek seductive jargon not 'at the cutting edge of chemistry'. Nevertheless it is clearly of importance in the longer term to the understanding of a number of problems of fundamental significance in the plant and animal kingdoms. In the more parochial shorter

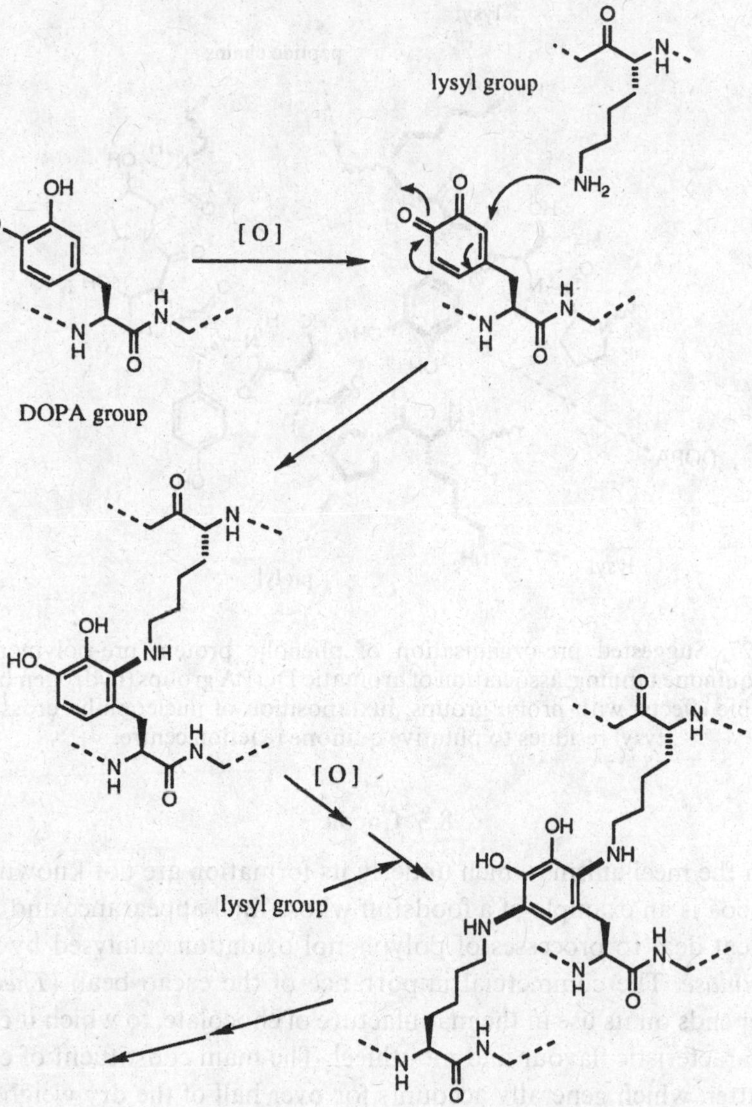

Figure 8.26. Oxidative cross-linking of polyphenolic protein chains in mussel byssus by quinone tanning: participation of pendant DOPA and lysyl residues.

term Waite has speculated, whilst noting the uncanny prescience of nature, that she is simply anticipating advances in industrial technology. Whether this be in the design of particles or to that of glues and varnishes to be applied under water, these anticipations will only be revealed by scientific 'observation, deduction and the application of relevant knowledge' (Waite, 1991; Doyle, 1981).

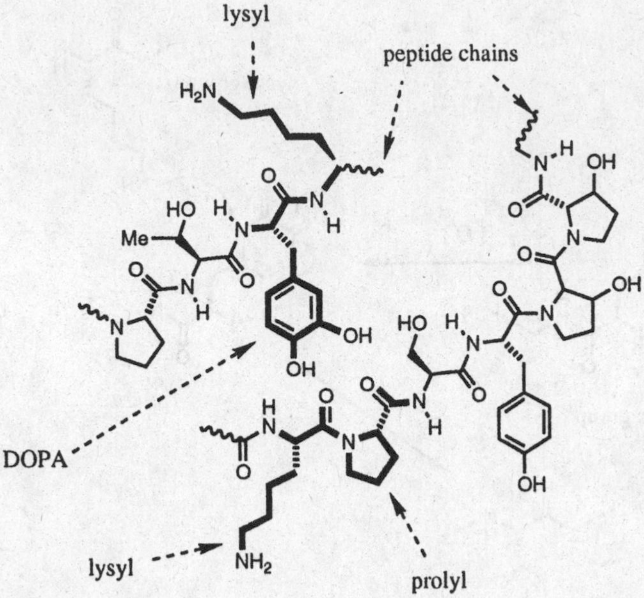

Figure 8.27. Suggested pre-organisation of phenolic protein pre-polymer before oxidative quinone tanning; association of aromatic DOPA groups (hydrogen bonding, hydrophobic effects) with prolyl groups; juxtaposition of nucleophilic cross-linking lysyl residues to putative quinone reaction centre.

8.5 Cocoa

Although the mechanisms which underly its formation are not known in any detail, cocoa is an example of a foodstuff whose final appearance and flavour owe a great deal to processes of polyphenol oxidation catalysed by a *polyphenol-oxidase*. The commercial importance of the cacao bean (*Theobroma cacao*) depends on its use in the manufacture of chocolate, to which it contributes a characteristic flavour and mouthfeel. The main constituent of cocoa is cocoa butter, which generally accounts for over half of the dry weight of the cacao bean. It has unique melting properties – it is solid at room temperature but melts in the mouth (∼ 37°C). The fatty acid composition of cocoa butter is also unique when compared to other plant fats. In particular it contains large amounts of stearate, which to a large measure influences its melting properties. Cocoa butter is mainly triacyl glycerol – (palmitate–oleate–stearate), (stearate–oleate–stearate) and (palmitate–oleate–palmitate) are the principal species.

The basic tastes of cocoa are sour (acidic), bitter and ***astringent***. These characteristics originate from particular components which are either present in the bean or are derived during fermentation, drying and roasting:

(i) *sour (acidic)* – due to organic acids, principally citric but also lactic, oxalic and succinic acids. The presence of significant quantities of acetic acid can be a sign of incorrect drying procedures;

(ii) *bitterness* – attributed to the methyl xanthines (purines) – theobromine (mainly) and caffeine;

(iii) *astringency* – the various phenolic metabolites present in the bean (phenolic flavan-3-ols (catechins) and their oligomers (procyanidins)) are usually held to be responsible for this particular mouthfeel (Haslam *et al.*, 1972; Porter *et al.*, 1991). The purple colour of beans is due to the presence of small amounts of anthocyanins (derivatives of cyanidin).

The uniqueness of cocoa flavour comes from the volatile compounds which contribute to its aroma – a total of 462 such compounds have been found from several sources of cocoa, variously comprising pyrazines, aldehydes, alcohols, volatile acids and esters (Macdonald, 1993).

The oval cacao pod has an embryo consisting of two irregularly formed cotyledons and a radicle; the whole is surrounded by a seed coat, which in turn is covered with a thick, white layer of endocarp – the so-called pulp. Only the cotyledons are used in chocolate manufacture; they are very rich in fat. Fresh cotyledons contain approximately one-third water and one-third fat (cocoa butter) and are largely composed of two types of parenchyma cells. The most common type of cell is small and contains protoplasm, starch granules, aleurone grains and fat droplets. The second type (~ 12% of the tissue) consists of larger storage cells. They range in colour from white to purple (dependent upon the anthocyanin content which may be ~ 3%) and are completely lacking in fat but contain all the phenolic metabolites (~ 60%) and the methyl xanthines (~ 15%).

Three processes are required for cocoa flavour development from the raw cacao bean, namely **fermentation**, **drying** and **roasting** (Forsyth and Quesnel, 1963; Wood, 1985).

8.5.1 Fermentation

The bulk of the world's cacao is fermented by either the 'heap' or the 'box' method. The essential requirement is a mass of beans large enough so that the heat generated by the process is not lost too rapidly but causes a rise in temperature of the mass substantially above ambient temperature. The beans together with the surrounding mucilaginous pulp are removed from the pods and placed in heaps or boxes. The pile of beans and pulp is then naturally inoculated with yeasts, followed by *Acetobacter* and lactic acid bacteria which dominate as conditions in the pile change. The yeasts favour the low pH (3.8) and the low oxygen levels in the pulp. They break down the sugars in the pulp to produce ethanol; the pulp rapidly disintegrates and an acid liquor drains off. As oxygen increases with aeration *Acetobacter* take over and convert the ethanol to give acetic acid which then permeates the bean lowering the internal pH to ~ 4.5. The combination of the effects of acid and heat eventually kills the beans.

In normal fermentation when the death of the seed occurs the cells lose their structural integrity and with it the cell membranes lose their semi-permeability. The phenolic and purine metabolites are then able to slowly diffuse away from their sites of storage within the cell (often first noticed as the diffusion of cellular pigments away from the polyphenol storage cells; Pettipher, 1986). The cellular disruption at the same time also allows enzymes to interact with their appropriate substrates to produce the precursors which are thought to be crucial to the development of cocoa flavour and appearance. These enzymic reactions principally include the following.

(i) Sucrose is converted to glucose and fructose by invertase. Glucose is then preferentially utilised so that fructose eventually predominates.
(ii) Cacao storage protein is hydrolysed by the cacao proteinases to give peptides and free amino acids.
(iii) *Polyphenol-oxidase* converts the various phenolic metabolites initially to quinones which may then react in self-polymerisation reactions or in cross-linking reactions with proteins (*quinone tanning*). In either case high molecular weight 'brown coloured', highly insoluble, polymers, of unknown structure, are formed.

8.5.2 Drying

Following fermentation (~ 7–8 days) the cacao beans are cleaned then dried (5–6 days, sun drying); the moisture content drops to around 7.5%. The type of drying can profoundly affect the final flavour of the cocoa. Thus forced hot air drying may cause the shell of the bean to lose moisture preferentially. As a result the shells harden before all the water is lost and thereby the diffusion and loss of volatile acids from the bean is inhibited. Sun drying allows the drying process to proceed much more gradually and the volatile acids are lost progressively with the water from the surface of the bean. Although a major proportion of the activity of the enzyme, *polyphenol-oxidase*, is lost during the fermentation its action on residual polyphenolic metabolites is prolonged under conditions of sun drying into the drying phase.

8.5.3 Roasting

Once the beans have been dried they possess many of the flavour volatiles – such as alcohols and esters – but their aroma is not usually chocolate-like. It is roasting (140°C, 30 min.) that converts the flavour precursors to give compounds which contribute to the distinctive flavour of cocoa. Almost all the reducing sugars and at least half of the amino acid content (either as free amino acids or small peptides) are lost in this roasting phase. The water content drops to around 1.5%. The beans are then decorticated to yield kernel fragments which are called nibs. The nibs are ground and milled to form **cocoa liquor** which, if it is then subject to hydraulic pressing, gives cocoa butter as the expelled fatty phase.

'Good' cocoa liquors are not generally derived from cacao beans which possess too high a polyphenol content. The reason, it is presumed, is closely connected with the fact that as the *fermentation* proceeds polyphenols gradually diffuse from the storage cells into the internal milieu of the bean. The internal pH is such that as they do so the polyphenols will certainly complex with other macromolecules, particularly proteins, with which they come into contact. There may be some selectivity in these associations but it is probably best to assume that these events are essentially indiscriminate in their nature. Amongst the proteins with which encounters may be envisaged are enzymes and those proteins which form structural elements in the cell. Polyphenol–enzyme complexation invariably leads to a substantial decrease and eventually loss of enzyme activity. It is therefore reasonable to suppose that at this point the rates of degradation of storage proteins and sucrose (*fermentation* (i and ii)) will steadily decrease (Forsyth *et al.*, 1958; Bracco *et al.*, 1969). Depending on the availability of oxygen to the disrupted cellular structure the polyphenols themselves as the fermentation proceeds begin to be oxidised to quinones which will either form covalent bonds with the proteins (*quinone tanning*) or self-polymerise (*fermentation* (iii)). With cacao beans which are rich in polyphenolic metabolites these reactions, essential to the development of 'good' cocoa liquors, will therefore be arrested prematurely. At the conclusion of a normal fermentation there will probably be present the following components in the system.

(a) Flavour precursors – principally amino acids and small peptides and reducing sugars, mainly fructose. These may well, in some cases, already be associated as Schiff's bases.
(b) Storage protein and fragments of partially degraded storage protein.
(c) Some free polyphenols (phenolic-flavan-3-ols and their oligomers (procyanidins)).
(d) Polyphenols which, through the action of polyphenol-oxidase, are self-poly-

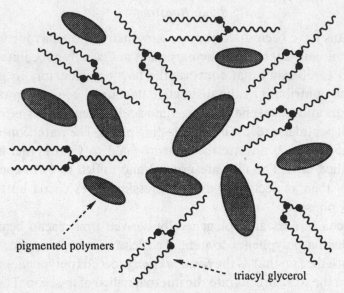

pigmented polymers

triacyl glycerol

Figure 8.28. Cocoa liquor: a butter fat medium in which there is a very, very fine suspension of particles including unreacted or partially degraded proteins and polymeric pigmented species derived by *quinone self-polymerisation* and *quinone tanning*.

merised or have become attached to proteins, cross-linking them to become part of a wide, complex structural network (*quinone tanning*).

Components (a)–(c) would remain relatively soluble in the predominantly aqueous phase of the medium, whilst (d) would be predominantly insoluble. During the *drying* phase, particularly if the pH slowly rises as volatile acids escape from the bean and whilst its water content remains moderately high, some residual enzyme activity would remain. Thus in this phase one would expect modest increases in categories (a) and (d) and corresponding decreases in (b) and (c).

The *roasting* phase is the final critical one in the production of the cocoa. At this point there is little water remaining in the system and one therefore envisages a partially structured environment in which butter fat abounds but one in which components (a)–(d) also occur. In the ideal situation the following events may be presumed to take place.

(e) The flavour precursors produce flavour volatiles. In addition – if one assumes that reactions of the Maillard type are also possible – then this system may also be responsible for some at least of the additional dark brown pigmentation, characteristic of a 'good' cocoa butter.

(f) The polyphenolic derived polymers (d), where quinone groups remained, would also under the influence of the increased temperature increase in complexity by further polymerisation and cross-linking. Where intact resorcinol nuclei were still present then these same polyphenols might also consume a fraction of the flavour precursors (particularly the reducing sugars) by additional covalent bond formation.

After milling the final cocoa liquor will be composed of a butter fat medium in which many of the flavour volatiles are dissolved and in which there is a very, very fine suspension of particles including unreacted or partially degraded proteins, polymeric pigmented species from a variety of sources and (possibly) some flavour precursors, Figure 8.28. The pigmented polymers would, because of their chemical derivation and origins, be expected to display both hydrophobic and hydrophilic ($-OH$; $-NH-CO-$; CO_2H groups, etc.) characteristics, both of which are important in the further processing which takes place in the formation of chocolate.

References

Andersen, S. O. (1971). Phenolic compounds isolated from insect hard cuticle and their relationship to sclerotization. *Insect Biochem.*, **1**, 157–70.

Andersen, S. O. (1974). Two mechanisms of sclerotization in insect cuticle. *Nature*, **251**, 507–8.

Bailey, R. G., Nursten, H. E. and McDowell, I. (1992). Isolation and analysis of a polymeric thearubigin fraction from tea. *J. Sci. Food Agric.*, **59**, 365–75.

Benasson, R. V., Land, E. J. and Truscott, T. G. (1993). Melanogenesis and melanoma. In *Excited States and Free Radicals in Biology*, chapter 8, Oxford University Press: Oxford, pp. 229–48.

Bracco, U., Graihle, N., Rostango, W. and Egli, R. H. (1969). Analytical evaluation of cocoa curing in the Ivory Coast. *J. Sci. Food Agric.*, **20**, 713–17.

Brunet, P. C. J. and Coles, B. C. (1974). Tanned Silks. *Proc. R. Soc. Lond.*, **187 B**, 133–70.

Cai, Y. and Haslam, E. (1994). Plant polyphenols (vegetable tannins): gallic acid metabolism. *Natural Product Reports*, **11**, 41–66.

Cheynier, V., Osse, C. and Rigaud, J. (1988). Oxidation of grape juice phenolics in model systems. *J. Food Sci.*, **53**, 1729–32 and 1760.

Cranwell, P. A. and Haworth, R. D. (1971). Humic acid. IV. The reaction of α-amino acid esters with quinones. *Tetrahedron*, **27**, 1831–7.

Doyle, A. C. (1981). The sign of four. In *The Complete Sherlock Holmes*, Doubleday & Co.: New York, pp. 89–94.

Fischer, E. and Schrader, H. (1910). Verbindungen von Chinon mit Aminosaure-estern. *Ber. dtsch. Chem. Ges.*, **43**, 525–9.

Forsyth, W. G. C., Quesnel, V. C. and Roberts, J. B. (1958). The interaction of polyphenols and proteins during cacao curing. *J. Sci. Food Agric.*, **9**, 181–4.

Forsyth, W. G. C. and Quesnel, V. C. (1963). The mechanism of cacao curing. *Adv. Enzymol.*, **25**, 457–92.

Haslam, E. (1965). Galloyl esters of the Aceraceae. *Phytochemistry*, **4**, 495.

Haslam, E., Thompson, R. S., Jacques, D. and Tanner, R. J. N. (1972). Plant proanthocyanidins. Part I Introduction; the isolation, structure and distribution in nature of plant procyanidins. *J. Chem. Soc.* (*Perkin Trans. I*), pp. 1387–99.

Haslam, E., Lilley, T. H., Warminski, E., Liao, H., Cai, Y., Martin, R., Gaffney, S. H., Goulding, P. N. and Luck, G. (1992). Polyphenol complexation – a study in molecular recognition. In *Phenolic Compounds in Food and Their Effects on Health. I. Analysis, Occurrence and Chemistry*. A.C.S. Symposium Series **506**, eds C.-T. Ho, C. Y. Lee and M.-T. Huang, American Chem. Soc.: Washington D.C., pp. 8–50.

Macdonald, H. (1993). Flavour development from cocoa bean to chocolate bar. *The Biochemist* (**April/May**), pp. 3–5.

Marumo, K. and Waite, J. H. (1986). Optimization of hydroxylation of tyrosine and tyrosine-containing peptides by mushroom tyrosinase. *Biochim. Biophys. Acta*, **872**, 98–103.

Mason, H. S. (1948). The chemistry of melanin – mechanism of oxidation of DOPA by tyrosinase. *J. Biol. Chem.*, **172**, 83–92.

Mason, H. S. (1956). Structures and functions of the phenolase complex. *Nature*, **177**, 79–81.

Mayer, A. M. and Harel, E. (1979). Polyphenol oxidases in plants. *Phytochemistry*, **18**, 193–215.

Nonaka, G.-I., Kawahara, O. and Nishioka, I. (1983). Tannins and related compounds. XV. A new class of dimeric flavan-3-ol gallates, theasinensins A and B, procyanidin gallates from green tea leaf. *Chem. Pharm. Bull.*, **31**, 3906–14.

Ollis, W. D., Eyton, W. B., Holmes, A. and Brown, A. G. (1969a). Identification of the thearubigins as polymeric proanthocyanidins. *Nature*, **221**, 742–4.

Ollis, W. D., Eyton, W. B., Holmes, A. and Brown, A. G. (1969b). The identification of the thearubigins as polymeric procyanidins. *Phytochemistry*, **8**, 2333–40.

Pettipher, G. L. (1986). An improved method for the extraction and quantitation of anthocyanins in cacao beans and its use as an index of the degree of fermentation. *J. Sci. Food Agric.*, **37**, 289–96.

Porter, L. J., Ma, Z. and Chan, B. G. (1991). Cacao procyanidins: major flavonoids and identification of some minor metabolites. *Phytochemistry*, **30**, 1657–63.

Prota, G. (1988). Progress in the chemistry of melanins and related metabolites. *Med. Res. Rev.*, **8**, 525–56.

Prout, C. K. and Wright, J. D. (1968). Observations on the crystal structures of electron donor–acceptor complexes. *Angew. Chem. Int. Ed.*, **7**, 659–67.

Pryor, G. M. (1940). On the hardening of the ootheca of *Blatta orientalis. Proc. R. Soc. Lond.*, **128B**, 378–92.

Pryor, G. M. (1962). Sclerotization. In *Comparative Biochemistry*, eds. M. Florkin and H. S. Mason, volume IV, Academic Press: New York, pp. 371–96.

Raper, H. S. (1928). The aerobic oxygenases. *Physiol. Rev.*, **8**, 245–82.

Roberts, E. A. H. (1962). Economic importance of flavonoid substances: tea fermentation. In *The Chemistry of Flavonoid Compounds*, editor T. A. Geissman, Pergamon Press: Oxford, London, New York, pp. 468–512.

Rorsman, H., Agrup, G., Hansson, C. and Rosengren, E. (1983). Biochemical recorders of malignant melanoma. In *Malignant Melanoma, Advances of a Decade*, editor R. M. McKie, S. Karger AG: Basel, pp. 93–115.

Sanderson, G. W. (1972). The chemistry of tea and tea manufacturing. In *Structural and Functional Aspects of Phytochemistry*, eds. V. C. Runeckles and T. C. Tso, *Recent Advances in Phytochemistry*, **5**, 247–316.

Sugumaran, M. (1987). Quinone-methide sclerotization: a revised mechanism for β-sclerotization of insect cuticle. *Bio-Org. Chem.*, **15**, 194–211.

Waite, J. H. (1983). Adhesion in bysally attached bivalves. *Biol. Rev.*, **58**, 209–31.

Waite, J. H. (1990). The phylogeny and chemical diversity of quinone-tanned glues and varnishes. *Comp. Biochem. Physiol.*, **97B**, 19–29.

Waite, J. H. (1991). Mussel beards: a coming of age. *Chem. and Ind.*, pp. 607–11.

Waite, J. H., Housley, T. J. and Tanzer, M. L. (1985). Peptide repeats in a mussel glue protein: theme and variations. *Biochemistry*, **24**, 5010–14.

Williams, T., Marumo, K., Waite, J. H. and Henkens, R. W. (1989). Mussel glue protein has an open conformation. *Arch. Biochem. Biophys.*, **269**, 415–22.

Wood, G. A. R. (1985). From harvest to store. In *Cocoa*, 4th edition, eds. G. A. R. Wood and R. A. Lass, Longmans: London and New York, pp. 444–504.

9

Polyphenols, collagen and leather

9.1 Introduction

The word tannin has a long and well established usage in the scientific literature which relates specifically to the application of plant extracts in the manufacture of leather from animal hides and skins. In this context it is an important etymological legacy. Despite the fact that the use of vegetable materials to convert skins to leather has declined remarkably over the past century the question of how these plant extracts (containing polyphenols as their major constituents) bring about this conversion has remained shrouded in mystery. It remains a question of legitimate scientific enquiry and importance, not least because of its relevance to the ageing and storage of leather-bound books and historical objects in which leather is an integral feature of ornament and design.

9.2 Vegetable tannage

Leather making is a craft and a trade of great antiquity and records exist relating to its operation in Mediterranean regions dating back to 1500 BC. The word tanner to describe a person who pursued this trade has probably been in use for a similar period of time, but it is doubtful if the corresponding term tannin to denote the substances responsible for the conversion of raw animal skins into leather was in common usage much before the end of the eighteenth century. Vegetable tannin chemistry can be said to have its origins around this time, for only then was it generally recognised that the tanning process involved a combination of substances – *tannins* – in the plant extract with the animal skins and was not merely a vaguely defined physical change in which the astringent tanning liquors caused the skin to harden and shrink. As befits an industry with this historical background much of the tanner's traditional

374

art has been gained by empirical methods, by countless years of observation, by hand and eye and frequently taste; Henry Procter in his Cantor lectures of 1899 was thus able to describe the science of leather making as 'still very young'.

9.2.1 Vegetable tannins of commerce

Tanning is the process whereby animal skins are turned into leathers – materials which have a greater stability to water, bacteria, heat and abrasion and which, in consequence, have a wide range of industrial and domestic applications. In Great Britain, up until the dawn of the Industrial Revolution, most leathers were produced by infusions of oak bark – the best was obtained from young trees and in the days of oak coppices this was readily available. The process was slow and took a period of some three to six months to complete. The rapidly increasing demands for more leather created by the Industrial Revolution meant that the British tanner had perforce to find new sources of vegetable tannin to supplement indigenous oak bark. These he discovered world-wide: some of the more important sources of vegetable tannins which were still widely employed commercially in leather production until comparatively late into the twentieth century are listed in Table 9.1. Substantial accumulations of vegetable tannins may be found in almost any part of a plant – seed, fruit, leaves, wood, bark, root. Increased tannin production is also often associated with a particular pathological condition. The most familiar example is that of plant galls caused by insect attack. Probably the most familiar insect gall found in Britain is the marble gall ('bullet gall', 'oak nut', 'Devonshire gall') found on pedunculate and sessile oak. It is plentiful on young hedgerow plants and on scrub-oaks in coppices and is a hard globular object (~ 2.5 cm in diameter) found singly or in groups. The galls arise as a result of the parasitic attack on terminal or axillary buds by the gall wasp *Andricus kollari* (*Cynips kollari*). However, British oak galls have much less tannin (~ 20% dry weight) than the commercially important Aleppo gall (*Quercus infectoria*) or the Chinese gall (*Rhus semialata*) which may contain in excess of 60–70% dry weight as tannin. Invariably the polyphenol composition (qualitatively and quantitatively) of the carapace of the gall is simply a reflection of the polyphenolic metabolic profile of the tissues of the host plant. Botanically speaking, vegetable tannins (polyphenols) occur in a substantial part of the plant kingdom and most notably in particular families of dicotyledons – the Leguminosae, Anacardiaceae, Combretaceae, Rhizophoraceae, Myrtaceae and Polygonaceae.

Despite the fact that by around the middle of the twentieth century the

Table 9.1. *Commercial vegetable tanning materials*

Barks of trees Wattle (*Acacia* sp.), Mangrove (*Rhizophora* sp.), Oak (*Quercus* sp.), Spruce (*Picea* sp.), Hemlock (*Tsuga* sp.), Eucalyptus (*Eucalyptus* sp.), Avaram (*Cassia auriculata*), Babul (*Acacia arabica*), Birch (*Betula* sp.), Willow (*Salix caprea*), Pine (*Pinus* sp.), Larch (*Larix* sp.), Alder (*Alnus* sp.).
Woods of trees Quebracho (*Schinopsis* sp.), Chestnut (*Castanea* sp.), Oak (*Quercus* sp.), Cutch (*Acacia catechu*), Wandoo (*Eucalyptus wandoo*), Urunday, Tizra.
Fruits and fruit pods Myrobalans (*syn.* Myrobolams, Myrabolans, Myrabalans (*Terminalia chebula*)), Valonea (*Quercus aegilops*), Divi-divi (*Caesalpinia coriaria*), Algarobilla (*Caesalpinia brevifolia*), Tara (*Caesalpinia spinosa*), Teripods, Sant pods.
Leaves Sumach, American Sumach (*Rhus* sp.), Gambier (*Uncaria gambir*), Dhawa or Country Sumach (*Anogeissus latifolia*), Badan (*Bergenia crassifolia*).
Roots Dock (*Rumex* sp.), Canaigre (*Rumex hymenosephalus*), Siberian saxifrage (*Saxifragia crassifolia*), Garouille (*Quercus coccifera*), Sea lavender.
Plant galls Oak (*Quercus* sp.), Chinese (*Rhus semialata*), Tamarisk (*Tamarix articulata*), Pistacia (*Pistacia* sp.), Aleppo (*Quercus infectoria*).

Information obtained from Howes (1953), Haslam (1966), White (1957) and Grimshaw (1986).

annual consumption of oak bark in this country had fallen to some 1% of the total consumption of vegetable tanning materials (~ 1000 tons per annum), oak bark tannage remains one of the few time-tested and traditional forms of leather manufacture still practised today – albeit on a very modest scale for the · making of speciality leathers used in fashion accessories and the manufacture of orthopaedic equipment. Chestnut extract (from the wood of *Castanea sativa*) is likewise still used to a limited extent in France and other European countries. These dramatic shifts in the commercial exploitation of the natural vegetable tannins for leather production are almost entirely due to the development of synthetic tannins (syntans) and the recognition of the enormous potential of mineral tannage (particularly that of basic chromium salts) for the manufacture of leathers. Processes are much faster and tannage of animal skins is complete with chromium salts when the skin has absorbed some 3% of the inorganic substrate (compared with the need to take up about half its own weight in the case of a vegetable tannin). Nevertheless, increasing concern has been expressed in recent years about the environmental problems associated with the widespread use of chromium salts in leather manufacture – not only those of usage but also those which concern the disposal of waste leathers themselves.

In the 21st Procter Memorial Lecture, R. L. Sykes (Sykes, 1986) made a very shrewd and accurate observation:

Even now with sophisticated separation procedures the composition of commercial vegetable tannins is not completely understood although we have a working hypothesis which is adequate for most purposes. Indeed, it is probably not unfair to say that, in commercial terms, the contribution of the physical chemists' approach ... to vegetable tanning systems ... has been much greater than that of the organic chemists' elucidation of the molecular structure of vegetable tannins.

Whilst some may consider the judgement harsh it nevertheless remains true that although the composition of a few extracts (e.g. Sumach, Chinese and Aleppo galls) is now well known, that of the vast majority of commercially important extracts, such as wattle, chestnut, quebracho, valonea, myrobalans (Table 9.1), cannot yet be defined in sufficient detail in molecular terms. There are a number of reasons for this state of affairs, not least of which is that these extracts are derived from plant materials after numerous post-mortal treatments such as storage and drying, fermentation and extraction under harsh conditions, all of which invariably multiply the complexities of the molecular composition. Nevertheless, as Sykes remarked, there is a working hypothesis which is adequate for most purposes – the broad outlines of the *general* chemical nature of most extracts (Table 9.1) is known and this must form the basis of a working hypothesis of the mechanism of vegetable tannage.

9.2.2 *Early ideas on the mechanism of vegetable tannage*

There is general agreement amongst the cognoscenti that chrome tannage occurs by intrafibrillar cross-linkage of carboxyl groups (aspartic and glutamic acid) which occur along the surface of tropocollagen molecules, as indicated in Figure 9.1. Since these charged groups are known to occur in clusters this explains the 'cluster-like' appearance of chrome tanned collagen (Heidemann, 1982).

By contrast the picture of vegetable tannage of collagen is much less clear; answers to the question 'where do the vegetable tannins go?' are correspondingly much more vague. Many workers believe that the deposition of the vegetable tanning materials also takes place analogously within the charged clusters along the collagen molecule. This concept was first expounded by Gustavson (1949) and Bear (1952) who suggested that vegetable tannins, as a

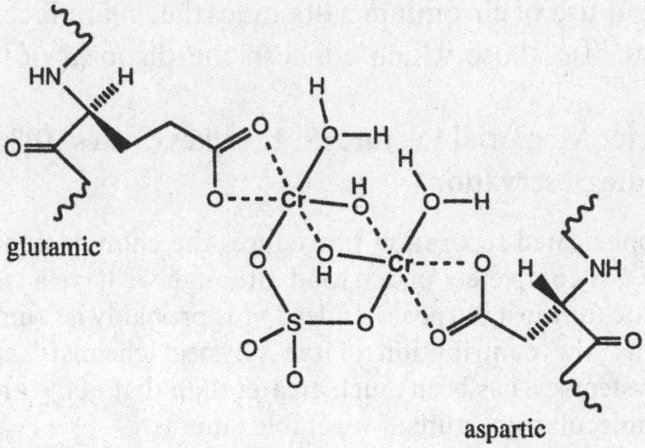

Figure 9.1. Intrafibrillar cross-linkage, by a binuclear chromium compound, of aspartic and glutamic acid groups in the charge clusters of collagen molecules.

result of their general chemical characteristics, impart hydrothermal stability to collagen fibrils predominantly by interaction with basic groups, e.g. lysine and arginine. Since these are also located at 'bands' in the collagen structure (*vide infra*) then tanning agents *generally* are presumed, in this theory, to largely enter the supramolecular structure at these same points (Bear, 1952).

This early work also suggested that the 'band' structure, by virtue of the presence of the sterically bulky acidic and basic amino acid side-chains, is more open and accessible to invasion by vegetable tannin molecules, Figure 9.2. In this interpretation the crystalline parts of the fibrillar structure (the 'inter-bands') do not require tanning since the compact and orderly arrangement of the amino acid side-chains at these points offers, in itself, protection against the penetration of water and attack by bacteria. During tannage the accessibility of these intrafibrillar regions to vegetable tannins would be increased by osmotic swelling in the aqueous acidic environment, Figure 9.2. Gustavson (1956) also strongly favoured the view that once the vegetable tannins had successfully penetrated these intrafibrillar spaces the peptide groups therein would provide additional reactive sites within the collagen structure at which the tannins could then combine.

These ideas concerning the mechanism of vegetable tannage were conceived and developed some 40–50 years ago and they have remained essentially unchanged in the intervening years. They reflect the status of the fundamental chemistry of collagen and the vegetable tannins at that time. It is therefore both timely and important to review these ideas in the light of the discoveries made over the past 40–50 year period, not only in relation to the structure of the vegetable tannins and the forces which drive their complexation with other

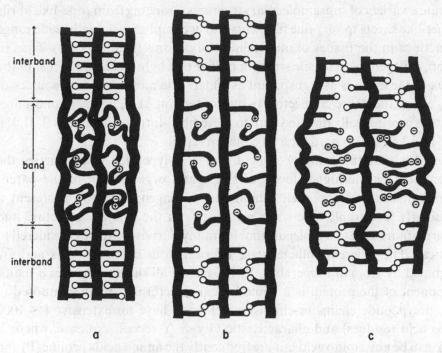

Figure 9.2. Diagrammatic representation of 'band–interband' structure of collagen fibrils showing (a) a dry fibril; (b) a fibril swelling with water at neutrality and (c) the result of acid swelling of a fibril (Bear, 1952).

molecular species, but also, and more particularly, against the background of the very detailed knowledge which has now emerged concerning the structure of the collagen molecule, collagen fibrils and collagen fibres. The essence of tannage is stabilisation of the protein structure by the incorporation into the collagen fibril structure and immobilisation therein of substances (tannins) which have an affinity for various functional groups in the protein structure. Very probably the most effective way in which this stabilisation is achieved is by some form of cross-linking of such groups by the tanning agent.

9.3 Collagen – primary structure and the triple helix

Collagen is arguably, probably, possibly (dependent on one's source of information) the most abundant protein in the animal kingdom. The word 'collagen' is derived from the Greek *kollagen*, meaning glue forming, and describes a family of structurally related proteins that are located in the extracellular matrix of connective tissue. Its properties are remarkably diverse and vividly illustrate how the basic motif of the collagen molecule has been utilised to generate a variety of supramolecular structures – ranging from rope-like fibrils and net-like sheets to very fine filaments – to accomplish an equally wide range of functions in the tissues of multicellular organisms (Kadler, 1994). Thus in tendon, collagen has a tensile strength of that of a light steel wire whilst in the cornea of the eye it is as transparent as water. The most abundant sources of collagen are the dense connective tissues of tendon, skin, ligament and bone; attention is principally directed here to these fibril-forming collagens (I, II, III, IV, V and XI), and in particular to collagen type I.

The idea that the primary structure of a polypeptide chain carries the information for the higher levels of shape and organisation of the parent protein molecule is now well established. Collagen may well represent a particularly favourable case since regularities in the primary structure and conformation result in a molecule that for many purposes can be considered to be linear. The different collagens are referred to as collagen types and are designated by Roman numerals I, II, III, etc. In all these molecules a major component of the protein is a triple helical structural domain composed of three polypeptide chains (α-chains) which each have an extensive (~ 1000 amino acid residues) and characteristic Gly–X–Y repeat sequence, where X and Y can be any amino acid but are frequently the amino acids proline (P) and hydroxyproline respectively. Other specific features of collagens are a high content of alanine (A) and lysine (K), post-translational modifications that encompass hydroxylation of proline and lysine residues, various glycosylations and the formation of intermolecular cross-links through lysine and hydroxylysine residues. For each collagen type the α-polypeptide chains are identified with Arabic numerals followed by Roman numerals, e.g. $\alpha 1(I)$, $\alpha 2(I)$, $\alpha 1(II)$, etc. The chain composition of the heterotrimeric type I collagen is thus written as $[\alpha 1(I)]_2 . \alpha 2(I)$.

Glycine, Gly, G

Alanine, Ala, A

Lysine, Lys, K

δ-Hydroxy-lysine
Hylys

Proline, Pro, P

γ-Hydroxy-proline
Hypro

The basis of the distinctive triple helical structure of a collagen molecule can be understood in terms of two key features of the primary structure of the

Figure 9.3. The tropocollagen molecule: three left-handed helices are twisted together to form a right-handed triple helix. Individual molecules may be regarded as stiff rod-like molecules (Kadler, 1994).

α-chains; the presence of glycine in every third position in the sequence and the high concentrations of proline and hydroxyproline (in collagen from verte-brates glycine, proline and hydroxyproline together account for slightly more than half of the some thousand or so amino acids in each of the three α-chains). Glycine in every third position is a prerequisite, a *sine qua non*, for the folding of the three chains into a triple helix. The imino acids (Pro, Hypro) direct the individual chains into a left-handed polyproline type II helix and place glycine residues into the centre of a triple helix in which adjacent chains are mutually staggered by one residue. The triple helix thus formed is a right-handed superhelix that repeats every 30 residues and is stabilised by intramolecular hydrogen bonding – (i) between the NH group of a glycyl residue and the CO group of the residue in the second position of the triplet in the neighbouring chain and (ii) via a water molecule and the hydroxyl group of hydroxyproline in the third position of the triplet (Gly–X–Y) (Ramachandran, 1967). Figure 9.3 depicts an outline representation of such a triple helix, viewed along its vertical axis. As the principal source of mechanical strength in the animal body, changes in the primary structure of the collagen molecule are often associated with diseases where mechanical support becomes defective. For example Os-teogenesis imperfecta (OI) is a heritable generalised connective tissue disorder

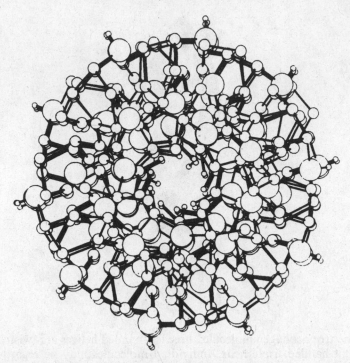

Figure 9.4. End-on view looking down the central long-axis of a molecular model of the triple helix formed by (Gly–Pro–Hypro)$_n$ – a prototype for the collagen molecule. Functional side-chains in the collagen molecule may be envisaged as projecting, approximately orthogonally, from the central triple helical core (Kadler, 1994).

characterised by brittleness of bones and weaknesses in other tissues rich in collagen I. In the majority of cases mutations have been found that alter the primary structure of the α1(I) and α2(I) chains of procollagen I. The most common mutations found to date have been single-base mutations that substitute amino acids with bulkier side-chains for glycine. This subtle steric change in the scaffold of the triple helix leads to diminished thermal stability, intracellular retention and enhanced degradation and reduced secretion (Raghunath et al., 1994).

Figure 9.4 shows an end-on view looking down the central long-axis of a molecular model of the triple helix formed by (Gly–Pro–Hypro)$_n$. In the collagen structure itself various other amino acids (with a range of functional side-chains – amino, carboxyl; hydrophobic and hydrophilic) are found in positions X and Y in the extended helical (Gly–X–Y)$_n$ sequence. These functional side-chains may be envisaged as projecting, approximately orthogonally, from the central triple helical core of the collagen molecule (Figure 9.3), much as the bristles on a test tube brush do so from its central spine. Although there is very good reason to believe that the identity and order of these amino acids is not critical to the molecular structure of collagen *itself*, it is known that the major collagen (type I) of higher animals contains two kinds of chain (two α1(I) and one α2(I)) and that certain features of the amino acid sequence have been well conserved throughout evolution. It therefore seems eminently reasonable to presume that this high degree of specificity is utilised in the molecular interactions which are responsible for the packing of collagen molecules into fibrils and fibres.

9.4 Biosynthesis of collagen (tropocollagen)

The α-polypeptide chains of fibrillar collagens are biosynthesised as precusor pro-α-chains with amino and carboxyl terminal extensions, generally referred to as N- and C-propeptides. The nascent pro-α-chains also undergo a series of post-translational modifications. Specific hydroxylases act on particular pep-tidyl lysine and proline residues converting them to hydroxylysyl and hy-droxylprolyl groups respectively. Vertebrate collagens similarly contain O-linked galactose and glucosylgalactose sugars that are covalently attached to specific hydroxylysyl residues in the triple helical domains of the pro-α-chains. The function of these carbohydrate groups is not clear but it has been pro-posed that these play a part in the subsequent regulation of the diameter of the collagen fibrils. Association of the C-propeptides of three individual pro-α-chains (in the case of collagen I, two pro-α1 and one pro-α2 chain) by interchain disulphide bond formation brings them into correct register and forms a single nucleation site for triple helix folding. Helix formation is then propagated towards the N-terminus following a mechanism like that of a zip-fastener. The molecule of procollagen thus formed is finally converted to collagen (tropocollagen) by enzymic cleavage of the N- and C-propeptides by procollagen N-peptidases and C-peptidases respectively, Figure 9.5 (Ward, 1978; Kadler, 1994).

Type I collagen (tropocollagen) $[\alpha1(I)]_2$. $\alpha2(I)$, is ~ 290 nm in length and ~ 1.5 nm in diameter. The main body of the molecule consists of a triple helical domain (~ 1000 amino acids; 337 Gly–X–Y units from each contributing polypeptide chain) with short telopeptide sequences at the N- (17 amino acids) and the C- (26 amino acids) termini. The telopeptides differ from the main body of the molecule by their inability to form a triple helix. High field NMR studies suggest that both the N- and the C-telopeptides are predominantly extended, with a small proportion of the N-telopeptides existing as a β-turn. Electron microscopic observations, however, suggest that the hydrophobic residues in the C-telopeptides are condensed into a hydrophobic cluster when collagen (tropocollagen) molecules are assembled into fibrils. Since these molecules are

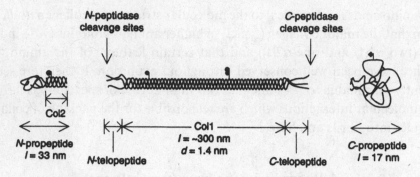

N-peptidase cleavage sites

C-peptidase cleavage sites

Col2

N-propeptide
l = 33 nm

N-telopeptide

Col1
l = ~300 nm
d = 1.4 nm

C-telopeptide

C-propeptide
l = 17 nm

Figure 9.5. Biosynthesis of tropocollagen molecule (Kadler, 1994).

evidently the fundamental sub-units of collagen structures as they are formed *in vivo*, they were named by Gross and others (Gross, 1961), tropocollagen – meaning 'turning into collagen'. In animals, tropocollagen molecules, once generated, assemble to form native collagen structures, Figure 9.6. Newly formed collagen structures are soluble in cold salt solution; simple warming yields reconstituted collagen whose organisation resembles the native form, Figure 9.6. Alternatively native collagen can be dissolved in acetic acid. Treatment of the resultant solution with ATP produces the 'Segment Long Spacing' (SLS) form of collagen in which the molecules aggregate in accurate transverse register, Figure 9.7. Alternative treatment of the same acetic acid solution with glycoprotein gives rise to the 'Fibrous Long Spacing' (FLS) form of collagen (Gross, 1961).

9.5 Fibrillogenesis

Once in the extracellular fluid outside the cells, tropocollagen molecules spontaneously aggregate first into fibrils and ultimately into fibres; they do so in a highly regular manner under the influence of forces directly dependent on the primary molecular structure, Figure 9.6. Assembly is thought by many to involve formation of a supercoiled helix of collagen (tropocollagen) molecules, the microfibril, followed by lateral association of the microfibrils into fibrils. Fibrils can vary greatly in length and diameter depending on the conditions involved in growth. Fibrils grown *in vitro*, however, are typically up to 500 nm in diameter and several micrometres in length and are thought to contain between 10^4 and 10^6 mature collagen (tropocollagen) molecules. Generation of these supramolecular structures is accompanied or followed by covalent inter- and intra-molecular cross-linking of the collagen (tropocollagen) molecules. This confers on the fibrils the high tensile and mechanical strength needed for

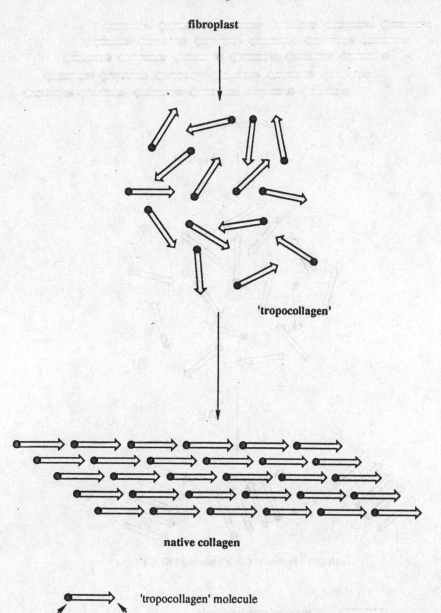

Figure 9.6. Synthesis and assembly of native collagen.

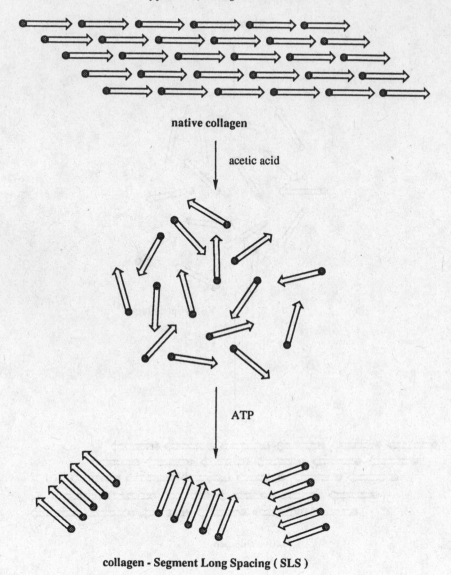

native collagen

acetic acid

ATP

collagen - Segment Long Spacing (SLS)

'tropocollagen' molecule

N-terminus   *C-terminus*

Figure 9.7. Reconstitution of native collagen in 'Segment Long Spacing' (SLS) form; dissolution in acetic acid followed by treatment with ATP (Gross, 1961).

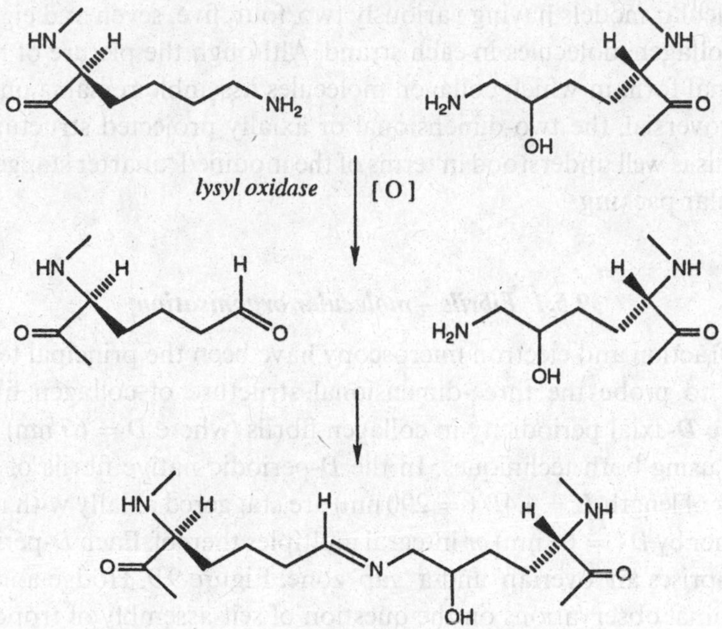

Figure 9.8. Oxidative deamination of lysyl (and hydroxylysyl) residues by lysyl oxidase and the spontaneous formation of covalent cross-links in collagen fibrils (Kadler, 1994).

tissue integrity. The process is catalysed by a single enzyme, the copper ion dependent lysyl oxidase, which oxidatively deaminates the ε-amino group of particular lysyl and hydroxylysyl residues in the telopeptide regions of the collagen (tropocollagen) molecules to form the corresponding highly reactive aldehydes. These then react spontaneously with the ε-amino group of other contiguous lysyl and hydroxylysyl residues to form a variety of covalent cross-links, Figure 9.8 (Kadler, 1994).

Thermodynamic studies indicate that the major driving force for fibril formation is the hydrophobic interaction between tropocollagen molecules leading to minimisation of the surface exposed to the aqueous environment (Parkinson *et al.*, 1994). Experimental data also show that both polar and hydrophobic interactions are involved in the ordered alignment and aggregation of molecules. Several different models have been proposed to describe the way in which collagen molecules are packed laterally into fibrils (Kadler, 1994). In one model it has been suggested that molecules are packed in a quasi-hexagonal molecular lattice without the need for discrete microfibrillar sub-structures (Hulmes and Miller, 1979; Hulmes *et al.*, 1981). Most other proposals embody the concept that fibrils are composed of sub-units, microfibrils, with particular models having variously two, four, five, seven and eight arrays of tropocollagen molecules in each strand. Although the picture of the three-dimensional form in which collagen molecules assemble remains uncertain if not controversial, the two-dimensional or axially projected structure of collagen fibrils is well understood in terms of the modified 'quarter stagger' theory of molecular packing.

9.5.1 Fibrils – molecular organisation

X-ray diffraction and electron microscopy have been the principal techniques exploited to probe the three-dimensional structure of collagen fibrils and fibres. The D-axial periodicity in collagen fibrils (where $D = 67$ nm) has been observed using both techniques. In the D-periodic native fibrils of collagen, molecules of length $L = 4.4D$ ($= 290$ nm) are staggered axially with respect to one another by D ($= 67$ nm) *or* integral multiples thereof. Each D-period in the fibril comprises an 'overlap' and a 'gap' zone, Figure 9.9. Hodge and Schmitt made seminal observations on the question of self-assembly of tropocollagen molecules into supramolecular structures in 1960 using electron microscopy to define the charge distribution along the molecule (Hodge and Schmitt, 1960; Bruns and Gross, 1973). In the 'Segment Long Spacing' (SLS) form of collagen the long triple helical, rod-like molecules align in perfect transverse register to form crystallites that are the same length as the tropocollagen molecule, Figure 9.7. When stained with electron dense compounds, such as phosphotungstic acid or uranyl acetate, and examined under the electron microscope such crystallites reveal a characteristic band pattern which reflects the distribution of clusters of charged and uncharged amino acid side-chains along the molecule. The individual bands on the SLS crystallite serve to identify loci (~ 2 nm) on the collagen molecule. The band pattern observed with the uranyl cation as stain is thought to reflect the distribution of aspartic (D) and glutamic acid (E) side-chains, whilst staining by phosphotungstic acid reveals principally arginine (R) and lysine (K) side-chains, Figure 9.10. Staining crystallites with phosphotungstic acid followed by uranyl acetate, in order to obtain the sharpest band patterns, Bruns and Gross (1973) identified a total of 58 dark bands in well formed crystallites. The SLS staining patterns indicated that the polar groups are localised in narrow defined clusters; whilst particular loci contain a predominance of basic groups, others contain a balance of both types of polar grouping, and others are largely acidic in character. Once the complete amino acid sequence data for the α-chains of type I collagen were known it was possible to compare the staining patterns with the charge distribution predicted from the primary structure. In SLS collagen the pattern

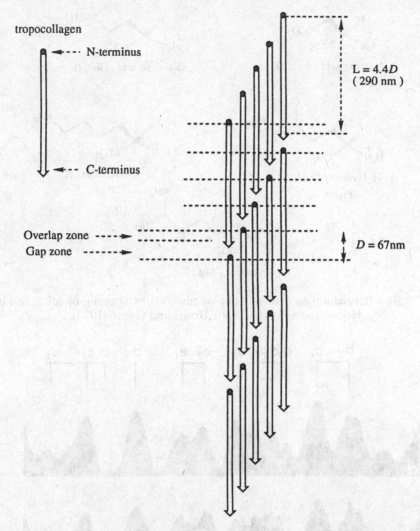

Figure 9.9. Packing of tropocollagen molecules in a collagen fibril (the Hodge–Petrushka model). Tropocollagen molecules are staggered by 1D (one of many possible forms of assembly with offsets as multiples of D). The gap and overlap zones represent 0.6 and 0.4D respectively.

compares directly with the predicted molecular charge profile, Figure 9.11.

Hulmes and his colleagues were the first to demonstrate that intermolecular electrostatic and hydrophobic interactions between two α-1 chains are maximal when the stagger is D (or a multiple of D), where $D = 234$ residues. The analysis was carried out by processing the triple helical region of ∼ 1000 amino acids of a tropocollagen molecule past itself; scoring for complementarity between opposing amino acids demonstrated that both charge–charge

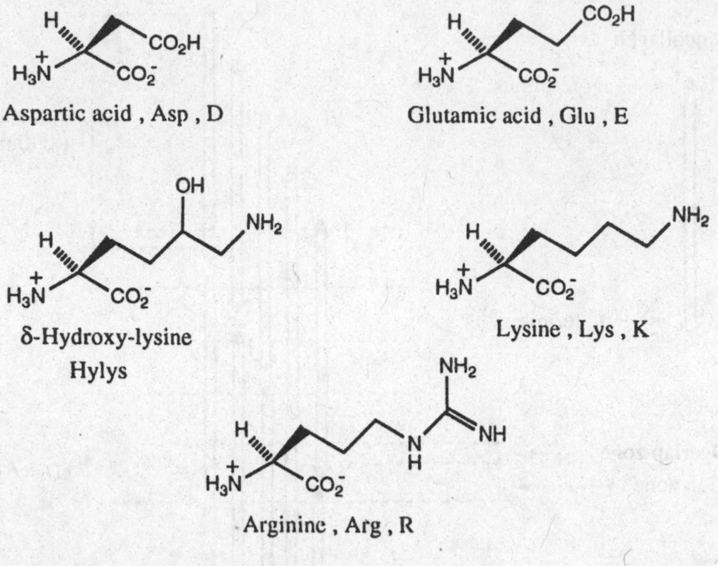

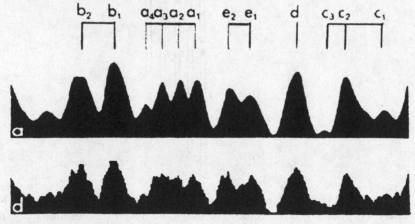

H
CO2H
H3N
CO2-
Aspartic acid , Asp , D
H
CO2H
H3N
CO2-
Glutamic acid , Glu , E
OH
NH2
H
H3N
CO2-
δ-Hydroxy-lysine
Hylys
NH2
H
H3N
CO2-
Lysine , Lys , K
NH2
H
NH
N
H
H3N
CO2-
Arginine , Arg , R

Figure 9.10. Charged amino acid side-chains revealed by staining of collagen fibrils, Hodge and Schmitt, (1960); Bruns and Gross, (1973).

b2 b1 a4a3a2a1 e2 e1 d c3 c2 c1
a
d

Figure 9.11. (a) Averaged densitometric tracing from the fibril staining pattern compared with (d) a 'smoothed' histogram generated from α1 and α2 sequence data showing predicted charge distribution along a complete *D*-segment (234 residues) with five triple chain sequences in the overlap zone and four in the gap zone, Figure 9.9. Information taken from Meek *et al.* (1979). The numering of the staining bands (a_1, b_1, c_1 etc.) follows that used by Bruns and Gross (1973).

interactions and those between large hydrophobic side-chains were major factors in determining the assembly and aggregation of tropocollagen molecules to form fibrils. Later workers have refined this form of analysis (Hulmes *et al.*, 1973; Hofmann *et al.*, 1978; Piez and Trus, 1978).

9.5.2 Microfibrils

These solutions to the packing problem are two dimensional; several attempts have been made to incorporate this two-dimensional structure of the collagen fibril into a three-dimensional one. The five-stranded microfibril in which the D-staggered tropocollagen molecules form a 5_1 or 5_4 helix, originally proposed by Smith, has been one of the most widely favoured models, Figure 9.12 (Smith, 1965, 1968). Smith argued that the only three-dimensional figure in which there is a constant $1D$ stagger between adjacent tropocollagen molecules is a monolayer in the form of a hollow cylindrical filament, Figure 9.12. Furthermore because the marginal molecules which are brought into contact by the formation of such a cylinder must also be staggered by $1D$, (strands A and E, Figure 9.12) the number of molecules in a cross section of a cylinder must be a multiple of five. That with the minimum number of five, Figure 9.12(b) would have the highest ratio of molecules to lumen. Successive $1D$ periods would contain one each of the five dissimilar tropocollagen segments plus one $0.6 D$ 'gap zone'. Furthermore, these 'gap zones' would be disposed in a discontinuous spiral around the tubular microfibril. The primary aggregation of tropocollagen molecules into this type of microfibril would necessarily be followed by their assembly into collagen fibres of mature collagen tissue. The concept of a D-periodic microfibril has grown in popularity and two, four, seven and eight-stranded microfibril structures have been suggested as discrete sub-units in collagen fibrils.

Although there are several lines of experimental evidence available which support the existence of microfibrillar units in collagen fibres, Hulmes and Miller (1979) have described a re-interpretation of the X-ray data which leads to an alternative model for the crystalline regions of the fibril, without the need to postulate microfibrillar sub-structures. These workers proposed that the collagen molecules are tilted to the fibril axis and packed on a quasi-hexagonal lattice. This model has the character of a molecular crystal. As a result of a search to reconcile this conflict of views, Trus and Piez (1980) have described a modified compressed microfibril model of the native collagen fibril. Trus and Piez retained the Smith concept of a five stranded microfibril but compressed it to place molecules on a pseudo-hexagonal lattice, as illustrated in Figure 9.13.

9.6 Connective tissue and the structure of skin

H. R. Procter was the founder of leather science and a truly remarkable man; he had no formal academic training and yet achieved great scientific distinc-

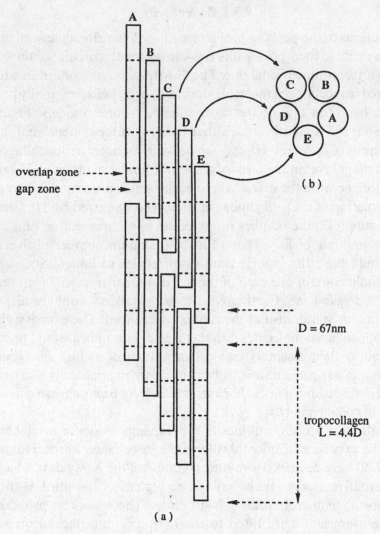

overlap zone
gap zone

(b)

D = 67nm

tropocollagen
L = 4.4D

(a)

Figure 9.12. Pictorial representation of the generation of the tubular Smith microfibril from the two-dimensional assembly of tropocollagen molecules (a); (b) is a diagram of a cross-section through the tubular microfibril (Smith, 1965, 1968).

tion and eminence through his intense interest in chemistry and his application of that interest to his family's association with the trade of leather manufacture. In his book *The Principles of Leather Manufacture*, published in 1922, Procter described skin structure in the following terms:

Mammalian skins all have the same basic structure, consisting of two distinct layers, the epidermis and the underlying dermis or corium. The corium is composed of white fibres – collagen, arranged in bundles of much finer fibrils cemented together by a substance, the nature of which is unknown but which is removed in liming. In the lower

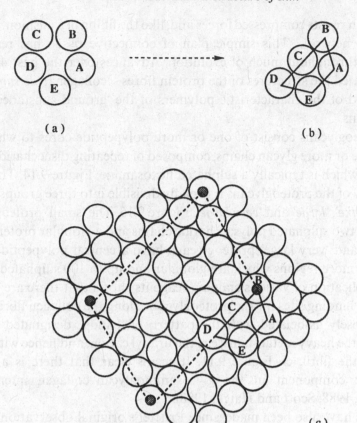

Figure 9.13. Schematic representation of the conversion of a Smith microfibril (a) to a compressed microfibril where molecules in cross-section lie on a pseudo-hexagonal lattice (b). A possible arrangement of the compressed microfibril in a native collagen fibril (c) (Trus and Piez, 1980).

corium the bundles are coarse, but in the grain layer the fibre bundles are finer and, at the surface they are separated into their elementary fibres.

With the development of both transmission and scanning electron microscopy knowledge of skin structure and the structure of collagen fibre bundles in particular has been greatly advanced since Procter's definition of skin structure. Procter's reference to a cementing substance, the nature of which was then unknown, has, for example, been superseded by the fairly detailed picture which is now available concerning the proteoglycan and its regular and highly organised distribution over the surface of the collagen fibrils.

There are two main elements in the structural make-up of connective tissue. These are the insoluble fibrils which resist tensile or pulling forces and the watery interfibrillar material (or 'ground substance', Procter's cementing substance) which resists compressed forces and, like the filling of a cushion, inflates the skeletal network. This simple plan of connective tissue has remained unchanged throughout much of evolution. Advances over the past 40 years have elucidated the structures of the protein fibres – collagen (*vide supra*) and elastin – and of the characteristic polymers of the 'ground substance' – the proteoglycans.

The proteoglycans consist of one or more polypeptide cores to which are attached one or more glycan chains, composed of repeating disaccharide units, one unit of which is typically a sulphated hexosamine, Figure 9.14. The gross morphology of the proteoglycans is broadly divisible into three groups: '*small*' or '*tadpole-like*', '*large*' and '*very large*', Figure 9.15. The 'small' proteoglycans have one or two sulphated polysaccharide chains and a globular protein head. The 'large' and 'very large' proteoglycans have a central polypeptide chain with one or more regions which are globular and up to 100 sulphated glycan chains. Application of various staining reagents shows that there are regular and specific binding sites for the proteoglycans along the collagen fibril which are very closely associated with the pattern of 'bands', designated a–e, as revealed by the heavy metal stains, e.g. the uranyl cation, and hence with polar groups on the fibril, cf. Figure 9.11. It seems clear that there is a strong electrostatic component in the glycosaminoglycan–collagen interactions (Scott, 1987, 1988; Scott and Haigh, 1988).

Advances have also been made, since Procter's original observations, using both transmission and scanning electron microscopy of the gross patterns of organisation of collagen within the skin structure (Haines, 1984). The collagen fibrils are remarkably consistent in size, except below the epidermis where the fibrils are smaller in average diameter. Data from electron microscopy using tissues fixed and embedded in resin suggest an average diameter of ∼ 100 nm, but results obtained with wet collagen by freezing the tissue, without prior fixation, suggested that the collagen fibrils were rather more densely packed with a diameter of ∼ 250–260 nm. These measurements also revealed an inner structure to the collagen fibrils – subfibrils the diameter of which were approximately one-tenth that of the individual fibril. The electron microscope has also shown the finest fibrillar units of collagen to be organised by helical spiralling into collagen fibrils, then into elementary fibres and bundles of fibres, and it has been suggested by Stirtz (1970) that the larger units have also, to some degree, a spiral organisation.

Although the collagen fibrils remain uniform in diameter through most of the skin thickness, the size of the bundles changes at different levels through the skin. In the centre of the corium, the fibre bundles are large and are encircled by a sheath of fine fibrils which do not form bundles. Approaching the junction

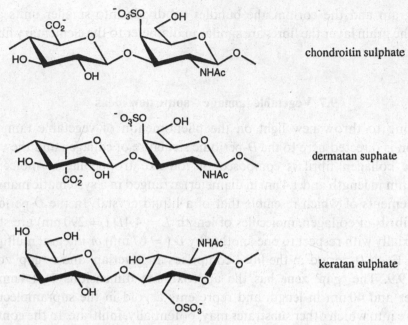

Figure 9.14. Structure of some typical repeating disaccharide units of the proteoglycans of connective tissue (Scott, 1987, 1988; Scott and Haigh, 1988).

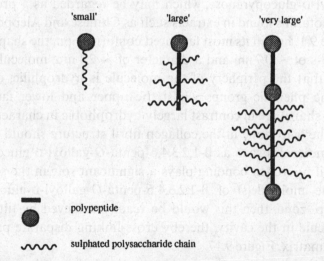

Figure 9.15. Diagrammatic illustration of the appearance of the proteoglycans of connective tissue. These macromolecules associate, often orthogonally, at specific sites along the collagen fibril (Scott, 1988).

of the grain and the corium, the bundles subdivide into smaller units and within the grain layer the fibres are similar in diameter to the elementary fibres in the corium.

9.7 Vegetable tannage – some new ideas

In seeking to throw new light on the phenomenon of vegetable tannage, attention is directed here to the D-periodic structure of collagen and the view that the collagen fibril is composed of rod-like tropocollagen molecules ($\sim 290\,$nm in length and 1.4 nm in diameter) arranged in a systematic manner, some elements of which resemble that of a liquid crystal. In the D-periodic native fibrils of collagen, molecules of length $L = 4.4D$ ($= 290\,$nm) are staggered axially with respect to one another by D ($= 67\,$nm) *or* integral multiples thereof. Each D-period in the fibril comprises an 'overlap' and a 'gap' zone, Figure 9.9. The **'gap' zone** has the approximate dimensions of 1.4 nm in diameter and 40 nm in length and represents a void in the supramolecular structure into which other substrates may, potentially, infiltrate. In the context of vegetable tannage the 'gap' zone therefore seems, *a priori*, a probable site for the binding of vegetable tannin molecules. Thus the molecule of β-1,2,3,4,6-penta-O-galloyl-D-glucopyranose, which may be regarded as a prototype of the typical gallotanins found in extracts such as Chinese and Aleppo galls and Sumach, Table 9.1, has, in its most favoured conformation, the shape of a disc with a thickness of $\sim 0.7\,$nm and a diameter of $\sim 2.1\,$nm; molecular models clearly reveal that the periphery of the molecule is hydrophilic, due to the presence of the phenolic groups whilst the upper and lower faces of the molecule (disc-shape) are in contrast largely hydrophobic in character, Figure 9.16. Clearly the 'gap' zone in the collagen fibril structure should be readily accessible to molecules such as β-1,2,3,4,6-penta-O-galloyl-D-glucopyranose. Furthermore, if hydrogen bonding plays a significant role in the subsequent binding of the molecule(s) of β-1,2,3,4,6-penta-O-galloyl-D-glucopyranose within the 'gap' zone, then this would be readily achieved by tilting of the disc-like molecule in the cavity, thereby cross-linking disparate parts of the collagen fibril matrix, Figure 9.17.

This view that the essence of vegetable tannage involves the interaction of polyphenols with protein functional groups within the 'gap' zones of the collagen fibril is also consistent with the knowledge that effective opening-up of the fibre structure of hide and skin, by the process of liming, is an essential feature to be accomplished before tannage is begun. Alexander and his colleagues (1986) have shown that, during liming, in addition to hyaluronic acid (which does not appear to be bound to the collagen), the proteoglycan derma-

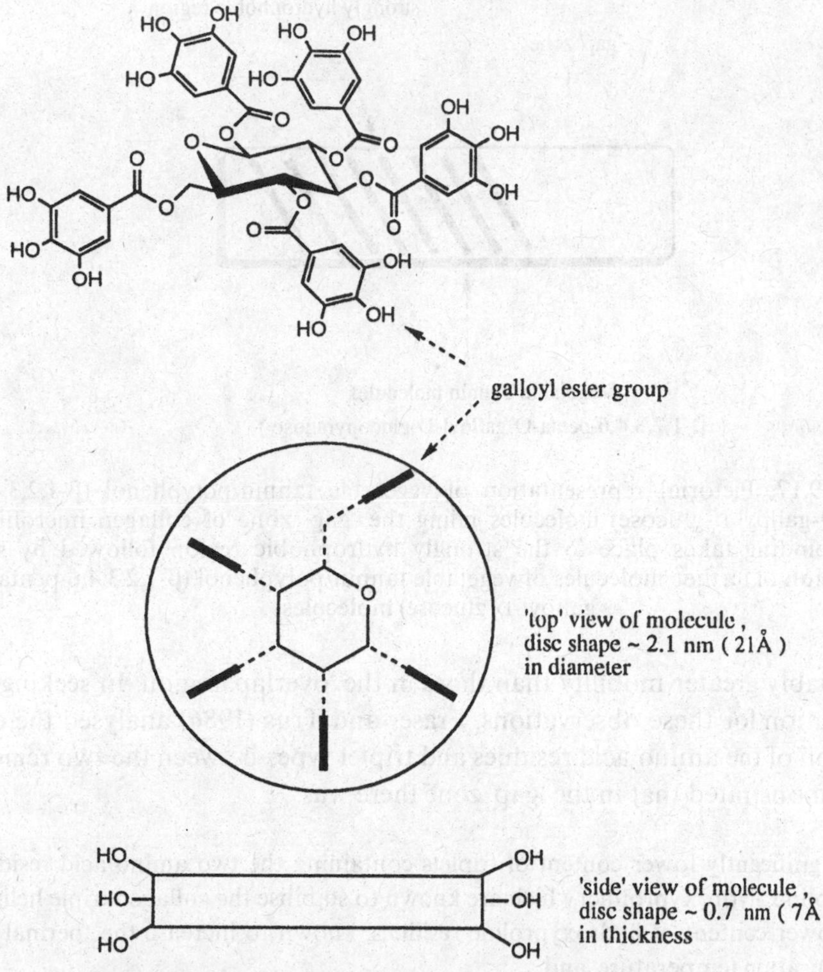

'top' view of molecule,
disc shape ~ 2.1 nm (21Å)
in diameter

'side' view of molecule,
disc shape ~ 0.7 nm (7Å)
in thickness

Figure 9.16. β-1,2,3,4,6-penta-*O*-galloyl-D-glucopyranose.

tan sulphate, Figures 9.14 and 9.15, is extensively removed from hides under the alkaline conditions that are employed. Dermatan sulphate is attached by strong electrostatic binding to the collagen fibrils at the 'd' bands in the 'gap' zones, Figure 9.11. Its removal has proved to be a very sensitive chemical marker of the opening-up process and the effectiveness of liming. This is entirely consistent with the need to remove non-collagenous materials from the 'gap' zones to make these positions directly accessible to the vegetable tannin substrates.

The 'overlap' region of the collagen fibril contains five molecular segments while the 'gap' zone has four molecular segments, Figure 9.9. Studies by Fraser *et al.* (1983) suggest that the molecular segments in the 'gap' zone have

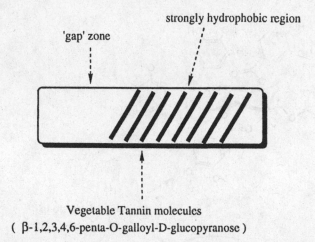

Vegetable Tannin molecules
(β-1,2,3,4,6-penta-O-galloyl-D-glucopyranose)

Figure 9.17. Pictorial representation of vegetable tannin/polyphenol (β-1,2,3,4,6-penta-*O*-galloyl-D-glucose) molecules filling the 'gap' zone of collagen microfibril. Initial binding takes place to the strongly hydrophobic region followed by self-association of further molecules of vegetable tannin/polyphenol (β-1,2,3,4,6-penta-*O*-galloyl-D-glucose) molecules.

appreciably greater mobility than those in the 'overlap' region. In seeking an explanation for these observations, Fraser and Trus (1986) analysed the distribution of the amino acid residues and triplet types between the two regions and demonstrated that in the 'gap' zone there was:

  (i) a significantly lower content of triplets containing the two amino acid residues (proline, hydroxyproline) which are known to stabilise the collagen triple helix,
 (ii) a lower content of hydroxyproline residues, known to increase the thermal denaturation temperature, and
(iii) a segment (∼ 10 amino acid residues in extent, ∼ 7% (3 nm) of the length of the 'gap' zone), which has a completely atypical composition compared to the remainder of this region. It has a very high hydropathy index (i.e. is strongly hydrophobic, in contrast to the rest of the collagen molecule), is devoid of charged amino acid residues, and contains very high concentrations of alanine, proline, hydroxyproline and phenylalanine (of the 10 aromatic residues in the 'gap' zone 9 are included in this short segment).

In the context of the search for a possible rationale to explain the mechanism of vegetable tannage the observations of Fraser and Trus (1986) are highly pertinent. The complexation of polyphenols (vegetable tannins) with peptides and proteins is time dependent and dynamic and is thought to be driven initially by hydrophobic effects with hydrogen bonding as a secondary effect serving to re-inforce these initial interactions. The short strongly hydrophobic peptide region (iii, above) therefore clearly has the potential to act as the principal site for the initial strong complexation with polyphenolic substrates, Figure 9.17. Thereafter further polyphenolic molecules may be envisaged as filling the remaining void volume created by the 'gap' zone by self-association with the initially bound polyphenol substrate and interaction, where feasible, with appropriate groups on the protein matrix. Each of these processes would be enhanced by the dynamic flexibility present in the four peptide regions which compose the 'gap' zone of the fibril.

The question of the relative disposition and distribution of these 'holes' or 'gap' zones in the three-dimensional matrix of the collagen fibril still remains a matter of conjecture. If one adopts the Smith microfibril model, Figure 9.12, for example, then the 'gap' zones would be disposed as a discontinuous spiral around the tubular microfibril. Alternative models of the collagen fibril have been developed by Hodge (1989) using short rods as models of the individual collagen (tropocollagen) molecules to construct extensive three-dimensional arrays, such as are found in the collagen fibril. Hodge assumed that the molecular packing was near hexagonal (Fraser, MacRae and Miller, 1987). By the manipulation of planar arrays of close-packed collagen molecules as depicted in Figure 9.9 (systematic D-stagger) Hodge (1989) demonstrated the possible distribution of 'holes' ('gap' zones) in native-type fibrils. He showed that two simple types of model were feasible using native-type close packing. In the first of these three-dimensional arrangements, 'holes' occur individually and are separated from one another by at least one molecular spacing, Figure 9.18a. In the other case contiguous 'holes' are located side by side and come together to form transverse channels or slots, one molecule thick and $\sim 0.6D$ in width, within the fibril, Figure 9.18b. These channels are separated by distances equal to five times the molecular spacing. This latter model is an especially interesting one in the context of the mechanism of vegetable tannage. If tannage were to occur by the filling of such extensive channels in the collagen fibril by vegetable tannin molecules then it would necessarily be a time dependent (due to the need for tannin molecules, of $R_M \sim 500$–3000, to permeate deeply into these intrafibrillar regions) and (to facilitate infiltration and subsequent binding) a dynamic process. This would thus account for the relative time taken to achieve tannage in the older traditional procedures (often 6 months to a year) and for the relatively large amounts of tannin often taken up by the collagen fibre (up to $\sim 50\%$ of its own weight). The whole question thus remains an intriguing one, but one ripe for further developments.

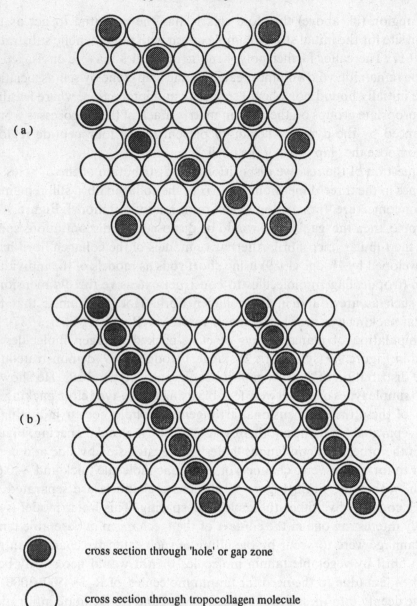

(a)

(b)

⬤   cross section through 'hole' or gap zone

◯   cross section through tropocollagen molecule

Figure 9.18. Possible distributions of 'holes'/'gap' zones in simple systematically staggered arrays of Type I collagen molecules in native type fibrils: schematic cross-section through the structure (Hodge, 1989).

9.8 Vegetable tannins and the durability of leather

The problem of the durability of leathers, particularly those which have been manufactured in the traditional ways using vegetable tannin extracts such as oak, mimosa and sumach, is an old one recognised well over a century ago. The natural ageing of leathers usually appears as a reddening of the leather (red rot), the leather surface cracks and flakes and the fibre structure begins to disintegrate. As a final stage the leather becomes a powder. In 1900 the (Royal) Society of Arts became so concerned about the effects of polluted industrial atmospheres on leather book bindings that the secretary, Sir Henry Trueman Wood, sent a circular letter to 40 of the most important libraries in the United Kingdom. The letter commenced:

Dear Sir,
In consequence of the widespread feeling of dissatisfaction, among those interested in the care of books, respecting the perishable nature of certain leathers used for book-bindings, the Council of the Society of Arts were moved to institute an investigation into the character of the evil, and the best means of remedying it. An influential Committee has been appointed to consider the whole question and report to the Council.

Identification of the chemical and physical changes which occur in naturally deteriorated vegetable tanned leather due to atmospheric pollutants and other environmental factors, however, remains as much a problem awaiting a solution today as it was 100 years ago. Its importance is of particular concern in respect of the protection of old books and archival materials as part of our cultural heritage. The main conclusions of the Society of Arts Committee, under the chairmanship of Lord Cobham, which reported in 1901, were as follows.

(i) Leathers tanned with 'pyrogallol' tans were less affected than those tanned with 'catechol' tans (*one presumes these tans were respectively hydrolysable tannins (e.g. Chinese, Sumach, Myrabolan) and non-hydrolysable (e.g. Mimosa, Quebracho, Gambier)*).
(ii) The fumes of burnt gas appeared to act more strongly than any other agent.
(iii) 'Over-tannage' was condemned as was the use of sulphuric acid in either tanning or dyeing.

Some 50 years later, in the Cantor Lectures of 1954 to the Royal Society of Arts, Dr. Henry Phillips, director of the B.L.M.R.A., concluded that the scientific evidence suggested the following pattern of events during the decay of vegetable tanned leather:

(a) the leather first absorbs sulphur dioxide from the atmosphere, which is oxidised to sulphuric acid, and

(b) when the leather becomes sufficiently acidic, a reaction takes place in which metallic impurities (e.g. iron), tan and the collagen molecular chains are involved leading to breakdown of the polypeptide chains by oxidation,

(c) the oxidation is probably a free radical one.

However, Roux and his colleagues later convincingly demonstrated that the atmospheric oxidation of tannins is at a minimum at strongly acidic pH values but increases rapidly after the pH rises to values of 4 and upwards. These observations are completely in line with chemical expectations and predictions which suggest that auto-oxidation of phenolic substrates takes place most rapidly via the extraordinarily reactive phenoxy radical, itself generated from the corresponding phenoxide anion. Very recently a multidisciplinary and multinational group has presented its first results on the natural and artificial ageing of vegetable tanned leather as part of the so-called STEP project, supported by the European Commission. The analytical data pointed to two principal mechanisms of breakdown in the form of an acid catalysed hydrolytic mechanism and an oxidatively catalysed fragmentation of the collagen molecule (Larsen, 1994). Contemporaneously Japanese workers (Uchida, Kato and Kawakishi, 1990) showed in model experiments that prolyl peptides were very susceptible to oxidatively promoted hydrolytic cleavage. Prolyl peptides exposed to Cu^{2+}/H_2O_2 generated considerable amounts of hydroxyproline and glutamic acid in the acid hydrolysates of oxidised peptides. In addition these workers detected significant quantities of γ-aminobutyric acid, which they suggested came specifically from the oxidative fragmentation of the prolyl peptides.

        phenol   phenoxide anion   phenoxy radical

Acid catalysed hydrolytic breakdown of the collagen structure is unexceptional since this would simply involve hydrolysis of the peptide bonds in the molecule. The analytical data obtained by the STEP group showed that the oxidation of collagen is inhibited by pollution (principally SO_2 and NO_x) and that inhibition increases proportionally to the amount of sulphate accumulated in the leather. Presumably this observation is related to those of Roux, namely that oxidation is inhibited at low pH values. Condensed tanned leather (e.g. Mimosa) accumulates more sulphate than hydrolysable tanned

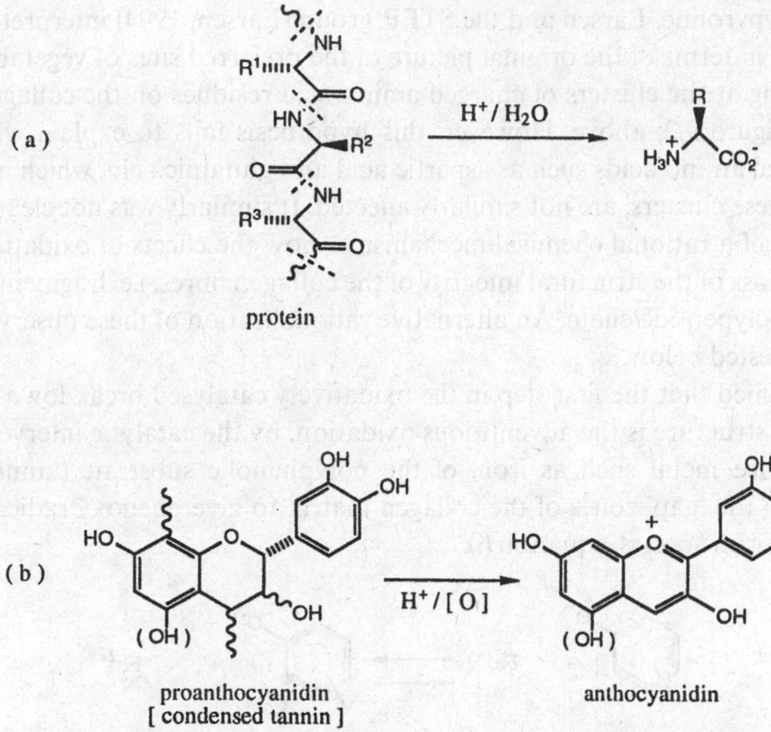

Figure 9.19. Ageing of leathers: the effects of oxides of sulphur and nitrogen. Acid catalysed breakdown of (a) protein structure and (b) condensed tannin (proanthocyanidin) structure.

types (e.g. Sumach) and hence the order of increasing inhibition of oxidative degradation by increasing pollution is *hydrolysable* tanned leather < *condensed* tanned leather. However, in these circumstances deterioration of the leather then takes place increasingly by acid hydrolysis. Furthermore not only will the increased concentrations of acid degrade the collagen but it will also lead to breakdown of the original tannins. In the case of the hydrolysable tannins this will give rise to the formation of gallic and ellagic acids; in the case of the condensed tannins (proanthocyanidin based) to the formation of anthocyanidins and hence to the familiar reddening which distingishes many aged leathers, Figure 9.19.

The results of the STEP group (Larsen, 1994) also showed that the rates and extent of oxidative degradation of collagen are greater for tanned as opposed to untanned skins and hence that oxidation takes place via the intermediacy of the polyphenolic molecules embedded in the collagen matrix. Amino acid analysis indicated that, under these conditions, the greatest loss of amino acids was for the basic amino acids lysine and arginine and the imino acids proline and hydroxypyroline. Larsen and the STEP group (Larsen, 1994) interpreted these results in terms of the original picture of the preferred sites of vegetable tannage being at the clusters of charged amino acid residues on the collagen structure, Figure 9.2, above. However, this hypothesis fails to explain why other charged amino acids such as aspartic acid and glutamic acid, which are located in these clusters, are not similarly affected. It similarly was not clear – on the basis of a rational chemical mechanism – how the effects of oxidation also lead to loss of the structural integrity of the collagen fibres, i.e. fragmentation of the polypeptide chain. An alternative rationalisation of these observations is suggested below.

It is presumed that the first step in the oxidatively catalysed breakdown of the collagen structure is the adventitious oxidation, by the catalytic intervention of a trace metal such as iron, of the polyphenolic substrate (tannin) embedded in the 'gap' zones of the collagen matrix to give phenoxy radicals and reduced iron species, equation (i).

(i)

$$\text{ArO}^- + Fe^{3+} \rightleftharpoons \text{ArO}^{\cdot} + Fe^{2+}$$

(ii)

$$Fe^{2+} + O_2 \rightleftharpoons Fe^{3+} + O_2^{\cdot -}$$

Iron ions are themselves free radicals and ferrous ions can then participate in electron transfer reactions with molecular oxygen to give the superoxide radical, equation (ii), which itself can protonate to give the hydroperoxyl radical or dismutate to give hydrogen peroxide, equations (iii) and (iv) below. Reaction of hydrogen peroxide with ferric iron could then finally give rise to the highly reactive hydroxyl radical, equation (v). These are thus all the essential ingredients for 'Fenton' chemistry with the polyphenol (ROH) acting as the surrogate ascorbate molecule. These species would perforce be present at highest concentrations in the regions where the polyphenol was located (i.e. the 'gap' zones) and it is in these same regions that the oxidation of the collagen molecule would preferentially occur. In this model therefore the role of the polyphenol (ROH) is to serve as a source of reducing power for ferric iron which is probably present, in very small amounts, throughout the tanned skin. Oxidative breakdown of the collagen is then assumed to take place via the agency of one or all the oxidising species – the superperoxide, hydroperoxyl, or hydroxyl radical. Reactions may then be formulated to account for the various experimental observations relating to the amino acids preferentially destroyed

(iii) $\quad O_2^- \ + \ ROH \ \longrightarrow \ HO_2^{\cdot} \ + \ RO^-$

$\quad\quad$ superoxide $\quad\quad\quad\quad\quad\quad$ superoxyl

(iv) $\quad O_2^- \ + \ HO_2^{\cdot} \ + \ ROH \longrightarrow \ H_2O_2 \ + \ O_2 \ + \ RO^-$

$\quad\quad$ superoxide $\quad\quad$ superoxyl

(v) $\quad Fe^{2+} \ + \ H_2O_2 \ \longrightarrow \ Fe^{3+} \ + \ HO^{\cdot} \ + \ H^+$

$\quad\quad\quad\quad\quad\quad\quad\quad\quad\quad\quad\quad\quad\quad\quad$ hydroxyl

in the oxidative degradation process and to the concomitant oxidatively catalysed fragmentation of the collagen molecule.

Because of their basic nature amines are generally oxidised with comparative ease. It is not therefore too surprising that both lysine and arginine appear to be relatively highly susceptible towards oxidation ($-CH_2NH-$ to $-COOH$) during the ageing process to give the corresponding carboxylic acids – α-aminoadipic acid and glutamic acid respectively.

$$\text{[O]}$$

$n = 3 ; R = H ;$ lysyl $\cdots\cdots\cdots\cdots\cdots\cdots\longrightarrow m = 2 ;$ α-aminoadipyl
$n = 2 ; R = -C(NH)NH_2 ;$ arginyl $\cdots\cdots\cdots\longrightarrow m = 1 ;$ glutamyl

Rupture of the polypeptide chain probably proceeds via the formation of a hydroperoxide at the α-centre of an amino acid residue. The process, it is postulated, is initiated by hydrogen abstraction, involving one of the reduced oxygen species, from the α-centre of a particular amino acid to produce a stabilised radical in the first step. Capture of this radical by the hydroperoxyl radical would lead to a hydroperoxide which, it is postulated, may break down hydrolytically to fragment the polypeptide chain at this point, Figure 9.20. The process is illustrated for the case of attack adjacent to proline but is equally applicable to attack at other centres; the mechanism outlined is similar to that originally suggested by Uchida *et al.* (1990) but differs in chemical detail.

Although the picture relating to the fate of collagen and its constituent amino acids during oxidative degradation is now steadily emerging, what happens to the polyphenol (tannin) in these same circumstances is much less clear. However, what is known is that whilst the majority of the polyphenol may be readily re-extracted from freshly tanned skin, as the leather is aged the

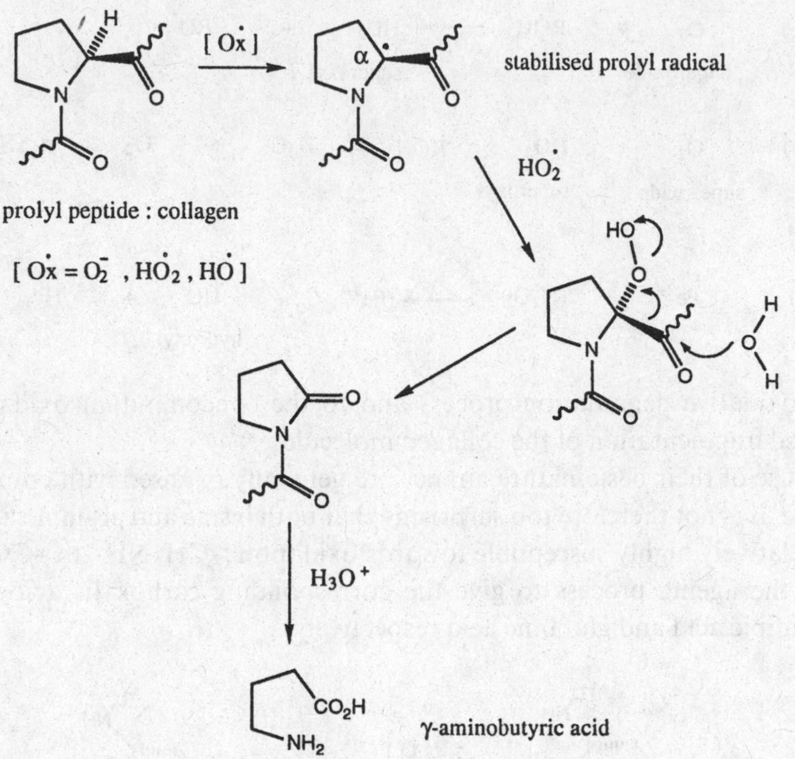

Figure 9.20. Postulated mechanism for the oxidatively catalysed fragmentation of the collagen molecule during ageing.

tannin becomes progressively more difficult to re-extract from the fibre. Several explanations appear possible. The hypothesis adumbrated above defines the initial stage as the multiple generation of phenoxy radicals from the polyphenol substrate (equation (i) above). Once formed such highly reactive species might be expected to polymerise by extensive C–O and/or C–C *intra*- and *inter*-molecular coupling, producing a highly condensed material with greatly reduced solubility. Alternatively the phenoxy radicals could undergo further oxidation to give *ortho*-quinones which again might polymerise or react with one or more nucleophilic groups (e.g. the amino group of lysyl residues) on the protein and thereby become irreversibly covalently bound to the collagen fibre (quinone tanning), Figure 9.21.

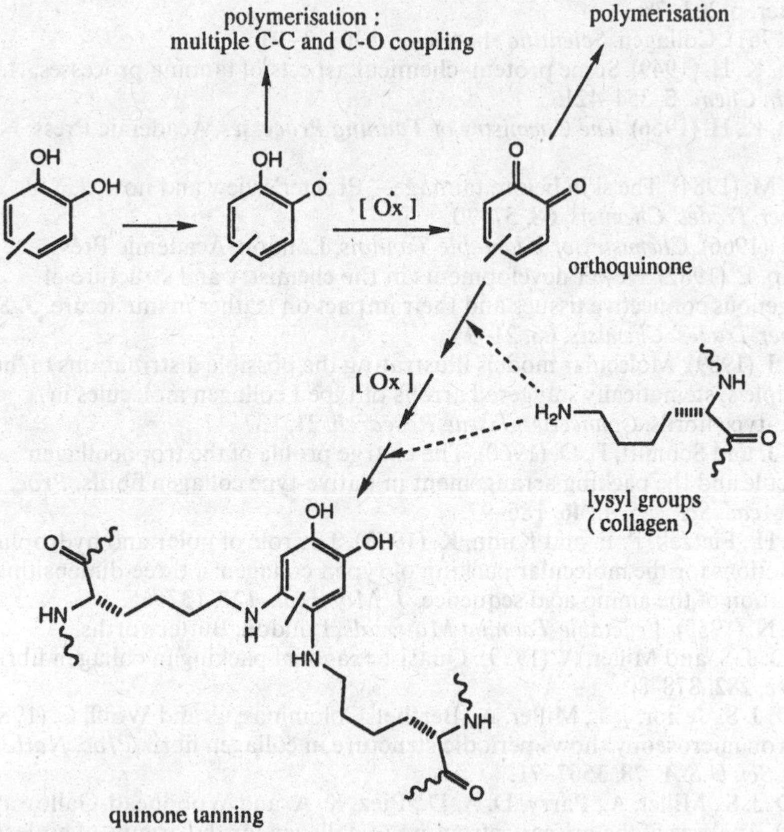

Figure 9.21. Ageing of vegetable tanned fibres: routes to the oxidative polymerisation (insolubilisation) of polyphenols (tannins) and cross-linking of the collagen fibres by quinone tanning.

References

Alexander, K. T. W., Haines, B. M. and Walker, M. P. (1986). Influence of proteoglycan removal on the opening-up in the beamhouse. *J. Amer. Leather Chemists Association*, **81**, 85–102.

Bear, R. S. (1952). The structure of collagen fibrils. *Adv. Protein Chem.*, **7**, 69–160.

Bruns, R. R. and Gross, J. (1973). Band pattern of the segment long spacing form of collagen. Its use in the analysis of primary structure. *Biochemistry*, **12**, 808–15.

Fraser, R. D. B. and Trus, B. L. (1986). Molecular mobility in the gap regions of type I collagen fibrils. *Bioscience Reports*, **6**, 221–6.

Fraser, R. D. B., MacRae, T. P. and Miller, A. (1987). Molecular packing in type I collagen fibrils. *J. Mol. Biol.*, **193**, 115–25.

Fraser, R. D. B., MacRae, T. P., Miller, A. and Suzuki, E. (1983). Molecular conformation and packing in collagen fibrils. *J. Mol. Biol.* **167**, 497–521.

Grimshaw, J. (1986). Depsides, hydrolysable tannins, lignans, lignin and humic acid. *Rodd's Chemistry of Carbon Compounds*, *3D*, Second Edition, Amsterdam:

Elsevier, p. 203–78.

Gross, J. (1961). Collagen. *Scientific American*, 1961, 3–10.

Gustavson, K. H. (1949). Some protein–chemical aspects of tanning processes. *Adv. Protein Chem.*, **5**, 354–421.

Gustavson, K. H. (1956). *The Chemistry of Tanning Processes*, Academic Press: New York.

Haines, B. M. (1984). The skin before tannage – Procter's view and now. *J. Soc. Leather Trades' Chemists*, **68**, 57–70.

Haslam, E. (1966). *Chemistry of Vegetable Tannins*, London: Academic Press.

Heidemann, E. (1982). Newer developments in the chemistry and structure of collagenous connective tissues and their impact on leather manufacture. *J. Soc. Leather Trades' Chemists*, **66**, 21–9.

Hodge, A. J. (1989). Molecular models illustrating the possible distributions of 'holes' in simple systematically staggered arrays of type I collagen molecules in native-type fibrils. *Connective Tissue Research*, **21**, 137–47.

Hodge, A. J. and Schmitt, F. O. (1960). The charge profile of the tropocollagen molecule and the packing arrangement in native-type collagen fibrils. *Proc. Natl. Acad. Sci. U.S.A.*, **46**, 186–97.

Hofmann, H., Fietzek, P. P. and Kuhn, K. (1978). The role of polar and hydrophobic interactions for the molecular packing of type I collagen: a three-dimensional evaluation of the amino acid sequence. *J. Mol. Biol.*, **125**, 137–65.

Howes, F. N. (1953). *Vegetable Tanning Materials*, London: Butterworths.

H ulmes, D. J. S. and Miller, A. (1979). Quasi-hexagonal packing in collagen fibrils. *Nature*, **282**, 878–80.

Hulmes, D. J. S., Jesior, J. J., Miller, A., Berthet-Colominas, C. and Wolff, C. (1981). Electron microscopy shows periodic structure in collagen fibril. *Proc. Natl. Acad. Sci. U.S.A.*, **78**, 3567–71.

Hulmes, D. J. S., Miller, A., Parry, D. A. D., Piez, K. A. and Woodhead-Galloway, J. (1973). Analysis of the primary structure of collagen for the origins of molecular packing. *J. Mol. Biol.*, **79**, 137–48.

Kadler, K. E. (1994). Extracellular matrix I: fibril-forming collagens. *Protein Profiles*, **1**, 519–638.

Larsen, R., editor. (1994). *STEP Leather Project: Evaluation of the Correlation between Natural and Artificial Ageing of Vegetable Tanned Leather and Determination of Parameters for Standardisation of an Artificial Ageing Method*, Copenhagen, Royal Danish Academy of Fine Arts.

Meek, K. A., Chapman, J. A. and Hardcastle, R. A. (1979). The staining patterns of collagen fibrils. *J. Biol. Chem.*, **254**, 10 710–14.

Parkinson, J., Kadler, K. E. and Brass, A. (1994). Simple physical model of collagen fibrillogenesis based on diffusion limited aggregation. *J. Mol. Biol.*, **247**, 823–31.

Piez, K. A. and Trus, B. L. (1978). Sequence regularities and packing of collagen molecules. *J. Mol. Biol.*, **122**, 419–32.

Raghunath, M., Bruckner, P. and Steinmann, B. (1994). Delayed triple helix formation from patients with Osteogenesis imperfecta. *J. Mol. Biol.*, **236**, 940–9.

Ramachandran, G. N. (1967). In *Treatise on Collagen*, Volume I, ed. Ramachandran, G. N., Academic Press, New York, pp. 103–83.

Scott, J. E. (1987). Molecules for strength and shape: our fibre-reinforced composite bodies. *Trends in Biochem. Sci.*, **12**, 318–21.

Scott, J. E. (1988). Proteoglycan – fibrillar collagen interactions. *Biochem. J.*, **252**, 313–23.

Scott, J. E. and Haigh, M. (1988). Identification of specific binding sites for keratan sulphate and chondroitin–dermatan sulphate proteoglycans on collagen fibrils in cornea by use of Cupromeronic Blue in 'critical-electrolyte-concentration' techniques. *Biochem. J.*, **253**, 607–10.

Smith, J. W. (1965). Packing arrangement of tropocollagen molecules. *Nature*, **205**, 356–8.

Smith, J. W. (1968). Molecular pattern in native collagen. *Nature*, **219**, 157–8.

Stirtz, T. (1970). Structure and interweaving of collagen fibres in hide on the basis of light and electron microscope photographs. *Leder*, **21**, 218–28.

Sykes, R. L. (1986). Procter's Textbook of Tanning – 1885. A retrospective view in the centennial year. *J. Soc. Leather Trades' Chemists*, **70**, 1–10.

Trus, B. L. and Piez, K. A. (1980). Compressed microfibril models of the native collagen fibril. *Nature*, **286**, 300–1.

Uchida, K., Kato, Y. and Kawakishi, S. (1990). A novel mechanism for oxidative cleavage of prolyl peptides induced by hydroxyl radicals. *Biochem. Biophys. Research Communications*, **169**, 265–71.

Ward, A. G. (1978). Procter Memorial Lecture – Collagen, 1891–1977: Retrospect and Prospect. *J. Soc. Leather Trades' Chemists*, **92**, 1–13.

White, T. (1957). Tannins – their occurrence and significance. *J. Sci. Food Agric.*, **8**, 377–85.

Biological index

General index

Printed in the United States
By Bookmasters